INDUSTRIAL LOCATION

Location is vital to the efficiency and profitability of industrial activity. *Industrial Location* presents a comprehensive introduction to and critical review of this field of growing academic and business interest.

Using case studies of actual firm location decisions as well as several key industrial sectors, the book offers both neoclassical microeconomic and political economy based macroeconomic perspectives on issues such as why firms locate where they do, models of industrial location and the impact on local development. Taking both cost-minimizing and revenue-maximizing approaches, the authors discuss the incentives and constraints under which firms make location decisions, and stress the importance of public policy in shaping the resulting economic landscape.

Focusing chiefly on the United States, but drawing on an international range of examples, *Industrial Location* explains the economic, social, and political forces which have shaped contemporary patterns of industrialization and examines the changing nature of production systems.

J.W. Harrington is Associate Professor of Public Policy and Geography at the Institute of Public Policy at George Mason University; **Barney Warf** is Associate Professor of Geography, Florida State University.

INDUSTRIAL LOCATION

Principles, Practice, and Policy

J.W. Harrington and Barney Warf

London and New York

First published 1995
by Routledge
11 New Fetter Lane, London EC4P 4EE

Transferred to Digital Printing 2004

Simultaneously published in the USA and Canada
by Routledge
29 West 35th Street, New York, NY 10001

© 1995 J.W. Harrington and Barney Warf

Typeset in Garamond by
Florencetype Limited, Stoodleigh, Devon

British Library Cataloguing in Publication Data
A catalogue record for this book is available from the British Library

Library of Congress Cataloguing in Publication Data
A catalogue record for this book has been requested

ISBN 0–415–10479–3
0–415–11951–0 (pbk)

CONTENTS

CONTENTS

FIGURES

TABLES

PREFACE AND ACKNOWLEDGEMENTS

This text represents several years of work by each of us to pull together lecture notes, research experience, massive reading, and our somewhat complementary interests in industrial location into a book usable in university courses. The primary audience of the text are students in an introductory course in industrial location, typically at the upper-division undergraduate level in the U.S. and Canada, or in the second or third year in the U.K. The text covers issues broader than the microeconomic concerns of industrial location, including industrialization and the regional and national consequences of industrialization and industrial change.

This volume should serve as the primary text for a semester-long, undergraduate course on industrial location or industrial geography. In such a graduate course, it might serve as a basic text, augmented by more specialized readings (including those suggested at the end of each chapter). In an undergraduate business-administration program, this text might be augmented with a set of corporate case studies. For use in an economic geography course, it may be supplemented by texts that cover agriculture, urbanization, retail location and Third World development.

We are grateful for support from the universities from which we wrote the bulk of this book: SUNY Buffalo, George Mason University, Florida State University, and Kent State University. Ms. Debbie Angerman and Ms. Sonia Vargas assisted with the typing and organizing. Assistance with figures and tables came from Ms. Jane derBoghossian, Ms. Ana McGuinness, J. Bradford Hunter, Ms. Sonia Vargas, and Ms. Ferdouse Sultana. Dr. Ian MacLachlan, University of Lethbridge, and Dr. John Lombard, Connecticut Department of Economic Development, gave useful comments on portions of the text. Mr. Tristan Palmer, our editor, and two anonymous reviewers have given very helpful suggestions on the entire text, at more than one juncture. Finally, we thank John Burkhardt and Annette Schwabe for their continual support of our efforts.

James W. Harrington, Jr.   Barney Warf
George Mason University   Florida State University

1

WHAT IS INDUSTRIAL LOCATION?

Look around the city or town where you live. Think about the forces that structure your world and everyday life. What processes yielded the built environment? Why are companies and jobs located where they are? Why are skyscrapers, filled with banks and insurance companies, clustered downtown? Where did the steel or automobile factories go? Why is your shirt likely to be made in China? Go to the grocery store: where did all of these goods come from? What types of jobs do your parents have? How are they different from the ones their parents had? How will your job future resemble or differ from theirs? Why are you likely to work in the service sector? Do you know anyone who works for a foreign company? Why is it here? Why do television networks and newspapers give so much attention to matters like the European Union or the North American Free Trade Agreement (NAFTA)? How will these international communities affect you?

These are the sorts of questions that concern economic geographers, especially those interested in industrial location. To some, the word "industrial" connotes dirty smokestacks or boring lists of exports and imports. Yet, as we shall see, "industries" include service firms and office work, as well as many other types of jobs. You do, or will, work in an "industry." How this industry is structured, how it came to be the way it is, the competitive and regulatory environment it faces, the various technologies it can adopt, and other factors that influence why it decided to locate where it did, all have an immense influence on your everyday life and future prospects. For those

of you who hope to get a job in the business community, start your own firm, or are worried about your career prospects, industrial location is an essential topic. People who ignore the dynamics of industry do so at their peril. Obviously this doesn't mean that reading this book will get you a job, but it does mean that you will have an appreciation of the context, forces, and issues that structure your future job opportunities. This sort of knowledge is particularly important given that we live in a world of constant, incessant, breathtaking change.

For most of you, change has become one of the few things upon which you can rely. Constant change characterizes your personal lives, careers and career prospects, and the look and feel of the world you see around you and in the news media. This book provides narrow, focused lenses on one important aspect of those changes. *We are concerned with organizations' decisions about the location of production.* The following paragraphs describe the ways in which these decisions affect the direction of your personal prospects and world changes. This one aspect of change seems pretty narrow, but if you look at the range of offices, warehouses, and factories around you, and take note of the daily announcements of openings, closings, expansions, and layoffs, you get a sense of how big and unwieldy a topic it is.

We will use two lenses to address these topics, to make some sense of the daily announcements. One lens is fairly narrow and largely economic. The other lens is broader and more historical. We'll use the economic lens to pull into focus location requirements, relate them to the varied nature of production activities, and understand changes in the requirements. We'll focus on government influences on these activities and on the relevant characteristics of places. We'll present ways in which the day-to-day production of commodities (goods and services) for sale changes the characteristics of the production process and of the places where production is carried out, yielding yet further change in the economic landscape. This same lens can help us to understand the way that organization managers decide where to expand, open, or close facilities, as well as the way that these changes affect those of us who are not organization managers. Thus, our view is useful to managers, workers, and planners.

This lens allows us to focus and thereby make sense of a seeming jumble of changes. However, any lens distorts other issues. We will present a view of the social causes and results of industrial change, but we will view these social characteristics through our economic lens, when we discuss the roles of people in production: employees or labor. We will then discuss these social changes from a historical perspective, still focusing on their economic nature. We will not explicitly focus on the political process through which governments create the rules that affect organizational behavior, though we will note some historical changes in political organization. We will not focus on the physical environmental results of production, and the effects those results have on our lives and on subsequent decisions by organizations. We will stick with our two lenses to yield a view that is partial, but clear and simple (simple, at least, compared to the full view from many different lenses).

After an introduction to what and why we are studying, this chapter presents some suggestions for how we can study industrial location or any other economic and social activity. Again, the purpose of a careful approach to how we study is to provide a clearer focus on our topic, to make sense of the jumble of casual observation.

Industrial Location as Commitment

This book is concerned with the location of firms dedicated to commodity production – activity which results in a product or service to be sold in a marketplace. The dedication of a facility for particular use such as a bank or a manufacturing operation represents a double commitment on the part of the operator of the facility. One commitment is to the operation itself – can the company provide banking services that clients will want and that clients will value at a price sufficient to cover costs? Can the manufacturing operation produce goods of a particular type that will find a sufficient price in some market? This operation also represents a second commitment: to a particular place and a particular form of provision of service or production of goods. For example, the bank branch can only obtain customers if it is conveniently located for the customers, and can only maintain a work force if its workers can conveniently get to and from the facility. The manufacturing plant represents a commitment to being able to obtain supplies and inputs of high quality and in good time; a commitment to ship these products to their market locations in reasonable time, good quality, and at a low cost; and a commitment to maintaining a work force that is productive and able to get to and from the workplace. These are very substantial commitments, and given the speed and unpredictability with which technology, markets, and people change in the late twentieth century, the commitment becomes very difficult to make.

The commitment or lack of commitment to particular places helps define the organization, helps distinguish it from other, similar organizations. One of the principal ways that retail customers select a bank company is to compare the locations of company's branches and automated teller machines (ATMs) to the customers' own daily travel patterns to and from work, school, and shopping. Let's suppose that a particular bank owns no branches and no ATMs, but contracts with other banks to allow its customers to use the ATMs of all banks, at no charge to the customers. For customers, the bank is distinguished by its vast network of ATMs but also by the difficulty of getting to the bank's only building. The bank will find that certain customers cost a great deal to service: the ones who withdraw five pounds or ten dollars every day from an ATM that the bank pays to be using. Eventually, the bank will work out a pricing scheme that encourages the kinds of retail and business customers it can serve best, based in part on its locational strategy.

Similarly, the mix of manufacturing and distributive locations developed by a manufacturer over the years influences the costs that it faces and the markets that it can serve. One competitor may be better suited to produce very low-cost products, and another competitor can design, produce, and deliver a new product more quickly to a particular market. Our first manufacturer must understand what its mix of locations, personnel, and equipment allow it to do better than its competitors. Perhaps its strength is the design, manufacture, and marketing of trade-marked items for a large consumer market in the midwestern United States. After reaching that understanding, the company must predict how profitable that competitive advantage will be in the future. Decisions to modify that strength must include a plan to modify the locations in which the company operates facilities. These important investment decisions are based on analysis of current strengths and prediction of future conditions.

There are ways of reducing the commitment, of course. The space for the facility can be rented or leased rather than purchased. Workers can be hired on a specifically temporary basis. Production for a manufacturing firm can be contracted to some supplying firm, so that the original manufacturer finds itself in a much more flexible, less-committed position. Nonetheless, the provision of services and the manufacture of products do take place, and take place at particular locations. It is the study of these locations – no matter how flexible, no matter how temporary – that is the topic of this book.

The Study of Industrial Location

Given the increased level of complexity and of fluctuation in ownership and location of production and services, how can we study the location of these operations? Why should we care about their physical location, given the increasing transience of these

locations? One answer is obvious: the people who make their living by working in these operations are raising families, buying houses, and establishing their lives in the expectation that they will continue to be able to make their living. People, in general, are not terribly mobile. Productive facilities are somewhat mobile, insofar as they can be constructed, sold, or purchased, but these operations must take place at some fixed point. Financial capital, the flow of credits and debits that help us keep track of economic activity, is extremely, nearly perfectly mobile. This book will be concerned primarily with the interplay between imperfectly mobile productive facilities on the one hand, and imperfectly mobile individuals and households, on the other hand.

There are other reasons for us to care about the location, however fleeting, of productive activity. These productive activities form the backdrop for our lives. Where are there jobs? Where are taxes paid? Where is there pollution from economic activity? Where is there congestion, given the need for a constant flow of people and materials to and from these productive activities? From which productive activities and what locations are we likely to see charitable contributions, many of which are aimed at needs in the local areas of firms' productive activities? So, we can see that there are important macroeconomic results of the essentially microeconomic act of committing a productive facility to a particular location. Indeed, *microeconomics (the individual actions of producers and consumers, coming together through some market or control mechanism) yield macroeconomic phenomena (levels of output and employment) when these producers and consumers locate themselves in a particular region or nation.* This is why the study of industrial location, along with the study of consumption and residential location, is so vitally important. It forms one part of the juncture between microeconomic action and macroeconomic situations. Finally, an inappropriate location can cost a company dearly, in terms of high freight bills, unavailability of suitable workers, time lost to traffic congestion, or difficulty in responding to rapid changes in the product market. The activity may be able to afford the cost, if it is underwritten by a large, profitable corporation or by tax monies. A poorly located activity can attempt to find closer suppliers, or change its operations to use available workers, or increase its supply inventory, or establish better distribution networks. The option of relocation is often present, if funds permit. Concerted efforts by businesses and citizens can even change the characteristics of places, by improving transport infrastructure and educational systems. Nonetheless, the perils of poor locations for investment haunt managers, and hurt the general productivity of capital and labor in the economy.

Having seen some of the ways in which the location of productive activity matters, we must ask again how we can study and understand these locations. Should we assume that any location is equally good, given a profitable enough operation? Shouldn't we recognize the importance of the market, looking at the target market location as the preferred

location for a productive facility? In the paragraphs above, we have given some suggestion of the factors that come into play in locating facilities. These include the *accessibility to chief inputs*, *the location of the target market*, and *the availability of a quality work force*. This suggests that locations – cities, regions, or countries – which combine a wealth of material inputs with a large internal market and a large labor force are good potential locations for most economic activity. Indeed, we can see the prevalence of economic activity in large cities, large metropolitan areas, and the large countries of the world. Table 1.1 compares the international distribution of population, income, and manufacturing output. But what of locations within these countries or in the rest of the world? Why has there been such a concentration of manufacturing across the English Midlands, or in the southeastern United States? How likely is it that the rapid industrialization of the "Four Tigers" of

Table 1.1 Population, GNP per capita, and percent GDP in manufacturing and service industries

Country	Population (millions)	GNP per capita (US$)	Percent GDP in manufacturing[A]	Percent GDP in service[A]
Australia*	17,045	16,670	17.09	73.97
Canada[1]	25,950	16,870	19.78	63.20
China*	1,134,000	370	N/A	N/A
France*	56,735	19,420	24.20	78.17
Germany[2]	78,752	20,520	34.66	65.80
Ghana*	14,870	390	N/A	N/A
Hong Kong*	5,705	11,890	16.88	74.57
India*	850,000	360	18.58	39.46
Japan*	124,000	25,840	31.11	60.00
Korea*	42,869	5,440	32.75	52.48
Mexico*	81,724	2,610	25.51	67.49
New Zealand[2]	3,330	11,970	20.72	75.80
Nigeria[1]	90,866	360	7.52	28.57
Singapore*	2,705	12,430	32.03	67.91
South Africa*	37,959	2,450	25.45	51.20
Thailand*	56,303	1,410	30.48	56.55
United Kingdom[3]	56,930	10,560	23.22	62.01
United States[3]	244,000	18,480	20.91	74.14

Source: World Tables 1993, World Bank
Notes: * Based on 1990 data
[1] Based on 1988 data
[2] Based on 1989 data
[3] Based on 1987 data
[A] Calculated thus: GDP at factor cost in current price/GDP at factor cost

East Asia (Hong Kong, Singapore, South Korea, and Taiwan) could be replicated in central Africa, or central Europe?

DISTINGUISHING AMONG ACTIVITIES AND LOCATIONS

To begin to answer these questions, we will rely on two assumptions: that productive activities can be distinguished from one another in ways that are important for their locational needs, and that potential locations can be distinguished from one another in equally important ways. This provides the ability to match particular kinds of productive activities to particular kinds of regions. This basic set of assumptions is only a starting point for location analysis. It ignores **locational inertia**: the large extent to which the current location of an establishment or of its founder determines that subsequent investment will be in that same location. Most business entrepreneurs start their businesses within commuting distances of their homes, if not in their homes. Most business investment is put into existing operations, where they are. This set of simple assumptions also ignores the fact that the characteristics of locations are not static; indeed, these characteristics are influenced by the decisions of enterprises to found, expand, and locate establishments in the location. However, most new businesses fail, and existing businesses vary in their profitability and longevity. Over time and in aggregate, particular places are better suited for particular activities, and we will spend some time understanding this matching process.

Characterizing Activities

How, then, can we distinguish productive activities from one another? One important way is to rely on the industry of the productive operation. **Industry**, in this sense, is a complex term, defined in two ways. An industry can be defined as a group of operations that share similar inputs and technologies. An industry is further defined as a set of operations that

share similar market characteristics. For example, we can define pens and pencils as products of the same or very similar industries, because they are both manufactured items that find use in the manual recording of information. However, the inputs to a typical ballpoint pen differ substantially from those to a typical wooden pencil. The definition and delineation of industries is an imperfect science. However, we can see that, given the definition of industry, it is an important concept in an attempt to distinguish the potential locational needs of productive activities.

Take a look at Tables 1.2 to 1.4. Note the way in which several different classification schemes have divided productive activity into industries. The Standard Industrial Classification (SIC) system, used in the United States, comprises 80 two-digit categories, or sectors. Within these sectors are the three-digit and four-digit levels, which can be considered individual industries. The United Kingdom's industrial classification system is composed of industry groups which are subdivided into minimum list headings. Other countries have their own industry classification systems. In addition, international trade is classified and recorded using the Standard International Trade Classification (SITC) system. This system is encapsulated in Table 1.4. Imbedded within these various systems is the tension between the two bases for distinguishing industries. For example, wooden roof trusses and steel roof trusses are substitutable products in smaller commercial buildings. Yet they are made from different materials, and their manufacture is accounted for in very different industries in the U.S. classification. On Table 1.2, wooden roof support members are considered a part of SIC 2439, while the manufacture of steel members is included in SIC 3441. A more recent problem occurs because of the importance of basic software, called operating systems, to the functioning of computers. Software development is considered a service, in SIC 7372, while computer manufacturing is included in SIC 3573. These are merely two examples of the problems we encounter when we try to use these industrial classifications for industrial location analysis. We continue to use these classification schemes, however, because these are the bases from which empirical information is gathered. Our study of industrial location requires both theoretical understanding of businesses' locational commitment of operations, and empirical information about those operations and their locations. If we rely on secondary, widely available information, we must use widely available categories.

Just what are these operations with which we are concerned? We can simplify the answer to this question by defining the plant, facility, or **establishment** as a productive operation with a particular location. This is to say that a particular company, or government, or organization may operate facilities in more than one location. Each facility is considered a separate plant or establishment. The organizational unit which operates these facilities or plants can be called a company, a firm, an organization, or an enterprise. In this text, the word **company** will be used to imply or to name a particular enterprise, such as XEROX, or General Motors, or ICI. We may use this word interchangeably with the word corporation, though there are legal differences in the two forms of private organizations. In this text, the word **firm** represents an abstraction of neoclassical economics. We will use this term when we are dealing with an idea of a company. This idea is defined by its desire to maximize profit or minimize cost. It does not have specific managers or shareholders or employees, but is merely an instrument for production and for profit maximization. The term **organization** focuses on the internal complexity of these entities, companies, or firms. We will use this term when we want to emphasize the potentially conflicting goals of the individuals within the company or firm. Studies of organization are also concerned with the partial ignorance on the part of the organization of its external environment. Among the most commonly used terms in industrial geography is the term **enterprise**. The term enterprise is again a relatively abstract term, including the desire (if not the practice) for profit maximization, as well as the presence of conflicting goals and information within the enterprise. The words organization and enterprise may also be used for public agencies and other government-run operations.

So, the enterprise is the leading actor about which we will be concerned in this text. The enterprise is charged with creating or assembling input and with hiring and managing workers which will yield some good or service which the enterprise will then attempt to sell or for which the enterprise finds some demand. The assembly and the employment occurs at one or more specific sites, termed **plants** for manufacturing operations. A more general word is "establishment." An **establishment** is defined as a discrete operation at a specific location, owned by an enterprise. Because an enterprise can have many establishments, at many different locations, it is difficult to talk about the "location" of an enterprise or company. Each establishment, however, has a clearly identifiable location. An enterprise can also operate within many different industries: the same company can own an automobile assembly operation, automobile parts manufacturing facilities, a leasing company, and a consumer-finance company. However, individual establishments are usually identified according to a discrete operation: an automobile assembly plant, an automobile-seat cushion factory, the offices of the leasing operation, the retail outlets of the finance company. Establishments are very important to our study of industrial location, because each establishment has a specific location, a specific enterprise-owner, and a specific industry.

Characterizing Places

Having discussed the concept of industry, we need to discuss the concept of "location." What does it mean that a particular operation is located in a particular place? Speaking geographically, place can be a city block, or a city, or a given metropolitan area, or even a particular country – as opposed to another city block, or another city, or some rural area, or some other country. At what geographic scale do these location decisions matter? Of course the answer to that question depends on your interest in industrial location. If you are concerned about the well-being of your neighborhood, you may (or may not) want an industrial facility within your neighborhood. If you are concerned about the international trade balance

Table 1.2 U.S. Standard Industrial Classification codes

Code	Industry	Code	Industry
01	Agricultural Crops	45	Transportation By Air
02	Agricultural Livestock	46	Pipe Lines
07	Agricultural Services	47	Transportation Services
08	Forestry	48	Communication Services
09	Fishing, Hunting, Trapping	49	Electric, Gas, and Transportation
10	Metals Mining		Services
11, 12	Coal Mining	50, 51	Wholesale Trade
13	Oil and Gas Extraction	52–59	Retail Trade
14	Mineral, Stone, Clay Mining and	60	Banking Services
	Crushing	61	Credit Agencies
15,16,17	Construction	62	Security and Commodity Brokers
20	Food Product Manufacturing		and Services
21	Tobacco Manufacturing	63	Insurance Carriers
22	Textile Mill Products	64	Insurance Agents
23	Apparel	65	Real Estate Services
24	Lumber and Wood Products	66	Combined Real Estate and Insurance
25	Furniture and Fixtures		Services
26	Paper and Allied Products	67	Holding Companies and Investment
27	Printing and Publishing		Offices
28	Chemicals and Allied Products	70	Hotels and Other Lodging Places
29	Petroleum and Coal Products	72	Personal Services
30	Rubber and Miscellaneous Plastic	73	Business Services
	Products	75	Automotive Repair, Services, and
31	Leather and Leather Products		Garages
32	Stone, Clay, and Glass Products	76	Miscellaneous Repair Services
33	Primary Metal Manufacturing	78	Motion Picture Production,
34	Fabricated Metal Products		Distribution, and Services
35	Non-electrical Machinery	79	Amusement and Recreation Services
36	Electric and Electronic Equipment	80	Health Services
37	Transportation Equipment	81	Legal Services
38	Instruments and Related Products	82	Educational Services
39	Miscellaneous Manufacturing	83	Social Services
40	Railroad Transportation Services	84	Museums, Botanical, Zoological
41	Local and Interurban Passenger		Gardens
	Transit	86	Membership Organizations
42	Trucking and Warehousing	88	Private Households
43	U.S. Postal Service	89	Miscellaneous Services
44	Water Transportation Services		

between the United States and Japan, you are implicitly concerned with the distribution of productive facilities between those two countries. This text will consider a range of geographic scales, but will focus on the regional scale and the international scale. Some key differences, for example in productivity, total income, per capita income, and growth rates, are especially great among countries, as shown in Table 1.1 and Figure 1.1. The small-regional scale, or the metropolitan scale, is of particular importance because it is defined as the area within which individuals can commute daily from a fixed residential location. As was mentioned above, the range of employment or income opportunities for individuals

Table 1.3 U.K. Industrial Classification System, with selected detail

Division 0:	Agriculture, Forestry, and Fishing
Division 1:	Energy and Water Supply Industries
Division 2:	Extraction of Minerals and Ores Other than Fuels; Manufacture of Metals, Mineral Products, and Chemicals
	Class 25: Chemical Industry
	Group 257: Pharmaceutical Products
Division 3:	Metal Goods, Engineering, and Vehicles Industries
	Class 34: Electrical and Electronic Engineering
	Group 345: Other Electronic Equipment
	Activity 3454: Active components and electronic sub-assemblies
Division 4:	Other Manufacturing Industries
Division 5:	Construction
Division 6:	Distribution, Hotels, and Catering; Repairs
	Classes 64 and 65: Retail Distribution
	Group 652: Filling Stations (motor fuel and lubricants)
Division 7:	Transport and Communication
Division 8:	Banking, Finance, Insurance, Business Services, and Leasing
	Class 83: Business Services
	Group 835: Legal Services
Division 9:	Other Services
	Class 91: Public Administration, National Defence, and Compulsory Social Security
	Group 915: National Defence
	Class 94: Research and Development
	Class 95: Medical and Other Health Services; Veterinary Services

Table 1.4 Standard International Trade Classification codes

Code	Industry
00	Food and Live Animals Chiefly for Food
01	Beverages and Tobacco
02	Crude Materials, Inedible, except Fuels
03	Mineral Fuels, Lubricants, and related materials
04	Animal and Vegetable Oils, Fats and Waxes
05	Chemicals and related products, N.E.S.
06	Manufactured Goods classified chiefly by material
07	Machinery and Transport Equipment
08	Miscellaneous Manufactured Articles
09	Commodities and Transactions not classified elsewhere in the SITC

Source: ASI Annual Supplement, 1984

is one of the principal reasons for concern about industrial location. In addition, given the increased interaction between nations and the increased concern about international competitiveness, we will attempt to understand the patterns and causes of international redistribution of productive activities.

How do we draw distinctions among regions or locations, analogous to the distinctions drawn among industries (on the basis of inputs, technologies, and product markets) or among enterprises (on the basis of ownership)? There is no standardized grouping scheme for regions, like the SIC or SITC listings for

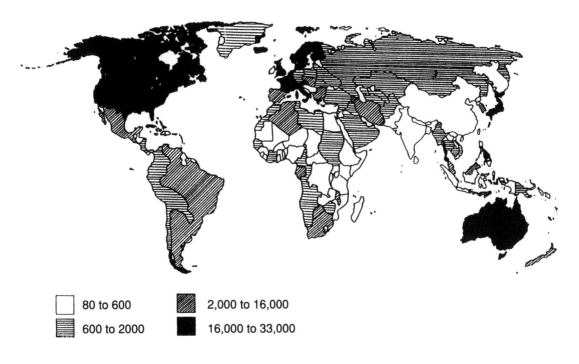

	80 to 600		2,000 to 16,000
	600 to 2000		16,000 to 33,000

Figure 1.1 GNP per capita, 1992

industries. Instead, location decisions are made and analyzed on the basis of measured characteristics of potential locations and their regions. We have already mentioned some important characteristics: availability of workers for hire, access to materials, and proximity to market demand. So, our agenda, so far, is to understand the spatial distribution of productive facilities, classified by industry, across geographic space at the interregional and international scales. These particular operations are located in regions or nations with identifiable characteristics. Our goal is to understand the varied ways in which a given enterprise can and does decide in what location to maintain an establishment.

STUDYING INDUSTRIAL LOCATION

What Should We Study?

As you might expect, abstract ideas of what firms should do is not always what actual companies do.

Which should we be studying? In fact, we must investigate industrial location from both points of view: the normative ways in which firms should make decisions, and the positive ways in which companies actually do decide to commit operations to particular locations. How do we carry out industrial location research? There are many possible ways to try to study why companies locate facilities in the places they do.

One way is purely **normative**: to understand what factors affect company profit, and to assume that the companies will locate facilities in places where those factors will maximize profit. One of the most basic locational models, the Weber model, assumes that facilities are located in such a way as to minimize the companies' total costs of transportation (of inputs to each plant and of output from each plant) and on-site production (for example, the costs of labor vary across places; facilities should locate to minimize their labor costs). According to this model, the reason that all facilities are not located in the same place is that different facilities have different inputs, serve different markets, have different logistical needs to be near an input or near a market, and use different types of labor. To use this model, you need to know the sources of a facility's inputs, the location of its markets, and the nature of its production process. Chapters 2 and 3 review normative approaches to firms' locations and Chapter 5 reviews a normative approach to regional and national industrial activity.

Two alternative ways of studying industrial location entail primary, empirical research. First, we can "simply" study actual location decisions. Roger Schmenner, in *Making Industrial Location Decisions*, researched the way in which companies make location decisions. Are decisions made by a committee? How many potential sites are inspected for each new facility? How do companies decide whether to relocate a facility, to establish a branch, or to expand a plant in its old location? Do these decisions vary according to the size of company? Do these decisions vary by industry? Do these decisions vary by reason for the location decision (capacity shortfall, versus cost increases, versus new product, etc.)? This approach is essentially **behavioral**, in that we are studying how people and groups make decisions. This is useful to study because it helps us improve the decision-making process and because it helps us know how to effect the decisions made. We will study location decisions in this way in Chapter 8.

A third way to study location, again using primary research, is to ask companies why they located facilities where they did. This is a common research approach, usually accomplished by personal interview or postal survey. A typical survey question is "Which of the following factors was most important in your company's decision to open this plant in this county?" These studies can be useful, but they have several inherent problems. The person who responds to the survey may not know, may not know accurately, or may not answer truthfully. For example, managers tend to exaggerate the importance of low taxes for their location decisions, because they think that the researcher may report

these findings to governments: "reduce taxes and you'll get much more industry." These studies are not always careful about the geographic scale about which they're concerned. If I tell you I located "here" because there was a nearby superhighway interchange, that may tell you why I selected a large town rather than a nearby, small village, but it does not tell you why I selected one region of the country over another region. Superhighway interchanges may exist in both regions, so that the highway variable is only important at the intra-regional scale. This approach is also essentially behavioral.

A fourth way to study industrial location is through secondary (published) empirical information: to note where facilities of different type are located. What do we mean by "facilities of different type"? Almost anything. One could compare the location of automobile assembly plants with computer assembly plants, or Japanese-owned automobile assembly plants with U.S.-owned automobile assembly plants, or factories of mass-market toy makers with factories of expensive toy makers. What does it mean to "compare the location of"? Is it important that one plant is in Amherst, New York, and another is in Amherst, Massachusetts? What matters about these locations?

This is where your **theoretical framework** is important: your sense of the causes of industrial location and locational change. What are the differences in facilities (or in locations) that interest you? What do you understand to be the important determinants of industrial location? From your understanding (drawn from your reading about and observation of industrial location), what characteristics of places should be important for the different facilities of interest? Alternatively, how should differences in locations affect the kinds of facilities across those locations?

There are any number of dimensions along which you might measure facility differences, such as industry, size, or type of product within an industry. Similarly, there are any number of dimensions along which you might measure locations and their differences, such as latitude and longitude, level of industrialization, or prevailing wage rate. This book will suggest specific reasons for you to expect particular differences among facilities and among locations to matter as you try to decide or analyze facility locations.

The interplay between what enterprises should do and what enterprises actually do will be a recurrent theme in this text. To suggest what an enterprise should do requires some sense of the enterprise's purpose. Most generally in economic study of market economies, we assume a desire of enterprises to maximize profit. For our current purposes, profit can be understood merely as the difference between total revenues and total costs. We will explore ways of understanding, and perhaps predicting, how the choice of location can reduce costs and increase revenues for the enterprise. However, companies do not necessarily behave in a profit-maximizing way. Given the long-term nature of the location commitment, and given the impossibility of knowing future costs and revenues, it is not surprising that industrial location decisions entail guesswork. We will explore ways of understanding these less optimal ways of reaching a location decision.

How Do We Describe What We're Studying?

Empirical knowledge is knowledge gained from observation. Empirical, then, means "observed" – as opposed to "imagined" or "deduced." Empirical does not necessarily mean "statistical," though observations can be described statistically (number of observations, mean, median, mode, standard deviation, etc.). Let's talk about empirical description, a good first step in scientific inquiry. The validity of a description depends on many difficult things. Take the following empirical statement, which seems simple enough. "In 1942, there were 15 steel plants in Western New York." This simple statement is unclear about its **categorization**. What's counted as a steel plant? Basic steel making? Making steel from scrap? Making basic steel shapes? The answer depends on your theoretical framework, your particular interest or purpose, and the available data. You cannot now observe what was present in 1942,

so you are dependent upon the definition of "steel plant" in the available records.

The **regionalization** used in this statement also needs to be clarified. What's Western New York? (See Figure 1.2.) Two counties? Eight counties? West of Syracuse? Why use political divisions when we're describing economic phenomena? (Data limitations, perhaps?) Regionalization is crucial to geography, obviously. Geographers have developed three bases for delimiting regions.

Homogeneous or **uniform regions** are contiguous areas that share important characteristics, which characteristics distinguish each region from others. What characteristics are important? That depends totally on your reason for distinguishing regions. A hydrologist would look first at the watersheds drained by particular rivers: each watershed is a separate region. An anthropologist might distinguish regions by the dominant language spoken or the dominant religion practiced. For our purposes, the area over which the same hourly or annual wage is required to hire a particular type of employee, or the area across which a facility's workers commute daily, would be ways of delimiting uniform regions. A **nodal** region is an area that shares a relationship to some key location. In a mid-twentieth-century metropolitan area, the central business district formed the core of the region. It contained the densest use of land and commanded the highest land prices and rental fees (Figure 1.3). Land-use density declined with distance from the center, until the metropolitan area gave way at its boundaries to rural uses. Rail and road lines converged in the center, adding to its centrality and congestion (Figure 1.4). While metropolitan areas in the United States stopped looking like this decades ago, this is still a powerful image of a nodal region. **Administrative regions** are very important for our purposes because they are the principal regional type for which government information and statistics are published. School districts, city boundaries, states or provinces, and nation-states all describe administrative regions. These regions may or may not coincide with important uniform regions (e.g., of wage rates) or nodal regions (e.g., of metropolitan areas). Nonetheless,

Figure 1.2 Western New York State

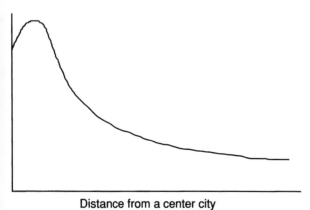

Distance from a center city

Figure 1.3 Hypothetical relationship between land values and metropolitan location

locational analyses often use them because of data availability.

Our simple statement requires defense of its **units of measurement**. Is the "plant" the best unit to use for your purposes? What about the number of companies? What about tons of output? Or employment? Again, the researcher or analyst must understand the purpose at hand, and the explanation being attempted, to be able to defend or change this unit of measurement. How do you tell when one "plant" ends and another begins? In Lackawanna, New York, the Bethlehem Steel Corporation operated many different mills and plants – all interconnected, all next to one another, all part of what we generally call "Bethlehem Steel." Finally, before using the information in our simple statement, we

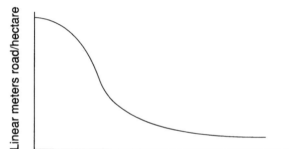

Distance from a center city

Figure 1.4 Hypothetical relationship between road density and metropolitan location

must subject it to **verification**. Is there some way you can check the observation? The observer might have misclassified or mismeasured. In surveys, the respondent might lie ("we may relocate because of the high taxes"). A next step after description of objects is the description of systems of objects. A **system** is a meaningfully bounded set of elements and the relationships among them. The steel-making industrial complex of Western New York included plants, transport networks, workers and their commuting patterns, law firms and shipping firms, etc.

Because it's difficult to keep all that detail in our minds, we abstract from the actual system, and develop models. A **model** is a simplified representation of a system. Think of a map, a globe. How do you decide what detail to omit? It depends on your purpose: think of your varied uses for road maps versus topographic maps versus political maps. Are you interested in the movement of water through the Western New York steel complex, or the financial linkages? How do you decide where the system "ends"? It depends on your purpose: a map of New York State versus the United States; the immediate material and financial linkages to steel plants, versus the material, financial, and governmental linkages of the things that are linked to steel making.

One approach is models. Models can be informal: logical, coherent, perhaps verbal, perhaps like your career plan. Alternately, models can be formal: clearly laid out, so that this connects to that at a particular point. A formal model can be verbal, graphical, or mathematical. Above, our description of a mid-

twentieth-century metropolitan area as a nodal region was a very simple verbal model of the metropolitan area. Two elements of that model, the relationship between land values and location and the convergence of railroads and highways in the central area, are depicted as graphic models in Figures 1.3 and 1.4. A very similar relationship, between land value and its location in the metropolitan area, can be described by the mathematical model

$$V = p - aD \qquad (1.1)$$

where V indicates the market value of an acre of land, p indicates the value of an acre of land in the densest, most expensive part of the central business district, a is the rate at which land values decline with distance from the center of the metropolitan area, and D is the distance from the central business district to the piece of land that interests us. V is the **dependent variable**, whose expected value is known as soon as we know the values of the other variables and parameters. The values of the **parameters** p and a will differ across metropolitan areas, and need to be measured empirically by discovering the value of central land, and (for a) by measuring the relationship among land values in different parts of the metropolitan area. The model's key contributions to our description are in naming these parameters, along with the **independent variable** D, the one piece of information we need to know about any piece of land whose value we want to estimate, and in specifying the relationships among these considerations. In this case, the minus sign specifies that we expect land values to decline away from the center. The multiplication of a and D specifies the way in which distance affects value – at a constant rate. Once we've done that and have specified these two parameters, then the model can be used to predict the value of any other acre of land in the metropolitan area. We only have to know how far the land is from the center of the area.

Such a simple mathematical model has many drawbacks. For one thing, the equation above describes a linear relationship between V and D, while the relationship may not be linear, as in Figure 1.3. Other considerations will affect V in addition to D, such as the acre's suitability for construction or agriculture, or the proximity of other activities that reduce or increase the value of nearby land. Finally, we must have some measurements of V and D for land across the metropolitan area we want to study, in order to find the relevant values of p and a. The strength of this simple model is twofold: the ease with which it and its interactions are understood, and the relatively small amount of data it demands to yield useful information.

How Do We Explain What We're Studying?

Note that a model, whether verbal, graphic, or mathematical, can be very useful in describing relationships among characteristics of places. However, the most general classes of models say relatively little about why these relationships hold. We need to know why, among other reasons, so that we will know when these relationships will not hold. For example, land values decline away from the center of a metropolitan area because the transportation system converges on the center, making the center the point of lowest transport costs. Entities in competition to rent or buy valuable land will bid up the prices of land, until the savings in transportation cost are built into the higher prices of land as we approach the center. With this understanding, we know that our simple models of Figure 1.3 and Equation 1.1 should not be used if land uses are allocated without market competition, or if the transport systems do not converge on the center of the region. It is possible to include these explanations of causality within our verbal presentation of a model, so that the model becomes a theory.

Let us define **theory** in very simple terms: a general explanation of why a set of events might occur. In the case of this text, theory entails the explanation for location of industrial activity at particular places, in such a way that the explanation can be used in a range of industrial and locational contexts. We will build two types of theory in this text. **Normative theory** tries to explain location on the assumption that some beneficial goal is desired

by the actor. Neoclassical economics assumes that producers are trying to maximize profit and that consumers are trying to maximize utility. Given this desire by firms to maximize the spread between their costs and their revenues, and given this expectation that purchasers of goods and services will attempt to purchase the best good or service at the lowest price, we will attempt to understand the interplay of input costs, transportation costs, and market accessibility. **Positive theory** attempts to explain without such clear-cut, optimal goals. We will build positive theory by understanding the ways in which enterprises react to stimuli such as competitors' actions, or changes in the labor market, or the product market for their goods.

We will build theory in a combination of ways. **Inductive theorizing** entails generalization from observation. For example, if we study 30 plant closures in New York State between 1970 and 1980, we may find that 24 of those plants faced declining sales for at least six months before closure. From a survey of 300 manufacturing plants that operated throughout the decade, we learn that 50 of them had periods of at least six months of declining sales. (See Table 1.5.) The proportion of closed plants that had a six-month decline before closing (80 percent) is much higher than the proportion of surviving plants that suffered six-month-long lean periods (16.7 percent). We can conclude from our observations that declining shipments are a cause of plant closures. We can also build theory from **deduction**. This entails a series of logical statements that lead us to an understanding or conclusion about the causes of an industrial location decision. For example, if we assume that enterprises want each plant to be profitable, there would be no reason to open a plant unless it is expected to be profitable. Similarly, there is no reason to close a plant unless it is unprofitable. Since profits can be simply defined as revenues minus costs, plants that have closed must have placed lower revenues over time or higher costs over time. We could then assume that the 24 plants that closed faced some combination of lower revenues and higher costs over time.

Table 1.5 Hypothetical example of empirical research findings

During 1970–80, did the plant face 6 or more consecutive months of declining sales?	Manufacturing plants operating during 1970–80		
	Still operating	No longer operating	Total
No	250 (83.3%)	6 (20%)	256 (77.6%)
Yes	50 (16.7%)	24 (80%)	74 (22.4%)
Total	300 (100%)	30 (100%)	330 (100%)

Positivist science combines both inductive and deductive theorizing. Positivist theorizing entails the deduction of hypotheses from basic logical statements. A hypothesis, such as decreased revenues or higher costs over time leading to plant closure, is then investigated using actual observation. Our current example would lead us to investigate the history of sales and of cost for all 30 of our New York State plant closures. If our search of plant records yields a finding that most of the closed plants maintained constant sales or shipments and constant costs in the year or two before closure, we cannot support our deductive expectation by the information at hand. We would also be interested in the sales and revenue figures of similar plants that remained in operation through the period. Note that this lack of support is meaningful only if the variables which we measured are reasonably drawn from the logical statements concerning profits and plant closures, and have been measured reasonably well in our investigation of plant records. If we can judge that the investigation of plant records yielded defendable measures of input costs and sales from the plant, and if indeed these variables are appropriate ways of measuring or defining profits, revenues, and costs, then our negative finding is a sign that our deductive logic is incorrect. Perhaps plants are closed for reasons other than insufficient profits. A whole host of other reasons are possible, and must now be investigated.

Structuralism is another way of developing theory. It relies on underlying, usually abstract structures as the causes of observed phenomena. Structuralist theory in the social sciences usually relies on historically created social structures – groups of people and the relations among the groups – as the cause for human behavior. Social structures and relations include economic classes, clans, castes, nationalities, gender relations, systems of ideology, the state, political alignments, the distribution of wealth, power, and property, and so forth. Economic structuralists generally trace behavior to economic structures. One dominant structure that is studied is the structure of people's relations to production: producers, workers, and consumers. More specifically, Marxist structuralists study the relation of people to the means (the machinery and technology) of production. The forces of capital (owners, top managers, shareholders) and the forces of labor (workers, who own neither the tools/offices/institutions in which they work nor own the output that they helped create) struggle. They struggle for work hours and work conditions, for wage levels, and for control of production. They struggle in the political arena (voting), economically (wage negotiations), and geographically (capital movement to areas where labor is more malleable, more controllable). Note that this structure, the capital–labor relationship, is abstract. Many actual people represent *both* capital (in that their wealth or retirement pension is bound up in the ownership of stock shares or some item that produces goods/services) *and* labor (in that they rent out their labor time for a wage or salary, and own neither the office/institution they work in nor the output that results from their work). It's not individual people we care about. Rather, it's the tendencies, the needs, of capital versus labor. Thus, one structuralist explanation of industrial locational change might entail the tendency of capital investment to reduce the power of (the returns given to) labor – by relocating part or all of operations to a labor-surplus location, or moving to a setting where labor is less accustomed to industrial relationships (and therefore less organized or militant), or by reducing industrial dependence on labor (via technological change).

Research entails a journey from a general understanding of relationships (a theoretical framework) to attention to a specific empirical or logical case, to the adding of specific insights to the general understanding. Without theoretical reasons for these expectations, without a theoretical framework, your investigation would have to entail ad hoc comparisons. Your findings would tell you nothing about situations other than the one you investigated, because you would not be able to understand or to defend the larger, more general importance of the differences you measured.

SUMMARY

This book is about the location of fixed, productive investment, especially in manufacturing facilities and service sites. One of the most challenging aspects of industrial location is the commitment that fixed investment represents to a given product, technology, and place, in the face of great uncertainty about the future of markets, products, and their geographic distribution.

One basis for our study of industrial location, which we have called the narrow and economic "lens" of investigation, will entail characterizing the activity studied: its labor market, production technology, markets or users, and perceived benefit from proximity or distance from suppliers, clients, competitors, sources of finance. Among the terms used in this characterization are "industry," "enterprise," and "strategy." In complementary fashion, we will characterize places: their sources of labor, materials, information, finance; the size of their markets; the cost of serving the markets; and the nature of government regulation. In this chapter, we have begun to clarify the many meanings of "place," by noting the different scales at which and reasons for which we might delimit one "place" (local region, larger region, country, or supra-national region) from another. This microeconomic study must recognize the extent to which the presence of an additional facility may modify the characteristics of a given place, and the extent to which a given facility could modify its operations to meet the exigencies of a given place.

Another, broader basis for our study will entail a historical overview of the economic, social, and political forces that have shaped contemporary patterns of industrial location. We will present some attempts to make sense of historical developments, relying on the changing nature of production and general production systems, but also on demographic change and on the effects of political and economic interaction across places.

Our concern for the determinants and patterns of industrial location stems from its importance in determining the efficiency and profitability of industrial activities, and from its importance in the social conditions and economic well-being of people where they live. We have emphasized that the macroeconomic results of microeconomic actions appear largely because these microeconomic actions – investment in a facility, migration of labor, orders from a producer – occur in particular places.

Because our goal of understanding industrial location is important and complex, we have ended the chapter with a vastly simplified set of guidelines for pursuing the kinds of research that have led to the generalizations and interpretations that follow.

SUGGESTED READING

Johnston, R.J. (1986) *Philosophy and Human Geography: An Introduction to Contemporary Approaches*. London: Edward Arnold.

Schmenner, R.W. (1982) *Making Industrial Location Decisions*. New Brunswick, New Jersey: Prentice-Hall.

2

LOCATING TO MINIMIZE COSTS

We will now turn to a very straightforward question – the location of a particular facility, in which a firm will invest for a specific productive activity. We will approach this question in a normative sense, making assumptions about the goals of the firm. In this chapter, the goal is assumed to be minimizing the cost of operating the facility and selling the product. In Chapter 3, we will assume the related but distinct goals of maximizing the amount of the product sold and of maximizing the profit from the fixed investment. Government-owned and -operated facilities may have yet other goals: to ensure that every citizen is within 100 miles of a particular type of government facility, to maximize interaction among government facilities, etc. Regardless of the goal, the firm or other enterprise is concerned with the investment and with its location only as a means toward that goal. Our question: how can the location of the investment affect the realization of the goal?

COMPETITIVE ENVIRONMENTS

In traditional microeconomic theory, the firm is defined as a productive unit which seeks to maximize profit through production and sales. How this is accomplished depends on the number and nature of the firm's competitors (competitors being defined here as the operations of other firms in the industry). Basic economics recognizes three types of competitive environments. To discuss the actions of a profit maximizing firm, including its locational actions, we

must investigate these three different competitive circumstances under which the firm can find itself.

Perfect Competition

The first circumstance is that of perfect competition. Under such conditions, the firm shares a production function, or a mix of inputs and products with all other firms in the industry. Indeed, the industry can be defined as that set of productive actors which share a technology and a set of products. Given this uniform set of inputs and products, the main differences in the costs faced by firms result from differences in input costs and logistics costs for each firm. Because the firms produce identical products, the market is indifferent to the products of any of the firms. As a result, there is a fixed price in the market which varies only with aggregate level of market demand. Each producer faces the same price for its output. No producer can greatly influence the level of output in the market. Therefore, no producer can increase price by restricting output, nor lower price by increasing output. However, an increase in the aggregate amount of products supplied to a given market, with no compensating increase in market demand, could result in a lowered price facing each producer. To maximize the difference between total revenues and total costs facing the firm, the firm produces as much product as it can sell until and unless its costs rise with increased production.

Why might costs rise with increased production? In the short run, a firm's facilities are established for a particular level of output. Once that level is neared,

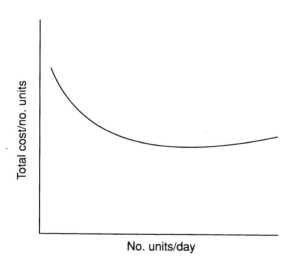

Figure 2.1 Short-run average cost as a function of output levels

capacity shortfalls arise. At output levels below that short-run capacity, however, a firm can benefit from increased production. Thus, Figure 2.1 can illustrate short-run average cost as a function of output level, for the individual firm (which operates one facility).

How does facility location fit into this picture? The firm that has located its facilities in such fashion as to minimize its total costs of production (where these costs include procurement and assembly of inputs, on-site production and manipulation of inputs, and dissemination or distribution of products to markets) finds its marginal cost curve steeper, or its minimal cost lower than its competitors which have not so minimized costs. This increases the level of profit received by the firm. Thus, under a constant and uniform price facing firms, their greatest competition is on the basis of cost. These costs are affected by location of the firms' productive facilities as we will see below.

Oligopolistic Competition

In some industries, and in some market settings, individual firms produce sufficient quantities, relative to the size of the market, that their level of output influences the price that the goods will receive in the

market. In this case the firm faces a demand curve that resembles the demand curve of the industry as a whole. As the firm increases output, the price it receives for the marginal units, the last units produced, falls. However, just as the firm's market price varies with its level of output, the firm's market price varies with the level of output of its large competitors. This sets up a fundamental uncertainty and instability on the part of the profit-maximizing firm. What are the likely actions of its competitors? What are the likely reactions of its competitors to any change in output level? Given the spatially circumscribed market areas facing many products (especially service products), and given in some cases the small level of aggregate demand in a limited market area, we find this oligopolistic competition very common in reality. Oligopolistic competition adds importance to the location of a firm's present facilities and the location of a firm's competitors' operations. If a competitor locates a facility near your operation, your price falls. If a competitor finds a lower cost location than you have, from which the competitor can serve the same market, your price falls, or your market share falls. This condition of oligopolistic competition leads to several potential locational outcomes when we are concerned with more than one facility belonging to more than one enterprise. We will inspect some of these location outcomes in Chapter 3.

Monopolistic Competition

A third condition in which a profit-maximizing firm can find itself is called monopolistic competition. In these circumstances the output of a given firm is not identical to the outputs of its competitors. This could be because of particular quality differences, size differences, trademark differences, or reputation differences. On the assumption that there are purchasers who prefer each of a varied set of product characteristics (some purchasers prefer large sizes, or particular trademarks, or particular levels of quality, or are willing to pay a premium for reputation), each firm can charge a premium for its product. Each purchaser has a preference for a given product's characteristics and will pay more to receive a product with those exact characteristics. However, these products are similar to one another; if the price facing the purchaser is too high the purchaser will indeed substitute some competitor's product. This fungible or limited monopoly is reflected in the oxymoronic phrase "monopolistic competition."

Under these circumstances, location affects competition in two ways. The monopolistically competitive firm has some leeway in its location. The price premium that it obtains for its slightly unique good allows it a wider range of operational options, including production locations. The production of trademarked or high-quality goods often occurs in more central and therefore more expensive locations, in part to allow a faster reaction to market needs and changes. Examples abound – from the lavish central business district offices of prestigious law firms to the Paris, Rome, and New York loci for high-fashion clothing production.

In addition, location itself may be a source of monopolistic advantage for a producer. One very strong reason for consumer preference of a particular product is convenience or cost in obtaining the product. Thus, a producer serving as the only source of products for a given geographic market has a monopoly within that market. It can thereby influence, even control the price charged for that product. However, if there are other potential competitors producing similar products but at a distance, price in the monopolist's market may rise to a point where purchasers acquire products from a distant producer at a higher transport cost. Whether the distant producer charges the purchasers directly for transport costs or embeds the additional transport cost in the total price charged, or whether the purchasers arrange transport themselves, this may be beneficial for the purchasers if the monopolist's price has been raised to a high enough level. Note that transport costs to market play a large role in this second way in which monopolistic competition affects location. This is a more important characteristic in industries where transport costs are a substantial part of the final purchase price, such as ready-mix concrete or bottled water.

A simple example of monopolistic spatial competition should suffice. The retail market for automobile fuel is a good example, because gasoline of a given grade is often assumed to be the same from most dispensers. Because gasoline is expended in the process of driving to buy gasoline, it makes little sense to drive far just to fuel one's car. Thus, each gasoline outlet has a small monopoly around itself: the market of passing cars that need fuel, that are closer to it than to a competitor. However, if the outlet's owner attempts to take advantage of this monopoly by raising its price above the level of other outlets, its effective market area will shrink as it becomes cheaper for some motorists to drive to another outlet. If the price-gouging outlet raises its price to make it worthwhile for a motorist to drive from that outlet to the nearest outlet offering a lower price, then the outlet loses all of its market except for emergencies.

In Chapters 2 and 3 we will investigate models of industrial location for profit-maximizing firms facing each of these three competitive circumstances. In subsequent chapters, we will relax the assumption of profit maximization, and pursue other normative or non-goal-oriented forms of enterprise operation and facility location.

A SIMPLE COST-MINIMIZING LOCATION MODEL

Our exposition of this model relies heavily on the writing of Alfred Weber, who wrote *Theory of the Location of Industries* in German in 1909. The work was translated into English in 1929, and has been an important starting point for industrial location study since then.

This is a normative model, based on an assumption of a particular goal. In this model, the goal is simple: the decision maker must choose a production location that will minimize the costs of transport and production. Implementing this simple goal entails several assumptions about the decision maker. First of all, production is defined as the conversion or modification of physical components, using mechanical equipment and energy, human labor, and known technology. No value is realized from this activity (and no actors are paid) until the products are obtained by purchasers. This definition is most useful in the production of manufactured goods, and this model is most widely used to explain manufacturing location. However, the parts of the model that portray the influence of labor availability and cost ("Step 2"), as well as the benefits to be gained from agglomeration of similar activities ("Step 3"), are very relevant to facilities whose output is a service rather than a good.

Given this definition of production, we can add four very strong assumptions, below. These assumptions make the model a simple, though somewhat unrealistic, way to understand the influences of cost minimization on facility location.

1 *The combination of inputs required is known and fixed.* In the simple model to be presented first, production is seen as a simple additive process. Production "recipes" are assumed to be unambiguous and to allow no deviation from the standard combination of factors.
2 *The locations of inputs are known and fixed.* We assume that the sources of the components and factors are also unambiguous. This includes the people who will work in the facility: we assume that the workers identified with each location remain there – or, if anyone relocates, (s)he takes on the characteristics of labor in the new location.
3 *The prices of inputs, including transport services, are known and fixed.* This way, cost minimization can proceed without uncertainty about the costs. Specifically, this assumption suggests that all the workers a facility could possibly need are easily available in one local area, so that industrial location has no effect on a region's wage rate. We could say that the supply of labor is inelastic, in that its amount does not respond to wage level changes, and wages do not respond to its scarcity.
4 *Consumption locations are known and fixed.* The amount of product demanded does not vary within the likely price range: thus, the producer can make

a location decision assuming a particular level of output (and purchases). If demand were price elastic, then the producer would have to know its costs (which vary by location) before it could know how much it can produce.

Step 1: Minimize Transport Costs

Given these assumptions, we attack the question of cost-minimizing location in three analytical steps. All three steps are required to accomplish the cost-minimizing good. In the first step, we want to select the location that minimizes the total costs of transport involved in assembly and distribution.

First, identify the relevant sources of inputs (mines, factories, seaports, or warehouses) and the final destination of the product ("the market"). "The market" is a tricky concept, for there is seldom only one purchaser of a facility's output. In most manufacturing industries, it is not economic to build a separate plant for each purchaser or each center of industry or population. The "market" is the center of a market area that is large enough to support an economically-scaled plant. If there are a few, mutually distant market destinations, each of them could be considered to be separate points in the discussion below.

What if there are alternative sources of a given input? Use the closest sites to the market center, unless a far-away source is so much cheaper that it will overcome the additional transport costs. Our goal is simple: find the production location where the total transport costs are minimized. This location is essentially the "center of gravity" of the material-input sources and the market center.

Figure 2.2 illustrates a case of a single market center C and two material inputs, each with a different source (M_1 and M_2). The center of gravity appears to be centrally located at P, our likely production location. However, each of these points (each material source and the market center) exerts a pull of a different strength. That strength is the proportion of the inputs in the production process multiplied by the actual cost of transporting the particular input or product.

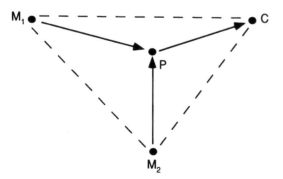

Figure 2.2 The "locational triangle" (broken lines) and the central production location P

Let's propose a very simple production process, combining inputs X_1 and X_2 to yield product Y. Equation 2.1 shows the proportional relationship of these inputs in a unit of Y, by including technical coefficients a_1 and a_2. Unlike a typical aspatial production function where the two inputs might signify capital and labor, we will assume that each of our inputs is a physical material that is used up in the production process, and that the coefficients a_1 and a_2 are in terms of weight of the inputs per weight unit of product. That way, we can compute the total transportation bill for this process as TT per unit of Y, as in Equation 2.2. However, the cost of transporting inputs X_1 and X_2 and product Y is not so simple: it is a product of specific transport rates and distances (see Equation 2.3). Equation 2.4 provides a fuller statement of the transportation costs involved.

$$Y = a_1 X_1 + a_2 X_2 \qquad (2.1)$$

where Y =   one unit of output,
a_1, a_2 =   units of inputs X_1 and X_2 required per unit of Y.

$$TT_y = T_1 a_1 + T_2 a_2 + T_y \qquad (2.2)$$

where TT_y =   the total costs of transportation (assembly and distribution),
T_i =   transport costs for i.

$$T_i = t_i d_{ip} \qquad (2.3)$$

where t_i =   charge to transport one unit of i one distance unit (freight rate),
d_{ip} =   distance from the source of input i or the destination of product Y to the production location p.

$$TT_y = t_1 d_{1p} a_1 + t_2 d_{2p} a_2 + t_y d_{yp}$$

or more generally

$$TT_y = \sum_i t_i d_{ip} a_i \qquad (2.4)$$

One critical complication remains: the d_{ip} variables in Equation 2.4 are interdependent, because each distance is measured from the same production location – the very location that we don't know! Another way of conceptualizing this interdependence is to recognize that the distances are substitutable. We can reduce d_{yp} to zero by locating production at the market center, but then we are farther from the sources of our disparate inputs. Therefore, we need to compute and compare the total costs of transport if the production facility is located at the market center or at the source of any of the material inputs. Locating production at such a beginning or ending point in the logistics stream eliminates all transport costs for one element in that stream, and may well produce an overall minimum. The resultant set of possible total transport costs must then be compared with any intermediate points that appear to minimize the total costs of transport.

The intermediate transport-cost-minimizing location can be approximated by the center of gravity of the relevant points, when each point i is given a weight that corresponds to $a_i t_i$, the amount of input transported per unit of output (equals 1 for the market point) multiplied by the transport rate per unit of i per unit of distance. If we give Cartesian coordinates (x_i, y_i) to each point i, the coordinates of the center of gravity (x^*, y^*) are

$$x^* = \frac{\sum_i (a_i t_i x_i)}{\sum_i (a_i t_i)}$$

$$y^* = \frac{\sum_i (a_i t_i y_i)}{\sum_i (a_i t_i)}$$

However, the center of gravity is the point that minimizes the sum of the squares of the weighted distances from the input sources and the market. Squaring these distances gives excessive weight to particularly remote inputs or market centers. Therefore, this procedure must call on a computer to search for the actual transport-cost-minimizing point in the vicinity of the center of gravity.

Kuhn and Kuenne developed an algorithm to test this point against the conditions for a transport cost minimum. Recall from the calculus that a curve reaches a minimum or maximum point, measured along a y-axis, for example, where its slope is zero (see Figure 2.3). This is the same as saying that a minimum or maximum is where an infinitesimal change in the x value brings no change in the y value, or that $dy/dx = 0$. The curve or function about which we are concerned is the change in total costs of transport as we change the x and the y coordinates of the potential production location.

If there is an intermediate minimum, other than one of the input sources or market centers, that minimum will be where the first derivatives of the total transportation costs with respect to the x-coordinate and the y-coordinate equal zero, in other

words, where $\partial TT_y/\partial x_p = 0$ and $\partial TT_y/\partial y_p = 0$ (see Equations 2.5). We can use these conditions to test for a transport-cost-minimizing point in the neighborhood of the center of gravity, and can have a computer converge on such a minimum.

When TT_y is at a minimum,

$$\frac{\partial TT_y = \sum_i a_i t_i}{\partial x_p \, d_i(x_i - x_p)} = 0$$

and

$$\frac{\partial TT_y = \sum_i a_i t_i}{\partial y_p \, d_i(y_i - y_p)} = 0 \qquad (2.5)$$

The presence of an actual transport network complicates this process, because Cartesian geometry cannot be used. Nonetheless, this simple exposition makes several conclusions possible.

1 If all inputs and final products face the same transport costs per unit of weight and distance, then material inputs that have large technical coefficients (i.e., that are needed in large quantities compared to the quantity of output) have a large pull on the transport-cost-minimizing production location.
   • If we consider only transport costs, we would expect cardboard boxes to be manufactured near the paper mills that produce fiber board. Some box facilities are located in this pattern. However, the cost of transporting finished boxes to clients may create more market-oriented patterns of box production, especially under conditions of location-driven monopolistic competition, described earlier in this chapter.

2 Material inputs or final output that face higher transport costs per unit of weight and distance have a pull on the cost-minimizing production location that is greater than the proportion of their weight in the production process. Material inputs or final output that face lower transport costs per unit of weight and distance have a pull on the cost-minimizing production location that is less than the proportion of their weight in the production process.

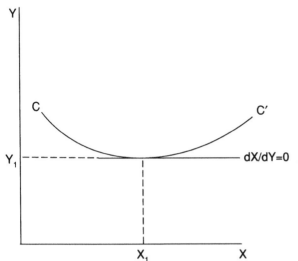

Figure 2.3 Marginal cost and marginal price curves
Note: The minimum Y value of CC′ is reached at (X_1, Y_1), where the slope of CC′ = dy/dx = 0

- Fragile and bulky inputs, such as the glass globes that are made into light bulbs, are produced very close to the process that turns out finished light bulbs. On the other hand, special trains devoted only to coal carriage substantially reduce the per-ton cost of coal transport. Therefore, this bulky, heavy input no longer provides a strong pull on the location of user industries, such as steel making.

3 A nearly ubiquitous material input like sand, water, or air does not have to be transported to the production location. Therefore, its d_i is zero, and it provides no locational pull.

- Industrial planners generally assume that air is available for cooling or chemical reaction in a production process, regardless of the location selected. Therefore, air seldom enters as a variable in location decisions. Any process requiring very clean air, or large amounts of fresh water, may find these resources in short supply, and may be driven to particular sites in pursuit of clean air or abundant water.

4 Material inputs that lose weight in the production process, yielding waste- or by-products, are not fully present in the product to be transported to market. Total transportation costs are reduced by locating production near or at their source, reducing their d_i perhaps to zero. Therefore, such gross inputs suggest a materials-oriented production location.

- Industrial activities that refine or process mined materials (ores, oil) often locate near the mines, or near a water port where the materials can be shipped cheaply.

5 Analogously, material inputs whose total weight remains in the final product – pure inputs such as sub-assemblies for automobiles or computers – provide relatively little locational pull, because their weight is present at both ends of the logistical journey. Market considerations or the availability and cost of immobile inputs will play a larger role in the siting of investment in these industries.

- Automobile assembly plants tend to be located near the centers of gravity of very large market areas, reflecting the relatively finished nature of the components to be assembled, and reflecting the large-scale economies in automobile assembly. Other assembly operations such as electronic goods, yielding products less bulky and less trade-barrier-protected than automobiles, often are given locations where labor wages are low, reflecting the greater spatial variation in wages than in transport costs of assembly and distribution.

6 A production process that entails ubiquitous and pure inputs suggests a transport-cost-minimizing location at the market center.

- The manufacture and bottling or canning of carbonated beverages combines flavor concentrate, water, carbon dioxide, and containers. The locational distribution of bottling plants across an industrialized country looks very much like the country's population distribution. This could not occur if scale economies in beverage production were huge relative to the size of the total market. However, the demand for carbonated beverages is so great that competing companies set up (or license) bottling plants in each large population center.

7 If final products are more fragile or otherwise expensive to transport than the material inputs, they pull the cost-minimizing production location toward the market.

- Production of perishable, fresh foodstuffs generally occurs at the outskirts of leading metropolitan areas.

Step 2: Consider Immobile Inputs

There are critical inputs to manufacturing that cannot be transported, at least, not in the ways considered above. Employees or other forms of labor are the most salient examples. Labor resources are relatively immobile across regions in two ways. First, while workers can commute daily, overcoming longer distances between home and work generally requires a change in residence. Moving is expensive and disruptive. Second, when people do move interregionally or internationally, they take on some of the characteristics of local labor (e.g., wage levels, work hours, work rules). Labor costs vary by location, based on a series of factors such as housing costs, number and rate of increase of potential workers, level of industrialization of the region, and changes in the local economies of regions. However, the assumption made above that labor is immobile implies that a manufacturing facility pays the prevailing wages for labor in a given location. To take advantage of lower wage labor in a particular location, the plant must locate in that low-wage area: "importing workers" is of limited help.

There are other elements of manufacturing costs that reflect the local environment, such as local taxes, general environmental pollution, and local public utility services and fees. These variables can be considered characteristics of places, and unlike physical inputs and products, their *delivered cost is not a clear function of distance from any point*. Recall that a critical feature of Step 1 is the way that the principal locational cost variable, transportation costs of assembly and distribution, increases with distance from each input and market location. If the proposed facility is located closer to one point, it is further from the other points. Because this is not the case for this next set of variables, our simple locational model has to deal with these variables differently from transportable inputs and final products.

The purpose of Step 2 is to determine whether or not it is worthwhile to locate in this low-wage (or otherwise low-cost) area. Because our overall objective is cost minimization, "worthwhile" means a net reduction in the unit costs of production and transportation. A low-cost location that does not coincide with the transport-cost-minimizing location reduces total unit costs if the difference in production costs exceeds the difference in total transportation costs. Using the example of a region of relatively low labor wages, and using symbols similar to those we used above, the low-wage location is advisable if

$$(w_p - w_b)\, a_l > tt_b - tt_p \qquad (2.6)$$

where w_p = the labor wage rate (per unit of labor) prevailing at the transport-cost-minimizing location p,

w_b = the labor wage rate prevailing in a relatively low-wage region b near the transport-cost-minimizing location,

a_l = the amount of labor required to produce a unit of output,

tt_b = the total costs of assembly and distribution (transportation) per unit of output, if production occurs in region b, and

tt_p = the total costs of transportation per unit of output if production occurs at p.

This simple inequality brings into focus the three key variables in determining the attractiveness of lower labor wages: the difference in the wage rates in the two potential locations, the difference in total transportation costs at the two locations, and the labor intensity of the production process. Especially low-wage regions provide a greater pull, everything else being equal. Low-wage regions that are proximate and well-connected to manufacturing inputs and market centers are especially attractive. Low-wage locations are particularly attractive for productive operations that rely heavily on wage labor.

If we recognize a few complications, we begin to understand the actual patterns of industrial location and locational change. First, everything else may not be equal in very low-wage regions. Other immobile inputs (in other words, other place characteristics) may represent higher production costs or may not be available in the low-wage location, such as utility infrastructures. Second, the close juxtaposition of low- and high-wage regions suggests some barrier to their interaction, such as poor transportation linkages or a trade barrier between countries. Third, some activities benefit tremendously from the wage differential, while it represents a small proportion of total costs for other activities, more easily outweighed by higher transport or other costs.

Of course, other types of immobile costs – such as electric power generated by a particular utility, or taxes levied by particular local, regional, or national governments – can be investigated in this same fashion, with a concern for the cost differential, the increased cost of transportation, and the degree to which the production process relies on the input or is affected by the tax. The crucial element of Step 2 is comparing reductions in immobile, region-specific costs to increases in the transport costs for mobile inputs and products.

Step 3: Consider the Potential Benefits of Agglomeration

The actions of other producers in the same industry have subtle but important effects on the costs facing a given producer. Some of these effects vary with the amount of distance among the producers' operations. **Agglomeration** is the term given to a grouping of similar operations. Such a grouping can benefit each operation, in ways that we will present in this section. Because the benefits of agglomeration depend on location within a particular place, these benefits can be weighed against additional transport costs simultaneously with our Step 2 concerns for labor wage rates. The individual decision maker can take the current location of similar operations as given, and can proceed on the assumption that these operations will continue. However, the presence of an agglomeration does depend on the joint actions of individual companies. For this reason, and from the industry-wide perspective (rather than the perspective of the individual company), Weber originally presented the consideration of agglomeration as a third step. (In addition, Weber distinguished between the dichotomous situation of being in or out of a low-wage region and the more spatially continuous variable of proximity to an agglomeration.)

Agglomeration economies represent unit cost reductions in production based on benefits accruing from the proximity of other, similar producers. Why would this proximity matter? Sources of agglomeration economies include shared, specialized infrastructure, such as roads, port facilities, and railroad sidings. A newer, smaller facility, requiring workers with skills specific to the industry, benefits from location near a larger, established facility in the same industry. Workers can be hired away from the larger facility, and/or workers in the area with some experience in the industry, currently unemployed, can be hired by the newer, smaller facility. This reduces

the expense and reduced productivity of training workers in those skills specific to the industry. Note that this motivation for location is not the only reason why facilities in the same industry might be located in the same region. Steps 1 and 2 may lead to this result, given the similarity in inputs and in labor intensity within the same industry. Indeed, the principal reason for dispersion of facilities in the same industry is the need to serve spatially dispersed markets. However, the possibility of agglomeration economies provides an additional reason for spatial concentration of a given industrial activity.

Agglomeration economies are the most difficult for a firm to estimate in advance or after a locational decision has been made. We can provide no neat inequality that has to be satisfied for an agglomeration to be the cost-minimizing location, in part because the concept of "agglomeration intensity," which would be symbolized as a_a, is impossible to measure and is nearly meaningless. Later in this chapter, we look more closely at the sources and limits of agglomeration economies. Much current research is now focused on this influence on production location.

Now we have a simple model of industrial location motivated by cost minimization. Before going on to other possible motivations for industrial location decisions, let's delve more deeply into some of the variables considered in this model.

More about Transport Costs

The costs involved in transportation have several components: packing, loading, the capital expense of the carrier (for vehicles or vessels as well as for right of way or routes), the operating expense of the journey, and unloading. Only operating expenses increase with the distance traveled. The other costs can be considered "fixed," in that they do not vary with distance. In predicting transport costs and in setting common-carrier fares, we can separate fixed from operating costs (so that a freight bill would separately list handling charges, capital-cost allocations, and actual carrying expense per mile), or we can establish cost schedules that are declining functions of distance (so that a freight bill would list the weight and distance, and charge per ton-mile, where this charge was greater for short distances than for large distances). Compare Figures 2.4 through 2.6; note why the presence of fixed costs is sometimes called "curvilinear" transport cost. It is customary to think of and to pay transport costs that are not constant multiples of distance. Truck or rail carriers charge much less per ton-mile for long-distance than for short-distance hauls, reflecting the amount of distance over which to amortize the fixed costs. Relative to rail and water transport, truck and air transport entail low fixed costs and high operating costs (see Figure 2.4), reflecting the smaller loads of trucks and aircraft, the public expenditure on highways, and the free use of air routes. Compare the fixed and operating costs of trucks (fixed cost = OT; operating costs as a function of distance represented by TT'), railway (fixed cost = OR; operating costs as a function of distance represented by RR'), and waterborne freight (fixed cost = OW; operating costs as a function of distance represented by WW'). Trucks are the cost-minimizing mode for shipping distances less than d_1, railways are the cost-minimizing mode for shipping distances between d_1 and d_2, and water transport (where available) is the cost-minimizing mode for shipping distances greater than d_2.

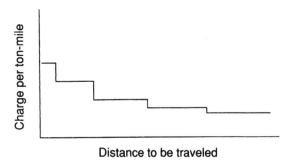

Figure 2.5 Typical freight rate schedule, expressed as the charge per ton-mile for shipping a load various distances
Note: This figure has been drawn to represent average charges rather than total charges, and this declines with greater distances, as the fixed costs of the transportation system are spread over increasing total freight bills

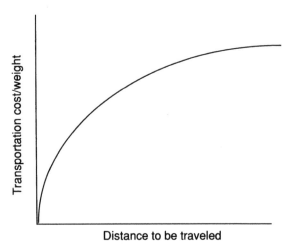

Figure 2.6 Curvilinear freight rates
Note: Idealized representation of total transport costs per weight unit, increasing with distance, but at a decreasing rate of increase: "curvilinear" freight rates

How does this affect production location decisions? If the Step 1 optimum is near one input source or market center, the fixed costs of transport suggest that it is cheaper to locate the production facility at the site of that input source or market center. The slight cost increase from shipping the other items a little farther should be much less than the fixed costs involved in a short-distance movement. Given curvilinear costs of transport, it is usually beneficial

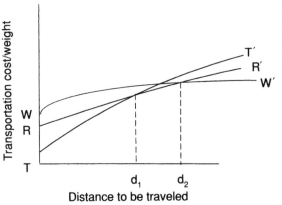

Figure 2.4 Fixed and operating costs by transport mode

to locate production at some (particularly gross) material input source or (given ubiquitous inputs or perishable products) at a market center. This tradeoff is especially important when rail or water transport is being used, because of the high fixed costs of these modes. The widespread use of trucks and aircraft in the post World War II period has decreased the industrial dominance of gross material sources and market centers, and has increased industrial investment in dispersed, intermediate locations.

Loading and handling costs are incurred when materials are transferred from one transport medium to another: ship to rail, rail to truck, etc. It is often cheaper to locate production at such a break-of-bulk point than to find the production location that literally minimizes all distances traveled. Again, the increased use of truck transport has reduced – but certainly not eliminated – the dominance of seaports and rail terminals in industrial investment.

The cost-minimizing model of the previous section implicitly assumes that the industrial producer pays the cost of shipping input materials to the production site (f.o.b. purchase of inputs) as well as paying for the shipping of product to markets (uniform delivered pricing of products). This combination motivates the producer to find the transport-cost-minimizing production location. If some transported inputs are bought at a uniform delivered price, they are ubiquitous inputs for the purposes of facility location analysis. If products are sold f.o.b., then the cost-minimizing firm does not care about its production location relative to the market center. However, the revenue- or profit-maximizing firm would be sensitive to the market's location, as we will see in Chapter 3.

PRODUCTION FACTORS: CAPITAL, LAND, LABOR

Other factors beyond labor wages and government taxes affect the possibility and costs of production in different places. These additional immobile inputs enter our Step 2 analysis of cost-minimizing production locations. Here, we'll present three such inputs: financial capital, costs of land, and aspects of labor in addition to wage rates.

Financial Capital as a Locational Factor

Financial capital – the crediting of accounts that allows individuals and businesses to order inventories, sign leases, and purchase equipment – epitomizes mobility. Financial credits and debits can be transferred, recorded, or sold electronically and immediately. A corporation can raise $20 million in several minutes' negotiation with banks and institutional investors in New York and London, and use that credit immediately to pay debts in Los Angeles and order equipment in Bangkok. The cost of capital, manifested in interest rates on loans or earnings expected to support the prices of stock, does vary across national credit markets, but the mobility of funds allows companies some choice in the credit markets they use for specific purposes. So, under what circumstances is financial capital a locational factor (a source of advantage from producing at one location rather than another)?

An important part of financial markets is information about the company or venture being financed. Information about large firms is widely available. Small firms and new firms, however, are sufficiently known by only a handful of owners, investors, and bankers. While you might buy stock in ICI, Northern Telecom, or Merck & Co., you're not likely to lend money to Davis Feedstocks in Dillon, South Carolina. In obtaining loans, Mr. Davis is somewhat limited to his local bank lending officer, who does have enough information to decide on terms for a loan. If the lending officer decides not to lend money, Davis's choices include borrowing on the equity of Mr. Davis's personal property, or borrowing from family members. In all these cases, long-term relationships, geographic proximity of borrower and lender, and (to a lesser extent) geographic proximity of lender and the actual fixed investment (the plant, shop, or office) determine whether a loan or stock purchase will even be considered. All these sources of funds – loans, savings, re-mortgaging personal property, family borrowing, and venture capital – are more likely to occur where the owner is settled and the investment is located. If the local area has low savings rates, depreciating real property values, and few risk-seeking financiers (venture capitalists), local small businesses find capital scarce and expensive.

Thus, in Weberian terms, financial capital is a Step 2 consideration for small companies: a characteristic of a place, an immobile factor that limits geographic expansion. For large companies, financial capital is infinitely mobile, becoming a ubiquitous factor. For new companies, financial capital is one of the factors that prevents a locational choice at all – entrepreneurial firms locate where the founder is settled.

Land Availability as a Locational Factor

Within any sub-national region, real estate can be purchased or leased at a great range of prices or rents. Prices increase with the land's suitability for construction, the public services available to it (public water, drainage sewers, sanitary sewers, electric power, natural-gas service, etc.), proximity to transportation networks of all modes, and proximity to transportation and employment nodes (urban and suburban commercial cores, superhighway interchanges). The variability of land prices across a region is at least as great as the variation across regions. Therefore, land is seldom a locational concern at the inter-regional scale. Once a firm has determined the basic region of its production operation, land availability and cost become important locational considerations. Again, a tradeoff is sought between the costs of not having access to public services, the costs of remoteness from railways, highways, or other activities on the one hand, and the cost of land parcels on the other hand. The tradeoff will vary with the particular activity, as well as with the scale of the activity: a large industrial operation can justify a large initial, private investment in building water delivery, draining, sanitation, or internal road systems.

In the simple model presented earlier in this chapter, land cost would be considered within Step 2, after the sub-national or metropolitan region is settled.

Labor as a Locational Factor

Labor is a complex and fundamental issue for industrial firms and industrial location. The importance of labor as a locational factor has increased historically for two reasons. First, transport costs worldwide and within countries have declined in real-money terms, because of technological changes (ranging from ever-larger ships and aircraft to containerized, intermodal shipping) and continual investment in infrastructure. This makes it increasingly likely that inter-regional and international differences in immobile costs (such as labor costs) will outweigh the costs of increased total transportation distances. Second, declining employment in agriculture across regions and in many countries has left increasing numbers of people looking for industrial work, and has led more countries to encourage capitalist industrial development. This spread of industrialization to formerly remote regions and countries has introduced very low wages to industrial firms' considerations.

At what geographic scale is labor a consideration? Adequate supply of labor and the prevailing wage or salary level are paramount considerations in international locational analyses, given the vast disparity in these characteristics across countries of the world. In addition, we can define a geographic region among which these labor force characteristics vary almost by definition. This region is the local labor market or LLM. Each LLM is the area over which an establishment could attract workers without the workers changing their place of residence. Within a LLM, wages for the same type of job should not vary, because workers could change jobs without great personal cost. Note that this generalization varies by type of labor (McDonald's restaurants do pay different wages across a metro area, because people don't commute across a metro area for a job in a McDonald's.) Variation across a local labor market for the same occupational category – for the same job – shows that the particular labor market is not truly unified.

Tables 2.1 through 2.3 indicate the disparity of average wages across regions in the U.S., the U.K., and Canada. Table 2.4 uses a group of countries to show the much greater international disparity in average wages. What causes wage-rate differentials across regional and national labor markets? The question has many answers, including supply, demand, and structural considerations.

The supply of labor – the proportion of people seeking employment – reflects cultural, demographic, and economic characteristics, each of which differs across regions and countries. The percentage of working-age residents who are employed (including self-employed) for money or who are looking for work is termed the **labor force participation rate** (LFPR). It is determined by the cultural acceptance of wage labor as opposed to domestic duties or other non-wage activities, by the proportion of households with children (especially young children who are not in school and therefore need supervision), and by the availability of adults past working age to assist with domestic and child-rearing duties. Economic

Table 2.1 Average weekly wage of production workers in manufacturing in selected states – United States

State	Weekly wage 1994 (US $)	Index US = 100
California	510.84	105
Colorado	507.96	105
Massachusetts	516.26	106
Mississippi	388.54	80
New York	492.05	101
North Carolina	405.41	84
Ohio	625.66	129
South Dakota	368.18	76
Texas	476.69	98
United States	485.32	100

Source: Employment and Earnings, U.S. DOL 41 (4), April, 1994 for the month of January, 1994

Table 2.2 Average weekly earnings (pounds sterling) in manufacturing by region – United Kingdom

Region	1989 Male	Index GB = 100	1989 Female	Index GB = 100
North	254.60	96.26	146.60	55.43
Yorkshire & Humberside	245.10	92.67	141.70	53.57
East Midlands	246.00	93.01	141.70	53.57
East Anglia	263.00	99.43	158.80	60.04
South East	303.60	114.78	189.80	71.76
South West	260.50	98.49	156.60	59.21
West Midlands	246.70	93.27	143.10	54.10
North West	253.10	95.69	150.10	56.75
England	266.80	100.87	160.70	60.76
Wales	246.10	93.04	155.10	58.64
Scotland	249.30	94.25	145.50	55.01
Northern Ireland	211.50	79.96	134.00	50.66
Great Britain	264.50	100.00	159.10	60.15

Source: Regional Trends 25, HMSO, 1990, Table 11.5 "All Mfg Ind"

Table 2.3 Average weekly earnings in manufacturing by province – Canada

Province	Average weekly earnings* 7/91 (Canadian $)	Index Canada = 100
Newfoundland	461.25	82
Prince Edward Island	332.19	59
Nova Scotia	484.39	86
New Brunswick	526.91	94
Quebec	518.53	92
Ontario	597.95	106
Manitoba	466.45	83
Saskatchewan	521.46	93
Alberta	484.75	86
British Columbia	614.44	109
Yukon	772.49	137
NW Territories	913.47	162
Canada	562.50	100

Source: STATS Can Employment Earnings and Hours Catalog 72–002, Table 2.2
Note: Canada figure equals average weekly hours multiplied by average hourly earnings, from summary statistics
* Employees paid by the hour including overtime

considerations influence the LFPR, as the availability of jobs – adequate labor demand – encourages people to join the labor force. The cost of living relative to average wages affects LFPR. Many middle-income households maintained or increased real income (income discounted for inflation) despite stagnant real wages in the 1970s and 1980s by increasing the number of jobs held by household members – one person holding multiple jobs or each adult in the household entering the labor force. Any of these considerations that increases the number of people looking for work in a labor market should depress the average wages in the labor market.

In addition, a low LFPR in a region or country with few cultural barriers to wage labor suggests that many people could be encouraged to join the work force if there were jobs available. This reserve labor force reduces the upward pressure on wages in times of increased labor demand: a small increase in average wage level (or, perhaps the mere availability of additional jobs at prevailing wages) will bring forth adequate additional supply. A region or country with low wages and low LFPR can keep its wages relatively low even as economic activity increases. This is what happened in the southeastern U.S. in

the 1960s and 1970s. Per capita incomes rose dramatically, while average manufacturing wages remained substantially lower than in the Northeast and Midwest, because increased industrial activity brought more people into the labor force.

In situations where the LFPR is not unusually low, sustained and increasing demand for labor puts upward pressure on wages. Increased demand for the region's production, or increased government employment can increase labor demand. A high-wage region can maintain and increase its economic activity if its products are unique or if the region provides unique benefits for industrial activity. From our cost-minimization model, these benefits include important material inputs, large internal markets, inexpensive services or utilities, or agglomeration economies. Of course, the region's relatively high wages makes it susceptible to slow rates of industrial investment because of Step 2 wage-rate considerations. Note, however, that expensive transportation services, relatively low proportion of labor expenses in the production process, or local agglomeration economies are all characteristics that reduce the pull of a low labor-wage location for industrial investment.

Table 2.4 Average weekly earnings in manufacturing for selected countries, 1990

Country	Average weekly earnings (U.S. $)
Germany	537.35
Greece	164.78
Hong Kong	138.06
Hungary	45.32
Ireland	380.11
Israel	328.74
Japan	654.80
Korea	206.16
United States	433.20

Source: for exchange rate: Monthly Bulletin of Stats, UN, Jan., 1992; for earnings: Monthly Bulletin of Stats, UN, Nov., 1991. Time factor estimated

In addition to being less pulled to low-wage locations, capital- or materials-intensive industrial activities (such as chemical manufacturing or petroleum refining) tend to pay higher wages. Because such activities use less labor than manufacturing or services on average, the economic return to an additional worker is greater, and the industry wage level is higher. Regions with an above-average proportion of their employment in such industries will have higher average wages, even if industry-specific wage levels are at national averages. Similarly, workers in different occupations earn different average wages and salaries, because of supply and demand, tradition, or unionization. A region with higher-than-average proportions of highly paid occupations will have a higher-than-average wage level, even if each occupation is paid at the national average. In the cases of regional (or national) wage disparities based on differences in industrial or occupational structure, the disparities are not necessarily causes for lower rates of industrial investment in the higher-wage locations.

Differences in regions' costs of living have a direct effect on regional wage disparities, in addition to their effect via the LFPR. Wages must cover culturally accepted subsistence for the individual (if not for the individual's dependents), or else the individual is unlikely to work for wages. A chief component of regional or national differences in cost of living is the cost of land, reflected in the cost of housing. Larger urban areas tend to have higher average land costs, which become manifested in housing costs. Average wage differentials tend to mirror this urban-size pattern.

Notwithstanding the simplistic assumptions of our cost-minimizing model, labor, in the person of individuals and families, is mobile. Inter-regional or international disparities in average wages should motivate migration to higher wage locations, increasing labor supply in those locations and reducing the disparity. The limitations to the motivation and the migration are substantial. Regional wage disparities that reflect cost-of-living disparities should not motivate migration. Disparities that reflect only differences in regional occupational structures should motivate migration only if some member of the migrating household can change occupation in the destination region. Regional (or national) disparities that reflect locational advantages (material inputs, capital availability, large markets, agglomeration) or industry structure (to the extent that workers' training and experiences allow them to change industries) can and do motivate migration.

Migration is limited by information about destinations, by the expense of moving, by the benefits of proximity to family, and by the desire to remain in familiar surroundings. Young adults are more likely to migrate than older working people, because they expect to reap the benefits of higher wages for a longer period. Migration is also limited by governments, particularly national governments which restrict immigration to particular countries, occupations, or absolute numbers. The limitations to migration and the wide variety of causes for inter-regional and international wage disparities combine to allow the disparities to continue.

SCALE, AGGLOMERATION, AND URBANIZATION

When we used the phrase "agglomeration economies" above, we could have used the more formal economic

Table 2.5 Circumstances affecting unit costs of production

Internal economies (or diseconomies) of plant scale: reductions (or increases) in unit costs resulting from operating a large facility at or near its designed capacity.

Internal economies of corporate scale: reductions in unit costs resulting from the competitive power, political influence, or organizational and financial resources of a large company or other organization.

Internal diseconomies of corporate scale: increases in unit costs incurred as organizations become larger, more complex, and less adaptable.

Internal economies of scope: reductions in unit costs resulting from the range of products or activities produced in one facility or one company.

External economies of agglomeration: reductions in unit costs resulting from a facility's proximity to other facilities of the same type.

External economies of localization: reductions in unit costs resulting from a facility's proximity to facilities from which it obtains inputs or services, or to which it sells products or services.

External economies of urbanization: reductions in unit costs resulting from a facility's location in an urban area with:

- general transportation, communication, and commercial facilities or infrastructure;
- wide range of potential employees; and
- wide range of educational, cultural, and residential choices for employees.

External diseconomies of urbanization: increases in unit costs resulting from a facility's location in an urban area with potential for congestion, high wages, high employee turnover.

---

phrase "external economies of agglomeration." Table 2.5 lists a set of similar phrases, each referring to a different phenomenon. Each phenomenon seeks to encapsulate reasons why characteristics of an establishment or its environment tend to increase or reduce its costs per unit of production. Most generally, each of these phenomena is considered to be characteristic of a narrowly defined industry, and thus helps us to understand the differences among industries and their locational tendencies. However, individual firms and facilities can configure themselves to be more or less dependent upon these various economies and diseconomies. We end this chapter with a review of some of these reasons, noting how these phenomena influence the location of cost-minimizing firms.

Internal Economies of Scale

Goods and services can often be produced at lower average cost when they are produced in larger quantities. Earlier, we presented the short-run average-cost curve, suggesting that most facilities have an optimum level of output that minimizes unit costs by taking full advantage of capacity without creating bottlenecks. In the present case, however, we are talking about reducing unit costs by increasing the capacity for which a facility is designed. Thus, the phrase "scale economies" refers to longer-run decisions than the output level of an existing facility. Scale economies are known as "internal" because their sources lie within the facility (or in some cases, within the firm as a whole).

What causes scale economies, and what are their limitations? If every industry exhibited no end to the possible scale economies in production, then the world would be served by one very large plant or office for each industry: one steel mill, one airline, one computer manufacturer, one bakery. Well, there are very large steel mills, airlines, computer manufacturers, and bakeries. These single facilities don't supply the entire world for several reasons: the degree of scale economies varies across these industries; the transportation costs involved in providing all the world's steel or biscuits from one plant overwhelm any unit-cost savings from huge mills and bakeries; national governments may insist on maintaining a national steel mill and airline. These are some of the limits on scale economies.

The reasons for scale economies include labor and equipment specialization, transactions costs, and market power. Let's investigate each in turn.

Specialization

Adam Smith wrote about the division of labor in British manufacturing in the late eighteenth century. Dividing complex tasks (like designing computers, running a company, or manufacturing an automobile) into smaller tasks (like defining customer needs, designing the function of a data storage device, designing the storage device, creating manufacturing guidelines, keeping track of new central-processing-unit technologies) provides several benefits. First, each worker gains greater experience and knowledge about a specific area or task than (s)he could possibly gain about the entire, complex task. This should yield a better end product and higher output per worker (labor productivity) than having each person become "a jack of all trades, yet a master of none." Second, each worker can be defined by her or his task – if that worker leaves, the characteristics or skills needed in a replacement are clearer than if the worker had been a jack of all trades. Third, some sub-tasks require more or rarer skills than others. By defining the employee according to the sub-task performed, employees with rare skills can be paid more than employees with more commonly found skills or knowledge. The firm can discriminate by occupation or skill level, spending less money on wages and salaries than if each worker were paid the amount necessary to obtain rare skills or knowledge.

This makes great sense, especially for the designing of computers. However, when simpler tasks (such as assembling the computers) are broken into fine sub-tasks in which the requisite skills are learned in hours, specialization can yield boring repetition. The results can be mind-numbing and even physically disabling, in the case of repetitive-motion disorders such as carpal tunnel syndrome in meat-cutters or keyboard operators. Every good idea has its limits.

Why, however, is the division of labor an element of scale economies? Task specialization requires a large number of workers, who can be paid only if the facility's sales are large enough to cover their wages, salaries, and overhead expenses. This requires a large facility, with a large marketing and distribution operation. Only with large-scale operations and demand near capacity does a minute division of labor reduce unit costs.

Specialized worker assignments are abetted by, and may be driven by, specialized equipment. One person can accomplish a great deal of carpentry with only three tools, perhaps a hand saw, simple screwdriver, and claw hammer. A skilled carpenter can be much more productive with a full assortment of specialized hand tools. Because of the carpenter's specialization in that livelihood, (s)he can afford to purchase such a full assortment of tools. In this case, task specialization (carpentry as a full-time occupation) allows the purchase of specialized equipment. If a company is established to build houses, its scale of output justifies hiring specialized carpenters, brick layers, plumbers, wall finishers, and the like. Each worker brings to and gains from the job his or her special expertise. In this case, increased output leads to greater division of labor. If the company becomes very successful and begins building whole residential neighborhoods, it will purchase or lease earth-moving equipment, concrete mixers, and computers for inventory and planning. At this point, even greater scale allows even more specialized equipment, requiring an additional set of specialized employees. The division of labor and the specialization of equipment go hand in hand, and each depends on the maintenance of adequate demand for the firm's output.

Why are there firms of different sizes? Specialization is carried further in some industries than in others, either because some industries' tasks are more complex or because the output of some industries can be transported farther to several different market centers, satisfying the requirement for large output. Within a given industry, the huge, low-cost production of large firms employing specialized labor and equipment may not satisfy the entire market. Smaller firms may satisfy demand in smaller geographic markets, or for less demanded specialty products, or for temporary peaks of demand that cannot be satisfied by large firms.

How do scale economies affect the location of fixed investment? Think about the nature of manufacturing: material and service inputs brought together with labor and capital equipment at one point of production, from which material products are shipped to one or more market centers. A large-scale operation requires transportation infrastructure sufficient to handle the large volume of shipments in and out. A small, wooden dock or a narrow highway will not suffice for a very large steel mill. Such an operation requires a large enough resident population that normal LFPRs yield enough employees. In addition, there are requirements for sufficient and stable demand for output that do not relate directly to cost minimization, and will therefore be presented in Chapter 6.

In addition, think about the determinants of cost-minimizing location, encapsulated in Equations 2.1 through 2.5. Recall the importance of the technical coefficients a_i in determining the weights attached to the transport of material inputs and products, and the importance of the labor coefficient a_l in determining the likelihood that a low-wage location would be better than the transport-cost-minimizing location. Now, think of two implications of scale economies for the relative sizes of these coefficients. First, if a larger facility running near capacity is more efficient in its use of materials than a smaller facility, then the locational pull of material sources is less for the larger facility than for the smaller one. To the extent that materials efficiency is one of the contributors to scale economies, we'd expect large-scale operations to be more market-oriented than smaller operations in the same industry. Second, the use of specialized labor and specialized equipment changes the technical coefficients of production. Greater reliance upon specialized or automated machinery

reduces the total labor component of production. A large facility may have more employees than a small facility producing the same product, but the large facility's wage bill is likely to be a smaller proportion of its total expenses. This should reduce the attractiveness of low-wage locations. However, Chapter 6 will present additional complexity regarding labor use, which will help explain the location of highly mechanized, large-scale production facilities in low-wage locations.

Transactions costs and market power

Two other sources of scale economies are very important, though they have less basis in the location of production. All firms require the negotiation of input prices and delivery schedules. All firms require capital financing occasionally, whether in the form of a commercial bank credit line, equipment leasing, construction finance, advance payment for monies receivable, or new issues of stock for major investment. All firms must comply with governmental requirements for reporting and for tax payments on inventory, payroll, and profits. These requirements of all forms loom larger as a proportion of total costs for small firms, because the costs involved in these transactions do not go up as fast as the size of the transactions. The costs are internal to the firms – the cost of a purchasing department does not rise linearly with the amount purchased, the attention that management must pay to secure a £1,000,000 credit line is not much different than that for a £50,000 line, and tax bills must be deciphered, checked, and paid regardless of the amount. In addition, the costs are charged to firms – common carriers pass on their higher unit costs to handle and ship small amounts, and banks typically require loan-processing fees that are a declining proportion of the amount borrowed. The large number of these transactions add up to a sizable penalty on small operations, and sizable economies or savings of large-scale production or firms. In addition to these quite real costs of transacting business, large firms that represent large proportions of a supplier's business, of a bank's lending, or of a municipality's tax base have the negotiating power to reduce their purchasing, capital, and tax costs even further. Each of the suppliers and banks needs a steady source of revenue to justify its own scale and specialized workers and equipment. The large corporate client represents such a source of stability, and it is worthwhile to reduce the prices and fees charged to that client, making more profit on other sales. Municipalities and national governments compete for additional investment by large firms (in part because of the lower transactions costs in trying to attract a single large investment rather than many small investments, and in part because of the rapidly vanishing assumption that large corporations' operations are more stable sources of employment and tax revenues). These large firms can negotiate tax reductions and tax holidays, reducing their unit costs. Such reductions in unit costs based on the negotiating power of large firms are sometimes called **pecuniary economies of scale**.

External Economies of Agglomeration

Table 2.5 emphasizes that agglomeration economies are externalities, or effects on the costs facing firms for which the firms do not pay. Similarly, agglomeration diseconomies are cost increases resulting from the proximity of other, similar producers, for which the firm is not compensated.

Every element of production costs can be affected by the proximity of similar producers. If the producers are willing to share transportation infrastructure or actual material shipments, the transport costs of assembly and distribution can be reduced for each individual shipper. These reductions stem from economies of scale in the provision of transport infrastructures (whether provided by a consortium of producers, or by a government on behalf of its powerful constituents) and in the provision of transportation services. Potential employees, already skilled in the industry, are more readily available in an agglomeration of the industry's activities. This ready availability of skilled employees comes at the additional cost of keeping employees who have other opportunities to sell their skills without incurring

the expense of migration. The net result is especially beneficial to small operations and to newer operations. It is also easier to attract employees to migrate to an industry agglomeration, because of the range of employment options that will face them there.

The agglomeration of electronic component design and manufacture exemplifies these and other characteristics of agglomeration. Highly trained and specialized electrical engineers and production engineers provide vital inputs to these operations. These engineers find great opportunity in agglomerations, and migrate to them after schooling. Job mobility across firms is common.

The enhanced job mobility in an agglomeration is a source of another benefit and cost of production in such a setting: sharing of technical and marketing information across firms. Despite the attempts of firms to protect their technical information and processes by patents and secrecy, basic elements of the information and processes are held by the employees who helped generate them. Employees are free to move on to other employers, with only limited restrictions on the knowledge they can take with them. These basic elements (if not an entire technology or marketing strategy) can be hired away from their original firm. Again, operations of newer and smaller firms are more likely to find these externalities to be net benefits.

If the agglomeration represents a large proportion of economic activity in the region, the policies and taxes of local governments will respond to the particular industry's needs, increasing the Step 2 attractiveness of the region for further investment in the industry. Financial capital may be more readily available in an industry agglomeration, insofar as local bank lending officers and venture capitalists have experience in judging risk in the industry. While very large firms may not be dependent on local bankers and venture capitalists, smaller and newer firms generally rely heavily on such sources of capital.

We've emphasized here that these various sources of agglomeration economies are greater for smaller operations and for smaller firms. In certain industries, it is uncommon for small, independently owned operations to be the first such facility in a large area, while large facilities owned by large companies may indeed survive in an area that has no similar activity. The latter facilities obtain highly skilled workers, technical information, and investment capital from elsewhere within the company.

What are these "certain industries" in which agglomeration externalities loom so large? From what we've presented here, these are industries in which technical information is important to the producers and to their investors. The information could relate to production methods, product development, or marketing methods. What's vital is that the information is not widely available, and is the source of much of the industry's profit. Compare basic cake and biscuit baking for a local market, in which the key inputs are flour, sugars, leavening agents, labor, and equipment, to the production and marketing efforts of a leading international food company. Then, compare the intensity of technology development that precedes and accompanies the development of ever-more powerful electronic components.

A final point: the sources of agglomeration externalities are distinct from the similarity of inputs and markets that can lead similar producers to the same cost-minimizing location. The latter similarity of locational needs and locational decisions can be understood without recourse to externalities, or uncompensated actions of other firms. The operations of an industry that requires large amounts of a bulky, gross, rare, raw material input are likely to gather around the few sources of that material. While agglomeration economies may develop, transport cost minimization led the operations to their locations. Industries that follow agglomeration economies are less likely to have weight-losing, gross inputs to pull them. Rather, agglomeration, as a Step 2 consideration, is more likely to pull an activity that has relatively low materials transportation costs.

External Economies and Diseconomies of Urbanization and Localization

Urbanization

Locating a production facility in an urban or metropolitan area brings net economies if the costs of using shared inputs are less than the costs of purchasing inputs specifically for the facility. Material inputs that are used up in the production process are not at issue here. Rather, general infrastructures such as railway termini, multiple highway interchanges, public port facilities, and services open to the public such as product testing laboratories and marketing consultants can be shared by many clients or provided within the production facility (or its larger company). Urban or metropolitan location also brings external (uncompensated) costs, such as time wasted by traffic congestion, or protecting reduction processes from environmental pollution not caused by the given operation.

Localization

Distinct from the benefits of the agglomeration of a specific industry and the benefits of location in a general metropolitan area are the benefits from having closely related activities nearby. These benefits are termed **localization economies**. These do not include the reduction in transport costs when a facility is near or at its source for an input good or service, or near its chief purchaser: those are basic Step 1 savings. If proximity to industry-specific consulting firms increases the management's awareness of trends or benefits of consultation, or if proximity to potential clients provides useful marketing information to the company, these benefits stem from localization of a set of related activities.

INPUT SUBSTITUTION AND THE INDETERMINACY OF LOCATION

Finally, we must complicate this chapter on cost minimization by noting a very real and very important source of cost reduction: substitution of other inputs for inputs that have become more expensive. Substitution is a fundamental component of economic theory, and of everyday life. If the cost of natural gas rises substantially, households substitute sweaters and home insulation for some of the natural gas they once used. If the price increase is permanent, new houses (and some older ones) will use oil- or electrically-powered heating. If additional demands for financial capital increase interest rates, businesses substitute other means of using existing equipment longer – including assigning more employees to the older equipment, or using more materials, or contracting out certain services (including repair).

This means that our first assumption, early in the chapter, is invalid. The combination of inputs required is not fixed, but depends upon the relative cost and availability of inputs. This poses a great difficulty for our cost-minimization model, because we weighted the transport rates of inputs and products by a_i, the proportion of inputs required for a unit of product. We also relied upon a_l, the labor coefficient, to weight the importance of region-specific wage-rate differentials. What technical coefficients should we assume? One easy response is to predict the average price level for each input over the life of the investment, and to assume a set of coefficients that minimizes costs at those input-cost levels. Then we are left "only" with the impossibility of good prediction, and the cost of basing a decision on the average of fluctuating values.

However, that solution, bad as it is, is insufficient. Part of the cost of using a material input is getting it to the production location. If the facility is far from a particular input source, it could substitute other inputs for the remote one, reducing its dependence on the expensively transported input. In other words, any location is as good as any other, because the input mix can be modified to minimize the total costs of transport and production. Thus, we need to study production processes, not production locations. We can argue each side of this point, below.

We can study location of productive investment. There are limits to substitution, and these limits help to distinguish industries from one another. Micro-computer production cannot be modified to use the same mix of inputs as wooden furniture, or as computer software development. Tendencies in input requirements remain, and these result in tendencies in investment locations. However, the range of possible production technologies within a single industry helps explain why all competitors serving similar markets do not produce in the same location. (Other explanations of spatial competition will be covered in the next chapter.) We must be able to propose locations for investment; otherwise, we should assume that successful industrial location is essentially random.

We must study the nature and determinants of production to understand the location of investments. Except for Chapters 2 and 3, most of this book focuses on the production process and the investment process, understanding the location of investments and the modification of places as a result of these two processes. Much of the most exciting work in current industrial geography begins with production and investment, studying how regional and national differences influence production technologies and investment decisions, and how corporate and governmental production of goods and services affect the regions in which we each live.

SUMMARY

The bulk of this chapter presents a straight-forward framework for dealing with a highly simplified problem: the optimal location of a productive facility to serve a market (or client base) whose location, size, and purchase price are known and fixed. In the process of presenting this framework, we have covered issues of production technology and substitution within a given technology, the nature and determinants of transportation costs, the influence of spatial wage differentials, and the sources and influence of scale and agglomeration economies.

The framework entails the consideration of materials logistics separately from the considera-tion of place characteristics. The materials and products can be transported by incurring a regular pattern of transport and time costs. Place characteristics, including the nature and expense of available workers, the level and type of government taxes and regulation, and the proximity of similar producers, do not necessarily vary in a regular, spatial fashion. In Chapter 8, we will explore ways of combining these varied considerations by creating a check list of characteristics at a limited number of potential locations.

Recognizing key differences across types of production (e.g., across industries and across functions within an industry), we developed a few generalizations about production location based on production characteristics. These generalizations reflect the variations in production technologies and market characteristics across industries and activities. However, once we recognize the possibility that production technologies allow substitution among material, labor, and capital inputs, we find our generalizations of less utility. Fortunately for us, the elasticity of substitution – the ability to respond to an input-price premium with a reduction in the use of that input – is not infinite, and our attempt to understand or propose investment locations based on production technology and the characteristics of places remains somewhat useful.

SUGGESTED READING

Hoover, E.M. (1948) *The Location of Economic Activity*. New York: McGraw-Hill.

Kuhn, H.W. and Kuenne, R.E. (1962) "An efficient algorithm for the numerical solution of the generalized Weber problem in space economics," *Journal of Regional Science* 4: 21–33.

Moses, L.N. (1958) "Location and the theory of production," *Quarterly Journal of Economics* 72: 259–72.

Weber, A. (1909) *Theory of the Location of Industries*. Translated (1929) by C.J. Friedrich. Reissued 1971, New York: Atheneum Publishers, Inc.

3

LOCATING TO MAXIMIZE REVENUES AND PROFITS

ere is a single, crucial difference between this
pter and the previous one. Recall that we began
 last chapter by assuming that each firm sold a
d amount to a particular market center. We
gested that if there were several (known and fixed)
rket locations, we would either consider them as
arate points with their separate pulls on the cost-
nimizing location, or combine them into a mean
rket center. The market was a point, and we were
king to serve the market from a point. In this
pter, we are explicitly concerned about the spatial
tribution of linear or areal markets, thereby
ving a little closer to typical circumstances.

While the analysis in the previous chapter got
newhat involved by the chapter's end, it clearly
ored some very important dimensions of economic
. We retained the early assumptions about invest-
nt decisions: that the level of output was known,
d that the location and the amount of demand
re known. These simplifying assumptions allowed
 to work around the uncertainty underlying all
estment decisions: uncertainty about the future.
 know that nothing, including demand, can really
 known in the future. All investment decisions
tail risks that conditions will change, and the
estor's decision to forgo expenditure (or profit)
w to have greater earnings in the future may bring
at disappointment.

The opaqueness of the future is not our only
oblem, however. In studying or contemplating
 location of productive investments, we have
 concern ourselves with additional demand
certainties. First and foremost, the minute we

recognize that facilities have unique locations and
that there are transport costs of distribution and
consumption, we must recognize that production
location affects the costs of distribution and
consumption. Facilities competing from different
locations relative to the market are not competing
as identical producers, because of their unique
production locations. Therefore, *perfect competition is
impossible as long as transport costs represent a significant
proportion of the costs of production and distribution.*
Without perfect competition, cost minimization, the
focus of the last chapter, cannot be equated with
profit maximization. We thus have to consider the
effects of location on revenues and market demand.
Investment decision makers and those who care
about the economic and social changes concomitant
with industrialization must study the sensitivity of
demand levels to prices, the spatial distribution of
demand, and the spatial distribution of competitors.

SPATIAL PRICING POLICIES

For now, we will focus on the pricing and demand
for goods or services sold across the regions within
a single country. Across national boundaries, pricing
and demand are heavily influenced by international
currency exchange, import tariffs, and international
differences in distribution systems. The on-going
trade discussions between the United States and
Japan provide a case in point, with American
producers arguing that the close relationships among
Japanese producers and distributors make it difficult
for non-Japanese goods to enter the market.

Across a single country, some goods are sold for the same price at retailers or distributors across the countryside, while other goods (and most services) are sold for a basic price plus delivery charges. In the case of services, the delivery charges may be billed by the service provider, may entail the service client (e.g., the retail shopper) to travel to consume the service, or may entail telecommunications charges for services that can be rendered in that format. When the producer absorbs the costs of market distribution, price is not a spatial variable. Then, does it make any difference where production is located?

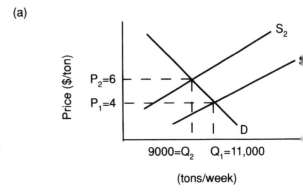

(a)

Uniform Delivered Pricing

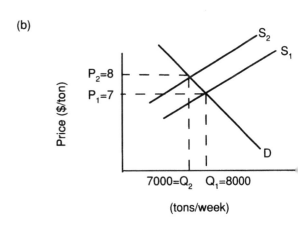

(b)

Yes, because the costs of distribution have to be paid whether or not they are charged explicitly to purchasers. Our first case, **uniform delivered pricing**, implies that the producer includes an average transport charge in the price charged to buyers (the buyers could be industrial companies, wholesale distributors, retail outlets, or individual consumers). Remote buyers are subsidized by buyers near the production location. Pricing is simple, and marketing can proceed on the assumption that the final users (perhaps after several intermediate sales) face similar prices. This pricing scheme is very common for consumer goods. What seems like a happy, if slightly unfair, situation becomes more worrisome for the producer once the producer recognizes that higher prices means fewer sales, even if everyone pays the same price.

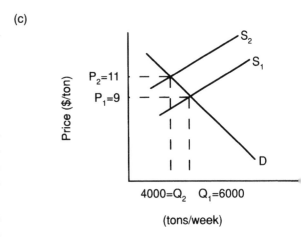

(c)

Price elasticity of demand

The market's response to higher prices is lower demand. Figure 3.1 illustrates this. This response is called the **price elasticity of demand**: the percentage change in demand resulting from a one percent change in price, multiplied by negative one. Equation 3.1 portrays this, by dividing the proportion change in quantity demanded (Q) by the proportion change in price (P). Note that proportional change is

Figure 3.1 Different price elasticities of demand

computed by dividing the absolute change by the average of the value before and after the change.

If $Q_a = (Q_1 + Q_2) / 2$ and $P_a = (P_1 + P_2) / 2$, then

$$E = -\frac{Q_2 - Q_1 / P_2 - P_1}{Q_a P_a} \qquad (3.1)$$

where Q_a, P_a = average values of Q and P,
Q_1, P_1 = original values of Q and P,
Q_2, P_2 = changed values of Q and P,
E = price elasticity of demand.

The negative sign in Equation 3.1 changes what should usually be a negative number (because price and quantity generally move in opposite directions) into a generally positive number. The degree of price elasticity of demand varies with the price level of the same product, and varies across products and markets. Values between 0 and 1 are said to describe low elasticity or *inelastic demand*. Values greater than 1 describe *elastic demand*, and elasticity of 1 is called *unitary elasticity*. Figure 3.1(a–c) illustrates these values. If a firm faces an inelastic market, the firm can raise prices slightly and gain more total revenue, because the negative demand response is smaller than the price increase (Figure 3.1a). If a firm faces an elastic market, a price increase reduces total revenue (Figure 3.1c), while a price reduction would increase total revenue. Figure 3.1 also shows that price elasticity of demand varies along a linear demand function, in which the quantity demanded is a fixed multiple of the price. This may or may not be realistic, and market analysts generally rely on other shapes of demand functions.

What affects the elasticity of demand facing a single producer? First and foremost, the degree of competition determines the elasticity. Under the hypothetical conditions of perfect competition, each producer faces a horizontal demand function. In other words, price is an externally determined value. Any increases in the individual producer's cost schedule (unit cost as a function of output level) or supply curve result in a reduced output level for the producer (see Figure 3.2). This extreme situation illustrates an important source of demand elasticity:

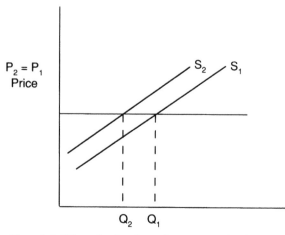

Figure 3.2 Effect of reduced supply on output level

the availability of substitutes for the buyers. Within an industry, competitors produce these substitutes. However, we can talk about the price elasticity of demand facing an industry as a whole. In this case, potential substitutes exist outside the industry. If the price of automobiles (the product of a manufacturing industry) rises very substantially, more people use public transportation (the product of a service industry).

Minimizing potential transport cost

How does elasticity affect the producer who sells under a uniform delivered price? Because transportation is not free, all the buyers are paying an average cost of transport, built into the general price. If the total market is elastic at the prevailing prices, the producer could increase revenues by lowering the price. One way to accomplish this while maintaining the uniform delivered pricing policy is to locate production in the center of the market area, so that the total transport costs of distribution are minimized. This can be expressed as a need to identify the point p that has the lowest *potential transport cost* to market (*PTC*), or

$$\min PTC_p = \sum_i M_i t_{pi} \qquad (3.2)$$

where PTC_p = potential transport cost of distributing product from p

M_i = size of market at location i

t_{pi} = transport cost from p to i, per unit of product shipped.

Various measures can be used to approximate M_i. For consumer products, the population of region i is a good proxy; even better is its population multiplied by its per capita income. For industrial products, M_i needs to be a measure of the industrial base of i, such as the total sales, employment, or value added in relevant industries in i. Note that t_{pi} is often not a fixed multiple d_{pi} of the distance from the potential production location to i, but may be a more complex function of distance, because of the nature of transportation costs and pricing (see Chapter 2). A firm considering a market-oriented facility location needs to calculate the PTC of many potential points p. The resultant calculations should have a pattern, with PTC reduced by the proximity of major markets. Such a pattern is developed in Figures 3.3 through 3.6. The distribution transport-cost-minimizing production location is near the densest market concentration, in the lower left of the figure, but is pulled slightly away by the other two dense markets in the lower right and upper right. Production locations in the upper left of the figure would incur the greatest transport costs of distribution.

This procedure is one part of cost-minimizing location analysis, the subject of Chapter 2. (Compared with Equation 2.4, what cost-minimizing considerations are omitted in the PTC calculation?) We arrived at this procedure via our concern for the impact of total transport costs on average transport costs, which are important under the policy of uniform delivered pricing. Price determines sales volume and total revenue via the price elasticity of demand. We recognize the influence of price level on total revenue (as in Figure 3.1).

F.o.b. Plant Pricing

An alternative pricing policy charges each buyer (which could be an industrial company, a wholesale

10	10	10	20	30
10	10	10	20	10
20	20	10	10	10
30	30	10	20	20
50	30	20	20	40

Figure 3.3 Local market sizes (M_i) across a market area

30	30	30	60	90
25	25	25	50	30
40	40	20	25	30
45	45	20	50	60
50	45	40	50	120

Figure 3.4 Cost of distributing product from local market P (shaded) to each cell i, or $C_t = M_i t_{Pi}$, where $t_{Pi} = 1 + .5d_{Pi}$

1245	1165	1130	1135	1195
1120	1015	975	1000	1115
1045	930	875	955	1080
1025	915	910	965	1090
1075	1005	995	1045	1145

Figure 3.5 Potential transport costs from each local market to all other local markets, or $PTC_p = \sum_i M_i t_{pi}$

distributor, or a retailer) the production cost (which includes the transport costs of assembling inputs as well as a return to all factors including capital) plus the cost of transporting the product to the buyer. This pricing scheme is prevalent in new-car sales: in that case, the final consumer generally pays the cost of transporting the car from factory to showroom. This is often referred to as f.o.b. pricing. F.o.b. is an abbreviation for "free on board," and designates the price charged for a product loaded on a transport

Figure 3.6 Isodapanes of potential total transportation costs, PTC

conveyance. Where is this conveyance? The convention is for the f.o.b. price to include the product and packing, ready to leave the production facility by some mode of transport. More fully, this should be called "f.o.b. plant" (or alternatively "f.o.b. mill" or "ex-factory"). The buyer is responsible for arranging and paying for transportation from the production site or other designated location.

F.o.b. pricing allows transport costs to have a direct effect on demand. Distant buyers face higher effective prices than proximate buyers. If the product's demand is at all elastic ($E > 0$), the amount sold to a distant market will be less than the amount sold to a proximate market with identical characteristics. A producer wishing to maximize sales should locate production at the densest market. This maximizes the effective demand from the richest market, while discounting the importance of demand from distant markets because their demand is dampened by the extra costs they are charged. In other words, the objective is to select the production location with the maximum **market potential**, which is the sum of the regional markets accessible from the production facility, where the size of each market is dampened by its distance from the facility. The rough mathematical expression of this measure is

$$\max MP_p = \sum_i M_i \left(1 - \frac{2E_i t_{pi})}{2P_p + t_{pi}}\right) \tag{3.3}$$

where MP_p is the market potential from production point p, and the other symbols are as defined in Equations 3.1 and 3.2. Equation 3.3 recognizes the need to add the effective demand from all the relevant market areas, and also recognizes the need to reduce the effective demand of distant regions based on the extent to which the costs of distribution will dampen demand. That dampening effect is price elasticity of demand, of course, so E appears in the equation. Note how difficult this equation is to implement, in that the elasticity can vary across market areas. This reflects the different competitive situation and opportunities for product substitution in each region. Estimating elasticity for all relevant market areas makes this a data-intensive operation. Nonetheless, the principle of market potential is important to revenue-maximizing locations under f.o.b. pricing.

In conclusion, if we ignore locational differences in production costs, we are left with marketing reasons for locating facilities. If the product is to be sold across the entire market area for a uniform delivered price, we want to find a production location in the center of the market area, allowing the largest markets to provide the greatest pull on this location. If the product is to be sold f.o.b. plant, we want to find a production location that is within the densest market, because the pull of large, distant markets is reduced by their distance.

SPATIAL COMPETITION

Competition entered the previous section only tangentially, as a potential explanation for price elasticity facing the individual producer. Once we recognize that the costs of transporting to a market vary according to location relative to the market, we understand that competition among firms can be based on production locations as well as on price.

One of the most famous statements of competitive location relative to a market was published by Harold Hotelling in 1929. It has been a point of departure for many discussions of this subject, including ours. Hotelling's concern was actually any form of monopolistic competition, not only that caused by the costs of transportation. For example, if tradition and campaign expenses result in a very limited number of political parties, a citizen will vote for the candidate whose political stance is "closest" to the voter, or will not vote. Hotelling developed his statement with a locational analogy, however, which will be our focus here.

Hotelling's assumptions are crucial to his analysis and its outcome. These assumptions were:

1 The market is a set of customers who are uniformly distributed.
2 The market is a one-dimensional line (length = 1 in Figure 3.7a).
3 Two producers serve the market (A and B in Figure 3.7a), though either producer is capable of serving the entire market.
4 The producers price their product f.o.b. at their facilities.
5 The cost of production is zero for each producer, regardless of location.
6 The market's demand is totally price inelastic ($E = 0$); each customer will buy x amount of the product daily, regardless of f.o.b. price or delivered price.
7 Each customer is ambivalent about the output of the two producers, and will buy x amount from the producer whose delivered price (f.o.b. price plus transport cost) is less. (Hotelling implicitly assumed that transport cost was the same function of distance everywhere on the linear market.)

Given these assumptions, the amount purchased from each producer (x_A and x_B) depends on the relative prices and locations of the producers. In Figure 3.7a, all customers on segment a will buy from A, and all customers on segment b will buy from B, unless A or B sets its prices so high as to lose preference by any customers, no matter how close. There is an equilibrium set of prices which maximize the profits ($p_A x_A$ and $p_B x_B$) of each producer. If we allow the producers to relocate freely, A will select a location near B, to increase the size of a. B will retaliate by moving to the left of A, creating a monopoly for itself that is larger than the monopoly A controls. The stable locations for the producers are very close to one another in the center of the market (Figure 3.7b).

Hotelling made two important comments about this result. First, given the assumptions, the total costs of transportation paid by all the customers would be minimized if the two producers were located at the quartiles of the market, so that no customer had to pay for transport over a distance longer than $l/4$. If the producers are in competition, however, this is not a stable condition. Therefore, under Hotelling's assumptions, the competitive equilibrium does not maximize the welfare of the consuming public. Second, the market l could represent any dimension instead of distance: the political ideology provided by two competing political parties or the sweetness of apple cider provided by two competing cider mills. He generalized that his model explained much of the sameness in everyday life.

Hotelling and many other writers have commented on the restrictiveness of the assumptions, and the effect of changing the assumptions. Indeed, Hotelling's statement is most useful for what it suggests about production location in situations that do not conform to its assumptions. In the following paragraphs, we present the counter-arguments to some of Hotelling's assumptions.

Concentrated market locations. If the market is not uniformly distributed, spatial competition and customers' preference for the lowest-delivered price will draw the producers toward the market

(a)

(b)

Figure 3.7 Hotelling problem: spatial competition in one dimension

concentrations. As we noted above, f.o.b. pricing will lead sellers to the densest part of the market. Note that the relevant measures of market size and density should reflect not only the location of households or establishments, but also the relative likelihood of each household's desire for the item and its relative ability to pay.

Two-dimensional market. Obviously, most human interaction occurs in two or three dimensions, two dimensions being the most relevant for industrial location concerns. The bounded linear market of the Hotelling model increases the isolation of peripheral locations. Note, in Figure 3.8, that a producer at point A (one end of a line segment) has adjacent market "area" in only one direction, while a producer at point B (one corner of a polygon) has adjacent market area in several directions. This reduces the penalty of peripheral locations in a two-dimensional market area, and may increase the tendency toward dispersion of producers when the number of producers serving a market region rises above two.

Greater number of producers. A new, third producer would not locate between A and B in Figure 3.7b. The interaction of a larger number of producers increases their dispersion from the duopolistic (two-seller) situation. An assumption that allowed new producers to enter the market (motivated by the excess profits being made by the duopolists facing totally inelastic demand) would quickly increase the dispersion of producers to approximate the market's distribution.

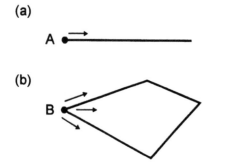

(a)

A

(b)

B

Figure 3.8 Hotelling problem: convergence of producers in market center

Locational variation in production costs. The entire focus of Chapter 2 was that production costs are not zero and do vary across production locations. The greater the locational variation in production costs (whether due to gross material inputs, high labor coefficients, or other factors), the greater the concentration of production at low-cost locations. However, an awareness of spatial competition suggests that some producers can overcome a cost disadvantage with a proximity advantage in certain parts of a distributed market. Such high-cost, market-oriented producers will be less common in a raw material dominated industry, like ore refining.

Elastic demand. Perhaps the most important generalization to be drawn from relaxing Hotelling's assumptions entails the usual case of price elasticity of demand, much more common than perfect inelasticity. This makes the profit-oriented producers much more concerned about the transport costs faced by the customers. The more elastic the demand, the more the producers will disperse to match the degree of market dispersion.

- Therefore, we would expect industries whose products have many substitutes to be more dispersed.

It follows from this that at any positive level of price elasticity of demand, higher rates of transport cost will cause greater dispersion of producers to match the market's dispersion.

- Therefore, we would expect industries whose transport costs are high, especially high as a proportion of total costs, to have relatively dispersed production locations.

Note that some buyers evaluate transport costs not in terms of money spent on transport, but on the time that elapses between placing an order and receiving a shipment. This time-sensitivity of demand increases the dispersion of profit-maximizing competitors across the market.

Differentiated products. Hotelling's seventh assumption means that delivered prices are the only basis of competition in the industry. If the producers' products are differentiated, customers have some reason in addition to delivered price to prefer one product over another. This influences producers' locations in two ways. First, if the two producers maintained dispersed locations (e.g., at the quartiles of the segment), some customers who prefer A's products but who are based much closer to B than to A may substitute B's products. Product differentiation is a form of monopolistic competition, so that customers will pay a premium for their preferred product, but there are limits to the premium (in this case, higher delivered price) that will be paid. Second, A, located at a quartile location, cannot be assured of all the customers who are closer to A than to B, because of the premium some customers are willing to pay for B's product. Both these influences increase the profitability of central production locations, for both producers. Furthermore, if new producers are allowed to enter the competition, the subsequent dispersion of producers is slower under a product-differentiation regime than under pure delivered-price competition.

Much of the foregoing discussion has been phrased to imply constant relocation of competitors' facilities. Given the real-world expense of facility relocation, the principal practical implication of these considerations is the siting of new facilities. Such investment decisions are often made by companies whose current facilities are aging or inadequately sized, and when market analysis suggests that the market could be better served from a different or additional location. In such cases, the location of competitors' facilities are noted. A decision to cluster near a competing facility versus maximize distance from existing facilities depends on the distribution of the targeted market, the market's elasticity, locational variations in production costs, the directness of the competition with other facilities, and the importance of scale in the design and operation of the new facility. Dynamic considerations appear, as well: a fast-growing region with promise to become an important market concentration may attract a company's new facility, even if it is not the lowest cost location, in an industry where transport costs to market are a substantial part of product costs.

MINIMIZING COSTS AND MAXIMIZING REVENUES

We noted at the beginning of this chapter that cost-minimizing location equals profit-maximizing location when competition is perfect. We noted that differential distances from individual producers to a spatially distributed market eliminate the possibility of pure price competition. By reducing the market to a point, the analysis in Chapter 2 eliminated the reality of a spatially distributed market. In addition, however, the analysis in Chapter 2 ignored competition altogether. We attempted to identify the cost-minimizing location for production investment of an individual facility: other facilities entered only in the inspection of agglomeration economies.

Combining the Concerns of Cost Minimization and Spatial Competition

In this chapter, we have focused on the interaction between the firm and its distributed market (spatial pricing policies) and on the interaction among firms competing for market share along the dimensions of price, proximity, and non-price factors (product differentiation). Thus far, we have ignored production costs, as well as the transportation costs of assembling localized material inputs. How can we combine these two kinds of concerns: competition for a distributed market and spatial variations in costs?

Recall the fifth assumption in the original Hotelling model, that production costs were zero for both producers and all potential production locations. Recall our conclusion about the relaxation of that assumption: the greater the locational variation in production costs, the greater the concentration of production at low-cost locations. This is the fundamental conclusion of combining our two concerns to create a profit-maximizing theory of industrial location. If the costs of assembly and production increase so dramatically away from the cost-minimizing location that any competitive benefit from market segment proximity is over-whelmed by increases in average production cost, then facilities serving a very extensive market area will cluster at the cost-minimizing point. This clustering usually has limits: at some distance from the cost-minimizing location, the costs of distribu-tion to the local market exceed the production cost advantage of that location. (However, think of the example of large commercial jet aircraft, in which there are very few competitors with truly global markets. The industry's extensive scale economies and the expense and transportability of its final product unify the international market.)

Now recall a major theme of this book and of industrial location study: we attempt to match the characteristics of facilities to the characteristics of potential locations. One important way of grouping facilities with similar characteristics is through our definition of industries. If production cost differen-tials are vitally important, we already know how to analyze them. Similarly, if spatial competition is the key consideration in location, we know how to analyze those forces towards concentration or dispersion relative to the market. So what remains is to understand industry and environmental character-istics that increase the salience of one or another set of considerations. What industry characteristics entail great locational variation in production costs? What characteristics of potential locations lead to such cost variation? What characteristics of indus-tries and of markets would exaggerate the tendencies toward dispersion, overwhelming the centralizing pull of production cost variation? How might these characteristics vary by size of facility, size of firm, or over time?

Industry characteristics

Industries vary in the intensity of cost variation across their potential locations. Activities that entail the refining of raw, gross, localized material inputs face very substantial penalties in locating away from these material sources. Activities for which a partic-ular type of worker is the chief input have little

choice but to locate within commuting distance of sufficient numbers of such workers: clerical workers for insurance payment and claims settlement, engineers for integrated-circuit design, mill workers for late nineteenth-century textile manufacturing. In addition, the penalty for locating where these workers are available at higher wages than elsewhere is substantial for such activities. These various penalties can be paid in the form of high unit costs, if other considerations mandate – such as a premium paid by customers for a proximate and prompt product. Activities that benefit greatly from agglom-eration face cost pressures to concentrate production. On the other hand, some industry characteristics encourage dispersion toward the market. (Note that spatial competition alone cannot lead to greater or more uniform distribution of production than of the market.) In Chapter 2 we indicated how the presence of a ubiquitous input within the product pulled the transport-cost-minimizing production location to the market. Now that we recognize areal markets rather than market centers, ubiquitous inputs pull competing facilities apart, distributed within the market. The presence or entry of multiple producers of the same product encourage their dispersion within the market area. The very quality of sameness of the product adds to this tendency, while industries that are highly differentiated tend toward more clustering in the densest part of the market.

Environmental characteristics

Especially varied landscapes provide more opportuni-ties for spatial differentiation in costs and revenue potentials, and thus for concentrations of particular industries based on their particular needs. The polar-ized nature of many countries, with a wealthy capital city and a large, poor, rural hinterland, provides firms with a clear locational choice: the capital for the major market, for skilled labor, for agglomeration, and the periphery for agricultural or extractive inputs or workers new to industrial employment. Relative rarity of a critical input – skilled or technical labor, seaport facilities, adequate water – across a particular environment motivates concentration of industry around that input. This suggests a relationship between the geographic scale being considered and the pull of locational cost differences: at the global scale, these differences are tremendous, and pull particularly labor-intensive or resource-intensive activities to certain countries despite the transportation costs to market. This also suggests a relative importance of locational cost considerations within spatially polarized (often newly industrializing) countries.

Characteristics of the market across the area being considered (the world, a country, a sub-area) influence the relative importance of cost versus revenue or competitive considerations. First and foremost, geographic dispersion of the relevant market affects the geography of the locational pull of the market: spatial competition cannot lead to greater dispersion than the markets being served. Second, we have seen that the price elasticity of demand by the market for a particular good or service at the relevant price levels severely affects the pull of dispersed market-oriented locations and the importance of market location to competitiveness. Markets differ in their price elasticity, because of different tastes, needs, income levels, or possible substitutes. If such differences exist across the area being considered, the market with the more elastic demand provides the stronger pull.

Size of facility

In an industry with sizable scale economies per facility, small, high-cost facilities play an interesting role. Their viability depends on some form of protection or subsidy. This could be a governmental subsidy from taxpayers or politicians who want to see a basic economic activity within their jurisdiction. There could be a cross-subsidy within a multi-establishment enterprise with a strategic reason to operate a costly small operation in a particular setting. The cost of transportation also provides protection for small operations. In Figure 3.9, facility

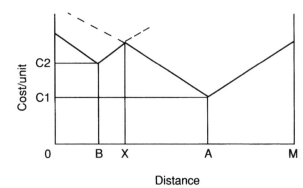

Figure 3.9 Effects of differing producer prices on market areas

A was established first, serving the entire market OM from a central location with an efficient, large-scale operation. Facility B can operate successfully near the edge of this market as a high-cost, small-scale operation, as long as its production costs (OC_2) are lower than the delivered cost of A's product in B's small, peripheral location. Thus, B can satisfy demand in the OX region, which is large enough for its small scale. This illustrates the ability of small facilities to locate in peripheral market areas, dispersing more finely within the market than can larger operations.

A final implication of this line of reasoning is the influence of changing transportation costs. Technological change along with public and private investment continue to reduce transport costs (in real, inflation-adjusted terms and in terms of time) per unit of distance and weight. This reduces the monopolistic advantage offered by distance, and reduces the importance of all our considerations that are functions of distance. These include the Step 1 considerations of our cost-minimizing model, and most of the rules of spatial competition. This leaves Step 2 and 3 considerations of our cost-minimizing model, as well as other considerations explored in upcoming chapters. Over time, we might expect increased production concentration according to industries' varied needs for skilled, unskilled, or technical labor, large amounts of localized energy sources, or agglomeration economies. (Other long-term temporal changes counteract this tendency: for example, the increasing time-cost of transportation in late-capitalist economies.)

Maximizing Profit

So far, we have combined the cost and revenue approaches by juxtaposing them, by noting the tendencies that increase spatial variation in production costs when the amount sold is held constant and the tendencies that increase dispersion of production to match the market when the costs of production are held constant. An alternative way of combining the cost and revenue approaches is to recognize that neither is the objective of the profit-maximizing firm. Such a firm desires a location where the difference between total costs and total revenues is maximized. Given the actual dependence of a facility's cost function (the schedule of cost as a function of output volume) and revenue function (the schedule of sales as a function of price) on location, this problem is intractable. The only way to break the circle is to simplify the problem, as follows.

1 **Decide production scale.** Decide the desired scale of operations first. This can be the scale at which long-run average costs stop declining with facility size. Long-run average costs are the schedule of unit costs for facilities of different sizes, each operating near capacity. The phrase *minimum economic scale* is sometimes used to

describe the size of facility beyond which unit costs do not fall (see Figure 3.10). Alternatively, the enterprise considering the investment may want to establish a smaller operation, because of capital constraints (lack of finances for a large operation), or because of a subsidy that will allow a smaller, higher cost facility to be viable. Our locational question is then, where can a facility of this size be located, given the location of inputs, markets, and competitors, so as to maximize the difference between total costs and total revenues?

2 **Identify a market area to satisfy the scale requirements.** The question can be answered by identifying the market first and then minimizing location-specific costs in the context of the market. If we know the scale of output and the likely average costs (which in fact depend on location), we have a sense of the context required. What is the range of locations from which a market of requisite size can be controlled at that price? This depends on the market area that can be served from a location, given transport costs and demand elasticity to delivered price, on the density of the market within that area, and on the share of the market that the facility can expect to control in the face of existing competition.

3 **Identify a cost-minimizing location within the market area, and make adjustments to scale and market.** Within that range, where are costs lowest? Does the consequent f.o.b. cost allow a larger market to be served at prevailing transport rates? This could result from lower delivered costs and a spread of the relevant market area, or an increase in market share because of a more beneficial cost structure than competitors. What would happen to total costs if the facility size were increased to serve that slightly larger market – would they increase more than revenues? Then do not attempt to serve the larger market. Would they increase less than revenues from the slightly expanded market? Then increase the proposed scale yet some more.

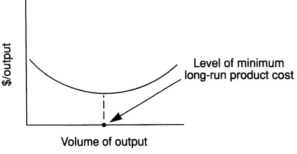

Figure 3.10 Long-run average costs

Combining Approaches

This is a difficult task, to combine the approaches into a truly general understanding of industrial location needs. The difficulties stem from the interlocking determinants of production costs and spatial competition. Perhaps more importantly, there are difficulties in combining considerations that are not strict functions of distance (availability and cost of required workers, tax rates, agglomeration economies) with considerations that we can simplify as distance-dependent (transport of inputs and products).

Even if we found a tractable, unified model of the considerations described in these two chapters, it would have limited applicability in the real world of incremental investment decisions made by complex organizations with limited information about inputs, costs, competitors, and markets. The next several chapters explore the evolution of technology and of investment decisions, and the way in which real companies tackle these intractable yet required questions.

SUMMARY

Chapter 2 made a set of heroic assumptions to equate cost minimization and profit maximization: specifically, that the facilities operate within a context of perfect competition, each producer facing the same product price, which it cannot affect by expanding or reducing output. At the beginning of this chapter, we noted that if some producers are closer to a given input source or a particular market and if transport costs are at all significant in the costs of production and distribution, then each producer was not in perfect competition with others in the same industry. If we add to these spatial concerns the reality of other forms of product differentiation, we recognize that we cannot go very far on the assumption of perfect competition. The rest of the chapter was devoted to generalizing about production locations under concerns for the influence of distance on costs and on quantity demanded.

If we recognize that market demand is sensitive to price and that market distribution entails costs, the producer's locational behavior should reflect its pricing policy. Assuming a spatially distributed market, a uniform delivered pricing policy yields an optimal distribution location that is near the densest market concentration, but is pulled by distant demand. An f.o.b. pricing policy yields an optimal distribution location that is more firmly influenced by an area of especially dense demand, as outlying markets are tempered by the expense incurred by remote purchasers.

However, this recognition that effective market size is influenced by the costs of distribution brings the recognition that location relative to the market(s) is an important element in competitive strategy – especially in industries where market access is important and varies with physical distance to the market. We engaged in a deductive exercise to illustrate the effects of various market and producer characteristics on the revenue-maximizing locations for competing producers.

Finally, we presented the difficulties of combining the cost-minimizing, market pricing, and revenue-maximizing approaches. As we approach the situation facing actual producers and their investment decisions, we recognized the importance of industry-specific characteristics such as number of producers, economies of scale, spatial extent of marketing areas, and the trend toward lower transportation costs, which allows other considerations (including geographic variation in costs and market density) to gain salience.

SUGGESTED READING

Greenhut, M.L. (1956) *Plant Location in Theory and in Practice*. Chapel Hill, NC: University of North Carolina Press.

Harris, C.D. (1954) "The market as a factor in the localization of industry in the U.S.," *Annals of the Association of American Geographers* 44: 315–48.

Hotelling, H. (1929) "Stability in competition," *Economic Journal* 39: 41–57.

Kennedy, P. and Copes, P. (1978) "Product differentiation and centralization of production," *Journal of Regional Science* 18: 323–36.

4

GROWTH AND LOCATION OF SERVICE ACTIVITIES

Three-quarters of the labor force of the industrialized world works in the service sector. Ninety percent of all new jobs generated in the U.S. are in services. When defined broadly to include activities besides extraction, agriculture, and manufacturing, services comprise the lion's share of advanced nations' economies (70 percent of U.S. GNP) and employment, as shown in Figure 4.1. Service activities dominate the formation of these countries' economic landscapes, and will be the source of the vast bulk of jobs of college graduates. As Figure 4.2 illustrates, many services doubled in employment in the U.S. between 1963 and 1991. Clearly, services have replaced manufacturing as the predominant form of

economic growth in many nations. One sign of the importance of services is that employment in McDonald's – the quintessential fast-food producer – is larger than that of U.S. Steel (now USX), once the largest manufacturing firm in the world. Long dismissed as an insignificant handmaiden to manufacturing, services have come into their own as a leading component of industrial location.

THE HETEROGENEITY OF SERVICES

What exactly are services? There is an enormous diversity encompassed by the term. Occupations

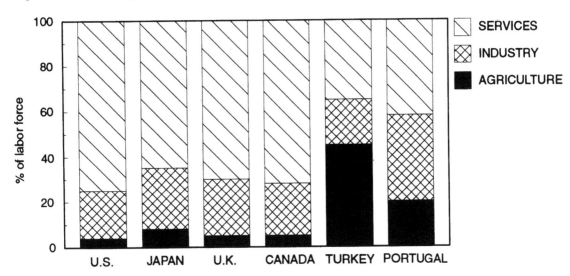

Figure 4.1 Labor force composition in five selected countries, 1990
Source: OECD Labour Force Statistics

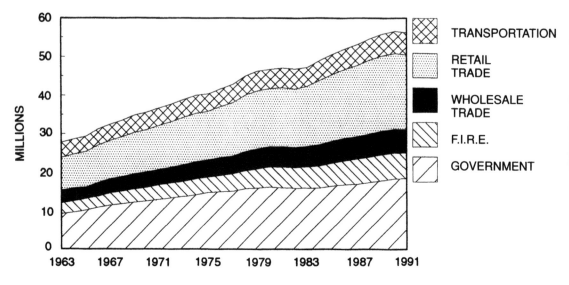

Figure 4.2 Employment in selected U.S. service industries, 1963–91

within the service sector range from prostitutes to professors, plumbers to philosophers. Indeed, so great is the variation among firms and occupations within services that the term threatens to lose any meaning whatsoever. Some observers argue that services include all information processing activities, including clerical activity, executive decision making, telecommunications, and the media. While many service jobs do involve the collection, processing, and transmission of large quantities of data, clearly others do not: the trash collector, hamburger flipper, security guard, and janitor all work in the service sector, but the degree to which these activities center around "information processing" is minimal.

Others emphasize that services involve the production of intangible outputs, in contrast to manufacturing, the result of which can be "dropped on your foot." What, for example, is the output of a lawyer? a teacher? a doctor? It is impossible to measure these outputs accurately and quantitatively, yet they are real. (Some approximations, including those of the U.S. government, use revenues as a proxy for output; yet revenues are the product of output and prices, so this measure is distorted by changes

in the relative prices of service outputs.) To complicate matters, many services generate both tangible and intangible outputs. Consider a fast-food franchise: the output is assuredly tangible, yet it is considered a service. The same is true for a computer software firm, in which the output is stored on disks. The fact that it is difficult, if not impossible, to measure output in services has enormous implications for their analysis. For example, some critics of the service sector argued that the slowdown in U.S. productivity growth in the late twentieth century reflected the growth of services. Yet if output in services cannot be adequately measured, how can one argue that output per employee in services is high or low, rising or falling? If wages are tied to productivity increases, what is the relation between services output and salaries?

If outputs do not differentiate services from manufacturing, then what about inputs? All forms of manufacturing as well as services involve inputs of *both* goods and services: an automobile producer must purchase legal, advertising, banking, and public relations inputs, just as an airline or securities firm must purchase hardware and equipment to do what it does. Clearly the lines between services and manufacturing are blurry and no simple definition will suffice.

Despite these problems, there is broad agreement as to the principal components of the service sector. These include the following:

1 **F.I.R.E.**, or finance, insurance, and real estate, including banking, securities, all forms of insurance, and real estate developers, brokers, and agents.
2 **Business services**, such as legal services, advertising, engineering and architecture, public relations, accounting, research and development, and consulting.
3 **Transportation and communications** including the printed and electronic media (newspapers, television, radio), trucking, shipping, railroads, and local transportation (taxis, buses, etc.).
4 **Wholesale and retail trade** including both "middlemen" involved in the intermediate steps between producers and consumers and retail outlets.
5 **Personal services and entertainment** including eating and drinking establishments, personal services (e.g., haircuts), repair and maintenance, private education and for-profit health care, and all forms of amusement (e.g., tourism, film production), including parts of the "informal" or "underground" economy that comprises a large part of the labor force in many Third World nations.
6 **Government** at the national, state, and local levels including public servants, the armed forces, and all those involved in the provision of public services (e.g., public education, health care, police, fire departments, etc.).
7 **Non-profit** agencies including charities, churches, museums, and private, non-profit health care agencies.

Because the definition of "services" is slippery, however, there are other components that escape these categories. For example, by "service," do we mean an *industry* or an *occupation*? (The U.S. government sidesteps this problem by adopting industries.) The difference, however, is critical. Many workers in manufacturing, for example, are in fact service sector workers, including personnel in headquarters, clerical functions, and research. Is the secretary who works for General Motors Corporation part of manufacturing, while the secretary who works for a bank part of the service sector? The use of industry versus occupational definitions is critical given the growth of many "nondirect" production workers within many manufacturing firms. In the ITT corporation, for example (officially, part of manufacturing), 90 percent of the labor force is not engaged in the immediate manufacture of electronics equipment, but in clerical, administrative, research, advertising, and maintenance functions. Clearly, different definitions of "services" will have significantly different implications for assessments of the size and composition of the service sector.

REASONS FOR SERVICE SECTOR GROWTH

Why have services grown so rapidly in the twentieth century? Services employment has increased steadily in the face of low rates of population growth, slowly rising rates of productivity and income, and significant manufacturing job loss. One common, but erroneous, perception is that services are the cause, or sometimes the consequence, of deindustrialization. Yet consider the logic behind this argument. Can service sector growth *cause* the decline in manufacturing? Such a hypothesis might be tenable if wages in services were significantly higher than in manufacturing, inducing workers to switch fields and causing wages to rise in industry. Clearly, in some services (e.g., securities) salaries are very high. Yet these opportunities are limited in scope, and the fact is that the bulk of service jobs pay lower wage rates than most manufacturing jobs. Working conditions in some services are obviously more comfortable and safer than is often found in manufacturing; yet not all service labor occurs in plush, air-conditioned offices; much of it takes place in near-sweatshop types of conditions. Further, the skill differences between many occupations in services and manufacturing are often so great that workers are unable to make a transition from one to the other. If not the cause, can the growth of services be the result of deindustrialization? Yet here too, the logic is faulty. Why should a law firm or hospital hire an additional secretary, for example, simply because a steel plant closed its doors? In fact, deindustrialization should decrease service sector growth through reductions in purchases and their multiplier effects, not increase it.

The "Post-Industrial Society" Thesis

A popular portrayal of the growth of services is that they herald the emergence of a so-called "post-industrial" society. (The term "post-industrial," however, tells us more what the new society is *not*.) Popularized by Utopian technocrats in the 1960s and 1970s, the post-industrial school of thought held that manufacturing created one type of society, characterized by a unique landscape, and that services are creating or will create a new type of society, with a different landscape. The essential argument of this school is that services are essentially information processing activities that require historically new levels of skill and education. Many postulated that services would generate a society populated by a "New Class" of educated professionals. The concerns of such a society, in which scarcity would be eliminated through enormous productivity gains, would center around "noneconomic" issues such as the quality of life, human needs, and social equity. Many post-industrial theorists emphasized the impact of telecommunications, which would allow the "New Class" to work at home (telecommuting) in the "electronic cottage," thus ending the need to commute to work on a daily basis. Why would anyone want or need to be physically present at a workplace when they could do their job at home? Such an arrangement would lead to flexible work hours and flatter, more egalitarian management structures. Telecommuting would generate the ultimate post-industrial landscape, a decentralized, low-density world in which workers would enjoy rural amenities while maintaining access to the world at a touch of a button.

In retrospect, the post-industrial thesis appears to be sadly mistaken. There is no denying that services are qualitatively different from manufacturing in many respects. However, services are also a form of commodity production, and firms in the service sector are bound by the same logic of profit maximization as those in manufacturing. The mistakes of the post-industrial school are easy to see in hindsight. First, services clearly embrace a wide diversity of activities, *some* — but not all — of which involve information processing. Many services (e.g., fast-food franchises) involve relatively little information processing, and involve the handling of tangible goods. Second, despite the rapid growth of skilled professional positions, the most rapid rate of employment growth has been in unskilled, low-paying jobs, hardly those predicted by the post-industrialists. Third, the landscape of services is more complex than the decentralized ones predicted by post-industrialist theorists: indeed, many services agglomerate together in downtown areas. In short, there is little evidence to support the post-industrialist vision of the future.

Alternative Explanations of Service Sector Growth

The fact is that the growth of services has relatively little to do with causing or resulting from the loss of industry in nations such as the U.S. In some parts of the U.S., deindustrialization *has* coincided with the growth of services (e.g., New York City); in others, the evacuation of manufacturing has seen virtually no service sector growth to take its place (e.g., Detroit); in yet other cases, both services and manufacturing have grown (e.g., southern California). While services and manufacturing clearly are interrelated, there is no simple, unidimensional causality at work. Indeed, growth in many services that are "bundled" to manufacturing (e.g., trucking) has probably been slowed by the decline in U.S. industrial employment. The key to explaining the growth of services, as with the loss of manufacturing, is the changing sources of demand. Simply put, services have increased rapidly because the demand for services has increased rapidly.

Six reasons for the increase in the demand for services are suggested below (one of these, however, is largely a statistical mirage).

First, service sector employment has increased in large part because of *rising average per capita incomes*. The demand for many services is income elastic, that is, small increases in income tend to generate proportionately larger increases in the consumption of these services (in contrast to most manufactured goods). Especially income-elastic services include entertainment and transportation (which is largely why the tourist industry is among the largest in the world today). In 1992, U.S. households, for the first time in their history, spent almost as much on services (43 percent of disposable income) as they did on durable and nondurable goods combined (Figure 4.3). An important reason contributing to this growth is the increasing value of time that accompanies rising incomes (especially with two income earners per family). As the value of time climbs relative to other commodities, many consumers will attempt to minimize the time costs of many ordinary tasks. While this phenomenon also explains the demand for washing machines, dishwashers, and automobiles, it is especially important for the growth of services. The explosion of fast-food restaurants, for example, has little to do with the quality of food (or even the price), and much to do with attempts by consumers to avoid cooking at home or shopping. Similarly, the growth of repair services reflects both increasingly sophisticated technologies (e.g., in automobiles or televisions) and a generalized unwillingness to spend precious recreation time doing such chores. Thus, the increasing value of time has led to a progressive *externalization of household functions*, so that which used to be done in-house becomes a commodity, or something performed through the market for a profit.

Second, the growth of services reflects the rapid *increase in health and education employment*. The provision and consumption of health care has increased steadily, in large part because of the changing demographic composition of the populations of industrialized nations. The most rapidly growing age groups today are the middle-aged and the elderly, precisely those groups that require relatively high per capita levels of medical care. Similarly, a changing labor market and increasing demand for literacy and numeracy at the workplace have driven the increasing demand for educational services at all levels.

Third, the growth of services reflects the *rising proportion of nondirect production workers* at the workplace, including many of those in manufacturing (recall the example of ITT). Most firms today must

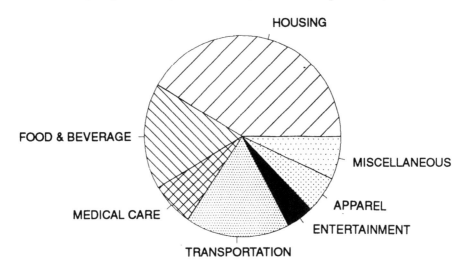

Figure 4.3 Consumption patterns of U.S. households, 1992

LOCATION OF SERVICE ACTIVITIES

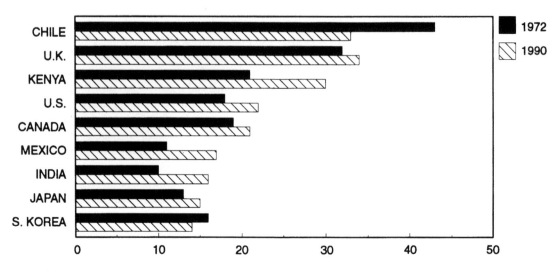

Figure 4.4 Variation and growth of government expenditures as a percent of GNP, selected countries, 1972 and 1990
Source: World Development Report 1992, The World Bank, Oxford University Press

devote considerable resources to dealing with a complex marketplace and legal environment, including many specialized clients, complex tax codes, environmental and labor restrictions, international competition, complicated financial deals, and sophisticated real estate markets. To do so, they require administrative bureaucracies to collect information and make decisions; clerical workers to assist with mountains of paperwork; sales people; researchers to study market demand and create new products; and legions of advertisers, public relations staff, accountants, lawyers, and financial experts to assist in a vital, but enormously complicated decision-making environment. Similarly, the introduction of sophisticated machinery requires maintenance and repair staff, while new offices or industrial plants require security and building maintenance staff – all nondirect production workers, and all services.

A fourth reason underpinning the growth of services reflects the increasing size and role of the *public sector*. The government contributes to the growth of services in two ways. First, government employment is largely service employment, at least in most mixed economies in which private ownership and operations prevail in most sectors. Government activity has been increasing in most

countries (Figure 4.4), for a range of reasons. In democratically governed countries, these reasons include the public demands for government services. One might argue that governments are inherently inefficient in providing these services, which may be more effectively provided through the private sector (i.e., with fewer workers per unit of output). Whether this is true or not, however, is not the issue: efficient or inefficient, government employment has increased steadily. Today, Uncle Sam is the largest single employer in the U.S., although federal employment, which has grown only slowly in the 1980s, is dwarfed by state and local government employment. A second way in which government contributes to the growth of services is indirectly: through a labyrinthine web of laws, rules, restrictions, and regulations, government has contributed to the growth in tax attorneys, accountants, consultants, and other specialists that assist firms in negotiating with the legal environment.

A fifth reason for the growth of services is rising levels of *global service exports*. Despite the widespread myth that all services cater to local demand, i.e., they are nonbasic activities and are thus of secondary importance to manufacturing, many cities, regions, and even nations derive a substantial portion of their nonlocal revenues from the sale of services to clients located elsewhere.

Internationally, the U.S. is a net exporter of services (but runs large trade deficits in manufactured goods), which is one reason services employment has expanded domestically. As Figure 4.5 reveals, U.S. service exports approached $140 billion in 1992. One hypothesis is that as the U.S. has lost its comparative advantage in manufacturing, it has gained a new one in services. However, the U.S. trade surplus in services does not begin to compensate for its deficits in manufacturing, and thus services are not likely to serve as an adequate long-term substitute for the loss of industrial jobs. The data on global services trade are poor, but the federal government estimates that services comprise roughly one-quarter of total U.S. exports. By industry, foreign sales account for approximately 25 percent of U.S. output in accounting, 23 percent in motion pictures, 21 percent in engineering, 15 percent in engineering, and 13 percent in data processing. These sales overseas take many forms, including tourism, fees and royalties, sales of business services, and profits from bank loans; unfortunately, the federal government includes repatriated profits from manufacturing investments overseas as a "service," which in reality it is not.

Geographically, most U.S. trade in services occurs with West European nations. It is important to note that foreign service exports do not generate revenues in all places of the U.S. equally: cities such as New York and Los Angeles, for example, which are critical to global capital markets, have benefitted the most.

Sixth, services have grown because a steady *contracting out of many functions from manufacturing firms* has occurred, i.e., vertical disintegration whereby many functions formerly performed in-house are subcontracted to suppliers, a process also known as externalization. For example, a steel company may lay off a janitor during a period of contraction, but then subcontract with a janitorial service (perhaps even the same person!); similarly, a ball bearing company may lay off a secretary and contract out to a clerical services company (e.g., Kelly). Because government statistics on employment rely upon a definition based on *industries* and not occupations, this process results in a reduction of manufacturing employment and an increase in service employment. Nothing has substantively changed, but the manufacturing sector in which the janitor or secretary used to work has registered a decrease in employment while the service sector has registered an increase.

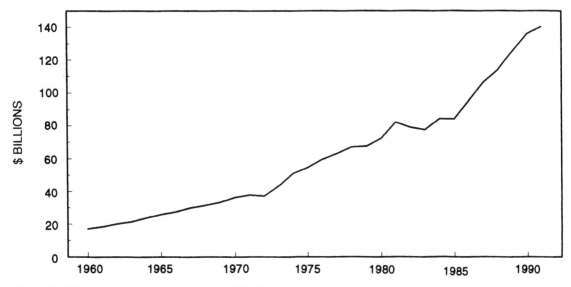

Figure 4.5 U.S. international service exports, 1960–91

The wave of vertical disintegration that swept many U.S. manufacturing firms in the 1980s resulted in substantial contracting out. Firms contract out as a means of cutting long-term costs (e.g., benefits) and coping with highly uncertain markets. In terms of services employment, therefore, this process is something of a statistical mirage; were the definition of services based on occupations and not industries, it would cease to be an explanation of services growth. Nonetheless, given the structure of our national employment accounting system, contracting out represents one reason why services employment has increased nationally.

SERVICE SECTOR LABOR MARKETS

A fundamental concern about the service sector is the way in which the jobs it generates differ from those typically found in manufacturing. Given the heterogeneity of both services and manufacturing firms, however, it is difficult to generalize about either sector. However, there are some broad characteristics found in many service sector firms that merit further consideration.

Compared to manufacturing, services are generally more labor intensive in nature, that is, more workers are required per unit of output (although, again, services output is extremely difficult to measure). Wages and salaries often comprise more than 90 percent of the total cost of production in many types of services, compared to a typical range of 5 percent (petroleum) to 50 percent (textiles) found in manufacturing. The degree of labor intensity, however, varies markedly among services. Some services are relatively capital intensive in nature (e.g., telecommunications). One of the principal reasons that services are generally more labor intensive than manufacturing is that they involve functions that are difficult to automate, including jobs in which there are considerable variations in routine, that involve the integration of diverse sources of information, or in which manual dexterity is involved, so that it is not cost-effective to mechanize them. Computerization, however, has begun to improve productivity

dramatically in many services, and has made some functions considerably more capital intensive.

A second important feature of service sector labor markets is that the income distribution within many such sectors tends to be less equal than in manufacturing. While there is much heated debate about this point, many observers argue that while manufacturing generated a relatively egalitarian income distribution, service employment tends to be bifurcated between two groups of occupations. On the high end, many service jobs are in the skilled professional and managerial category, involving occupations such as administration, attorneys, accountants, scientists, stock brokers, physicians, engineers, and so forth. At the other end, however, are innumerable low-paying, unskilled (or semi-skilled) jobs: data entry workers, janitors, security guards, cashiers, nurses' aides, boxboys, fast-food workers, etc. Indeed, the most rapidly growing occupations in the U.S. tend to be these low-paying ones, which are generally nonunionized, lack benefits, offer little upward mobility, and are largely filled by women. The concern about a bifurcated income structure is that there are relatively few jobs in between these two strata. Thus, the concern has been voiced that the deindustrialization of the U.S. may give rise to a polarized class structure, raising fears of the "disappearing middle class."

The shift from a manufacturing to service economy has undoubtedly reduced the standard of living for many people. For example, the average clerical job pays roughly 60 percent of the wage of the average manufacturing job, while the average position in retail trade pays only 50 percent of that in manufacturing. The unemployed automobile worker, therefore, who is forced to take a job in services is likely to suffer a significant drop in income. However, it is easy to exaggerate this phenomenon, leading to unrealistically pessimistic expectations of the future, particularly given the poor data available on many services jobs. Nonetheless, while many middle-class jobs are found in services, it is also true that the swelling numbers of the urban poor are in part due to the absence of well-paying jobs in many U.S. cities; many who try to sustain themselves and their families from the low incomes paid by low-end services find themselves caught in the ranks of the "working poor." This predicament is especially true for many minorities and women, who comprise the bulk of low-income urbanites.

A third dimension of labor markets in services concerns gender. The sex composition of workers in services tends to differ from that in manufacturing. Women have long worked in some forms of industry: many textile and garment workers, for example, were women. However, on the whole manufacturing was a predominantly male preserve; many jobs, particularly in heavy industry, were almost exclusively occupied by men, particularly when physical strength was an important criterion for employment. This gender composition had important repercussions for the structure of families, giving rise to the stereotypical, patriarchal nuclear family unit in which the father worked and the mother tended the home, a phenomenon increasingly rare in the U.S. The rise of services, in contrast, has seen large numbers of women enter the labor market. U.S. labor force participation rates of women now exceed 65 percent, and one-half of the labor force consists of women. The services types of jobs occupied by women, however, differ considerably from those filled by men. Most women are employed in semi-skilled, low-income services jobs, the so-called "pink collar" sector that includes health care (except physicians), clerical work, and teaching. Women comprise 99 percent of all secretaries, 97 percent of childcare workers, 88 percent of nonphysician health workers, 78 percent of personal services workers, and 66 percent of food service workers. This occupational distribution is largely, but not entirely, responsible for the fact that women earn, on average, only 60 percent of the income of men, giving rise to concerns over the "feminization of poverty." The growth of female labor force participation reflects both increased opportunities for women brought on by the reduction in physical strength necessary for service jobs and the women's movement and the gradual reduction in sex discrimination as well as the loss of manufacturing jobs; for many families,

two income earners is a necessity, not a luxury. This phenomenon has thrown constraints on women's labor market participation, such as the lack of affordable child care, into sharp contrast.

A fourth dimension of services labor markets concerns their general lack of unionization. In the post-World War II era, the proportion of U.S. workers in labor unions has dropped sharply, from 45 percent in 1950 to roughly 14 percent in 1990. While some of this decline is undoubtedly due to massive deindustrialization and efforts at union-busting by some firms, others point to the fact that services, which have grown quickly over the same period, are predominantly nonunionized. To be sure, there are unions in some services, including AFSCME (American Federation of State, County, and Municipal Employees), the American Federation of Teachers, and 1199, the union of medical workers. However, most services remain unorganized. The reasons for this are twofold. First, many workers in skilled professional and managerial jobs are employed in an individualized work climate not conducive to unions; many feel such organizations are unnecessary, counterproductive, or "beneath them." Workers in unskilled services jobs, in contrast, are very difficult to unionize; the abundant supplies of such workers, which holds down wages, gives them a decisive disadvantage relative to management and little ability to bargain for benefits such as vacations or health insurance.

A fifth and final facet of work in the service sector concerns its inputs of education. On average, jobs in services require more education than their counterparts in manufacturing. Obviously, skilled professional and managerial positions require a higher education, i.e., a college degree or advanced training. Many unskilled services jobs do not require this. However, over the last 20 years, particularly given the computerization of the office and the automation of many unskilled services, minimum skill requirements in services, including simple literacy, have risen steadily. It is difficult, for example, to find employment in any clerical position without some experience with computers. One consequence of this "upgrading" of skill requirements has been to increase the demand for educational services: few can find employment without a high school degree. Further, the proportion of high school students who go to college has increased steadily. For those without access to quality education, largely inner city minorities, this trend is devastating, leaving them permanently mired in a lifetime of poverty with few alternatives.

LOCATION OF SERVICE ACTIVITIES

As with all economic activities, the geographic distribution of services is highly uneven. To comprehend why service activities locate where they do, it is important to differentiate between two principal categories of services: consumer services (e.g., retail trade) and producer services (which primarily serve other firms, including manufacturing). The location of consumer services is relatively straightforward: because most such firms require close proximity to their market base, the distribution of such firms mirrors the distribution of population in general. Indeed, some consumer services (e.g., gas stations) mirror the distribution of population to the same extent as market-oriented manufacturing, such as soft-drink bottling. For this reason, consumer services have been abundantly studied through central place theory approaches, in which a hierarchy of services is distributed through a hierarchy of urban areas, each city serving a successively larger hinterland as one proceeds higher in the urban system. Although employment in consumer services is significant, this sector is relatively less important to explaining regional growth and change than are producer services, which can be exported from a region. Thus, attention below will be devoted entirely to locational considerations underpinning the geography of producer services.

By far the single most important consideration to take into account in studying the location of producer services is their tendency to agglomerate in large urban areas. In the U.S., financial and business services are most important to the large urban conurbations of the Northeast, particularly the

New York metropolitan region; this is reflected in the state-by-state distributions illustrated in Figures 4.6 and 4.7. Employment in business and financial services tends to be heavily concentrated in metropolitan areas for a variety of reasons. First, such locations allow service firms to maximize their accessibility to clients and suppliers. Although transport costs are negligible for producer services, other benefits of agglomeration are critical. Firms can minimize their input and output costs by locating near their market base, which generally consists of other firms, including many headquarter functions of manufacturing firms. Advertising firms, for example, must have quick access to a variety of clients (hence the formation of specialized districts such as Madison Avenue in Manhattan), while securities firms may need the rapid availability of specialized computer software firms or tax attorneys. Often such interconnections are mediated through face-to-face interaction (i.e., meetings of executives), which can be accomplished most profitably only in large urban areas. Face-to-face interaction allows firms to negotiate the details of particular transactions quickly and with security.

Second, in large cities, information often circulates in very narrow, specialized channels; by maximizing access to information, agglomeration reduces uncertainty. Third, locations in large cities maximize such firms' access to the urban labor pool; this criterion is important because producer services tend to be labor intensive and because they often have need of specialized, skilled personnel who are not readily available in smaller cities. A computer services firm that must train key personnel in other companies will often locate near its clients if its transactions involve repeated trips to other firms' workplaces. Fourth, large urban areas offer sufficiently large markets to allow the existence of specialized service firms that would fail to find a client base large enough to sustain them in small cities; by spreading their costs over many clients, such firms thrive when otherwise they would not.

Because of the tendency of producer services to agglomerate in large cities, the rapid growth of business and financial services in the 1980s was a

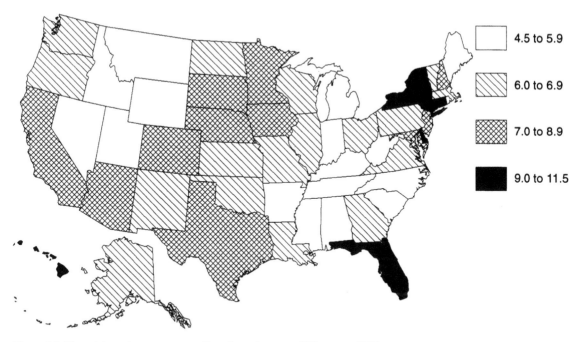

Figure 4.6 Financial services as percent of total employment, U.S. states, 1990

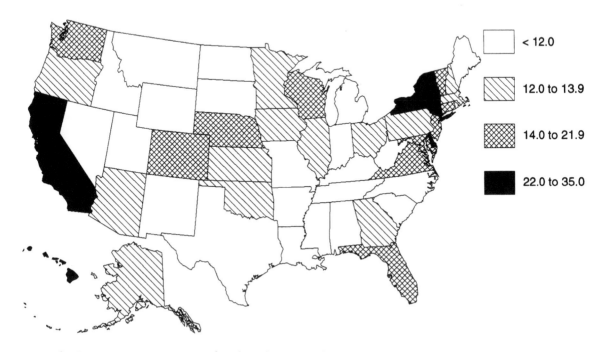

Figure 4.7 Business services as percent of total employment, U.S. states, 1990

primary factor in reconstructing central city landscapes, including the rapid growth of downtown office complexes and widespread gentrification of inner city neighborhoods and commercial areas. Like manufacturing firms, producer services also have need of an adequate urban infrastructure, largely to minimize transportation and communications costs. However, the infrastructural requirements of service firms are quite different from their industrial counterparts. For example, because many producer service companies must send their executive or sales and marketing representatives to other cities, air transportation is critical; airports are an important part of the service infrastructure. Similarly, many producer service firms require high capacity telecommunications systems, including telephone lines, fiber optics cables, satellite earth stations, and microwave facilities. Some newly constructed "smart" office buildings include these facilities in their design. The infrastructure of services tends to be most developed in large cities, which constitutes another reason for their tendency to agglomerate there.

From an historical perspective, it is important to note that the location of producer services is heavily influenced by the previous geographies created during earlier rounds of production. The clients and infrastructure of one region may reflect the types of production found there previously. For example, the complex of banking in Manhattan arose in large part because many merchant banking firms had close ties to the shipping industry and port functions; in London, these historical ties complement the trade–finance role of the city. Further, the locational needs of producer services vary from industry to industry. The geography of legal services, for example, is heavily affected by the presence of the federal government in Washington, D.C., where most large law firms have branch offices. The centralization of the U.K. legal system within the national government makes London even more of a legal center than Washington. The location of securities firms in Manhattan is accentuated by the presence of the New York Federal Reserve there, which is a leading player in the bond market. Finally, changes in the location of population, purchasing power, manufacturing, and international trade will shape the geography of producer services. Miami has become a leading banking center in large part because of its extensive ties to Latin American nations. Services in Vancouver, Seattle, and Los Angeles have grown quickly because of those cities' linkages to Pacific Rim trade and capital markets.

SUMMARY

Services comprise a vast, diverse, and critical set of industries that dominate much of the economic life and landscapes of many nations. Almost three-quarters of the jobs and output of the industrialized world, as well as that of many industrializing nations, consists of the production of intangibles. About 90 percent of all new jobs in Europe, North America, and Japan are in services. This growth in service activity must be attributed to a variety of factors, including a changing composition of final demand, changing technologies of material production, and changing organization of production. Service activities are very heterogeneous, encompassing a wide diversity of firms, skills, and occupations. We cannot reduce services to unskilled, low-paying type jobs such as in the fast-food industry or exaggerate the capability of services to raise incomes and productivity, the mistake of the post-industrial thesis. What is clear is that the growth of services has radically changed urban labor markets and economic landscapes, often in highly unpredictable ways.

Despite the growth and numerical dominance of service activities and service employment, there remains a stubborn perception among many planners, academics, and business people that services are somehow not as "real" or "important" as manufacturing. As this chapter has sought to demonstrate, services are in fact exported and traded among cities, regions, and nations, often forming the economic base of communities and entire countries. Service sectors such as tourism and finance are among the world's largest generators of profits, jobs, and output. There is no reason, therefore, to relegate services to a subordinate status with respect to manufacturing, or to regard services as simply extensions of the production of goods.

The geography of services is an important topic that deserves much more consideration than it has received. The locational logic of service firms can be understood as an outcome of commodity production, i.e., profit maximization, but service firms may emphasize different locational considerations than do industrial corporations. For example, access to labor and specialized information, i.e., agglomeration economies, is particularly important for most financial and business services, leading to their concentration in large metropolitan areas. Financial services exert an enormous influence over the formation and, at times, the obliteration, of economic landscapes by providing access to credit, without which large-scale production of any sort is impossible. The telecommunications sector, which embodies the electronic mediation of information, has had as profound an influence on production systems and the rhythms of everyday life as any other sector, including linking together national financial markets, the dispersal of back offices, and dispersing an infinite variety of information among social groups, leading to ever-changing experiences of time and space.

SUGGESTED READING

Brunn, B. and Leinbach, T. (eds.) (1991) *Collapsing Space and Time*. London: HarperCollins.

Daniels, P.W. (1985) *Service Industries: A Geographical Appraisal*. London: Methuen.

Daniels, P.W. (1993) *Service Industries in the World Economy*. Oxford: Blackwell.

Hepworth, Mark E. (1990) *Geography of the Information Economy*. New York: Guilford Press.

Moss, M. and Dunau, A. (1986) "Offices, information technology, and locational trends," *The Changing Office Workplace*. J. Black, K. Roark, and L. Schwartz (eds.) Washington, D.C.: Urban Land Institute.

Pred, A. (1977) *City-Systems in Advanced Economies*. New York: John Wiley and Sons.

5

COMPARATIVE ADVANTAGE AND INDUSTRIAL LOCATION

w that we've approached the individual location
sion within the context of firms' goals, the needs
markets of the activity at hand, and the charac-
stics of places, we're ready to pursue a different,
nplementary approach. The characteristics of
ces – availability of local resources, labor, and
rkets – influence the location and competitiveness
every activity located within the region. Thus,
can look at the aggregate pattern of regional
ustrial activity as a consequence of regional char-
cristics, as well as the reflection of all the industrial
ation decisions made to date in the region. This
croeconomic approach, aggregating decisions
oss actors, is the focus of this chapter.

An inherent part of capitalist economies is uneven
elopment over time and space, yielding highly
erentiated economic landscapes with rich and
r nations, containing growing and declining
ons. The formation (and destruction) of indus-
l regions is largely a product of the growth or
line of different economic sectors, which in turn
ects the locations where individual firms seek to
timize their profits. When a place offers the
ortunity to realize the highest rate of profit (e.g.,
ugh low labor costs), firms will tend to flock to
up facilities in it, generating jobs and incomes
having numerous effects on local housing
kets, land uses, public infrastructures, and so
h. When that same place becomes uncompetitive
tive to other places, be they cities, regions, or
ons, firms will tend to leave it for greener
ures elsewhere, causing a concomitant decline in
ployment, incomes, tax revenues, property values,

etc. In short, uneven development – the spatial
division of labor – is a necessary, inevitable outcome
of profit-maximizing production processes.

A central theme in understanding where and why
industries locate where they do is the notion of
comparative advantage, which may be broadly
defined as the ability of a place (or group) to produce
something more cheaply, and hence more profitably,
than other places. Comparative advantages don't just
"occur" spontaneously, they are constructed, and
reflect the historical development of regions,
including the costs of production, the prevalence of
specific natural resources, local government policies
and services, transportation systems, and many other
factors. In order to realize a comparative advantage,
an industry and a region must be able to export their
product; likewise, they must be able to import
the goods and services they need. In other words,
the formation of comparative advantage is closely
associated with trade, as different regions or nations
buy and sell their outputs and inputs to each other.
Because the advantages that a particular place offers
are most suited to the needs of one, or perhaps
a small handful of industries, the construction
of comparative advantages is useful in understanding the spatial distribution of industries and
the geographical structure of an economy. Thus, the
theme of comparative advantage allows us to
comprehend how uneven development occurs and
why it changes over time.

Comparative advantages are readily evident across
the globe, and both affect and reflect trade patterns:
Third World nations generally produce and sell

foodstuffs and mineral ores in return for manufactured goods, a legacy of their long colonial past; OPEC nations export oil in return for machinery, food, and consumer products; South Africa sells gold and uranium. Within Britain, the Northeast and Wales are known for the production of coal, Scotland for its output of whiskey and cloth, and London is the center of the nation's financial and business services. Within the U.S., individual areas have traded with one another and the rest of the world on the basis of comparative advantages that often reflect their natural resources: Appalachia has long been a leading producer of coal; the Southeast was, and still is, a leading producer of cotton and tobacco; the economies of Texas, Oklahoma, Louisiana, and Alaska are specialized around petroleum extraction and refining; the agricultural Midwest produces the nation's – and much of the world's – wheat, corn, soybeans, and meat.

Individual cities, too, often specialize in the manufacture of particular goods. For example, in the early twentieth century, New York City produced almost three-quarters of the nation's garments. Many towns in New England, and later in the Southeast, had economies that rose (and fell) around the textile industry. Detroit offers a famous example centered around the automobile industry. The steel industry made cities such as Buffalo, Pittsburgh, Jonestown, Youngstown, and Cleveland into major, wealthy centers of the U.S. economy during the first half of the twentieth century. Akron, Ohio, was for years the nation's "rubber capital," home of the four largest rubber firms. Often a comparative advantage reflects the presence of a single company with a disproportionately large influence over the local labor market (such as the infamous "company town" in many mining and lumber producing areas). Corning, New York, for example, is a significant producer of cut crystal; Rochester, New York, was long known as a shoemaking city; Memphis was the foremost U.S. cotton seed oil processing center; Seattle, Washington, prospered around the fortunes of one aircraft and aerospace producer. Many small agricultural towns depend heavily on local grain or meat processing. In order to understand comparative advantage, we must look at various ways it is theorized and how it is related to regional patterns of production, trade, and consumption.

THEORIES OF COMPARATIVE ADVANTAGE

There are several models of comparative advantage and its relation to production and trading patterns. Often, these models reflect the historical and economic circumstances of the theorists who invented them. While they differ in their assumptions and conclusions, some elements are common to all of them. This section explores the original theory, Ricardo's model, then examines the more sophisticated neoclassical economic approach, and finally the relations between comparative advantage and transport costs.

Ricardo's Model of Comparative Advantage

The first theorist of the nature of comparative advantage was the eighteenth-century British economist David Ricardo, who studied the interrelations between growing capitalist markets, trade, and regional or national patterns of economic specialization. In particular, Ricardo was interested in the effects of trade upon production in different places: when communities began to exchange goods, they gave up the necessity of producing everything they needed for themselves, purchasing them instead from other places (i.e., as imports). Consumers in one place generally only purchase goods from other places if they can obtain the same or better quality of goods for a price equal to or lower than the cost of producing it themselves. What happens to local or national economies, then, when the resources (such as land, labor, or capital) previously used to produce one set of goods are "released" i.e., made available for the production of something different, because the output they generated is being purchased by local consumers from producers located elsewhere in the form of imports?

To analyze this process, Ricardo made some simplifying assumptions, including the absence of barriers to trade (such as tariffs or quotas) and the presence of a competitive market (i.e., without oligopolies or monopolies, so that every producer is a price-taker unable to influence the market price). Ricardo controlled for the influences of different currencies (exchange rates) by framing his analysis without money, i.e., by setting the prices of goods in terms of one another. He also ignored demand. Like most early classical economists, Ricardo subscribed to the labor theory of value, believing that the value of goods reflected the labor time invested in their production. In his conception, comparative advantage reflects the distribution of natural resources, climate, labor costs, and differences in technology embodied in production.

To illustrate what happens when trade allows places to form comparative advantages, Ricardo invented a famous example modeling the relations between Portugal and England, two long-time allies with extensive trading ties. Before the emergence of capitalism, both nations had their own wine and clothing industries, producing both sets of products for themselves. However, shortly after they began to trade, Portugal's clothing industry disappeared, as did England's wine sector. What happened? Ricardo illustrated the underpinnings of this process as follows. Suppose that the costs to wine and clothing producers (in labor hours) in each nation before specialization were distributed as follows:

Before specialization

	wine	cloth	total
Labor used in England	120	100	220
Labor used in Portugal	80	90	170
Total units produced	2	2	390

Thus, to produce one unit of wine, England must use 120 labor hours; to produce one unit of cloth, Portugal must use 90 labor hours. To produce all the wine and cloth it consumes (one unit each), England must use 220 labor hours, while Portugal must use 170. Each nation produces one unit of wine

and one unit of cloth, indicating two units of each good are produced altogether.

A casual observer might think that since Portugal can produce both wine and cloth more efficiently (less expensively) than England (i.e., it enjoys an absolute advantage in the production of both goods), that it would have no reason to trade. In fact this is not the case. As we shall see, trade benefits both Portugal and England, although not to the same degree. In other words, *trade benefits less efficient producers as well as efficient ones.*

What would be the pattern of production assuming the two countries engaged in trade, as occurred with the growth of capitalist markets? Examine the distribution below of labor time involved in the production of wine and cloth, but in the case where England has specialized in the production of cloth and Portugal specialized in the production of wine:

After specialization

	wine	cloth	total	savings
Labor used in England	0	200	200	20
Labor used in Portugal	160	0	160	10
Total units produced	2	2	360	30

The same total volume of goods is produced with specialization (and trade) as before, i.e., two units each of wine and cloth. But in producing two units of cloth rather than one (at 100 labor hours each) – one for itself and one to sell to Portugal – England has invested only 200 labor hours total, a saving of 20 hours over the pre-specialization situation. In producing two units of wine – one for itself and one to sell to England – Portugal has invested only 160 labor hours, a saving of 10. In other words, *trade allows inefficient resources to be allocated at no cost*, generating savings in the process that may be invested to produce a yet larger output (and profit). For this reason, Ricardo and most economists advocate "free trade," and oppose barriers to trade such as tariffs and quotas frequently advocated by domestic pressure groups. Of course, not everyone benefits equally from this process: England realized a larger saving (20 hours) than did Portugal; and English wine producers and Portuguese cloth producers, who were both eliminated by trade, would scarcely find their post-specialization status appealing. The Ricardian model concludes that, given trade, regions will specialize completely in the production of one output, abandoning entirely the attempt to produce goods in which they do not enjoy a comparative advantage.

Ricardo's analysis illustrates the widespread specialization among nations and cities that began to occur with the advent of capitalism. British cities such as Manchester, Leeds, and Birmingham, began to produce textiles; Lyon, France, became a leading producer of silk; Italy became a leading shoe manufacturer; and so forth. However, Ricardo's analysis is incomplete, and, therefore, not altogether accurate. For a deeper understanding of comparative advantage, we must include the demand for goods, which may vary across places, and we must allow more than one input into production. Because complete specialization rarely occurs in practice, we must explain the common phenomenon of incomplete specialization.

The Neoclassical Model: the Heckscher–Ohlin Theory

By the late nineteenth century, economics changed from the classical approach, which was centered around the labor theory of value, to the neoclassical approach, which heavily emphasized the nature of demand and marginal costs and benefits. The neoclassical theory of comparative advantage, also known as the Heckscher–Ohlin theory after its two originators, differs from the Ricardian approach in two important ways. First, it includes more than one factor of production, i.e., capital as well as labor (which entails abandoning the labor theory of value). Second, it allows for the incorporation of demand, which Ricardo ignored, through the use of utility curves, a common tool of neoclassical analysis. Utility curves represent different levels of satisfaction among consumers given different tradeoffs that they must continually make. Just as producers seek to

maximize profits, consumers seek to maximize utility, or satisfaction, by purchasing a particular combination of goods.

Figure 5.1 illustrates the neoclassical theory of comparative advantage. In this example, a region produces two goods, guns and butter, each of which uses two inputs, capital and labor. The production of guns is assumed to be relatively capital intensive while the production of butter is assumed to be relatively labor intensive. The **production possibilities curve** (PP) indicates the maximum amount of both goods that can be produced using different combinations of labor and capital. Note that the PP curve is curved; unlike the Ricardian model, the Heckscher–Ohlin theory assumes the existence of **diminishing returns to production** (and nonlinear production functions). As a resource (say, labor) becomes increasingly scarce (e.g., as this economy begins to specialize in the production of guns), its price (wages) rises accordingly, and, to maintain the same level of profits, firms will reduce their output.

Utility lines above the production possibilities curve represent the aggregate consumption preferences for this society, i.e., the different tradeoffs between guns and butter that equally satisfy local consumers. Clearly, without trade a society cannot consume above its PP curve (ignoring the question of debt). To consume below the PP curve would be to consume less than the maximum amount, and hence not to maximize utility. Only one utility curve, U_1, is just tangent to the PP curve, and thus is not only technically possible but maximizes the society's satisfaction. The point of tangency between U_1 and PP indicates the optimal tradeoff for this society, given its resources, technological constraints, and preferences. Thus, *without trade*, this society will produce and consume G_1 amount of guns and B_1 amount of butter; note that without trade the amount of each good produced and consumed is identical. Domestic prices of guns and butter are represented by the line DD; the slope of this price line indicates how much of each good the society would sacrifice to obtain a given amount of the other good, i.e., their relative prices in terms of each other.

Now let us introduce trade, and specialization,

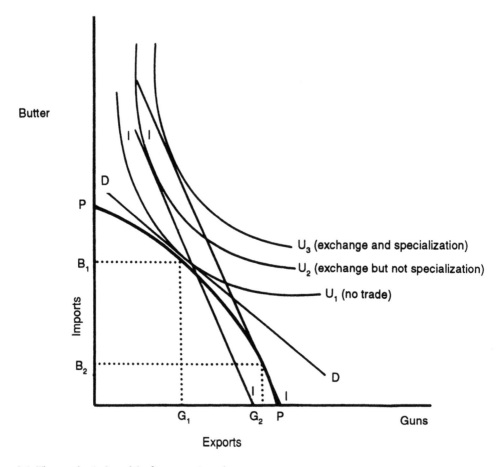

Figure 5.1 The neoclassical model of comparative advantage

into this analysis. If the international prices of guns and butter (in terms of each other) are different than the domestic prices, the slope of the international price line (II) will be different than that of the domestic price line (DD). Note that in this example, II has a steeper slope than DD, indicating that in order to purchase a given amount of guns on the international market, a larger amount of butter must be sacrificed than is necessary on the domestic market: internationally, therefore, guns are more expensive (and butter less so) than on the domestic market. In other words, this society can produce guns more efficiently (and butter less so) than producers elsewhere — it enjoys a comparative

advantage in guns. We noted that gun production is relatively capital intensive compared to butter production; because this society can produce guns more cheaply than other producers on the international market, it must have relatively inexpensive supplies of capital. This relation illustrates an important part of the Heckscher–Ohlin theorem: *regions specialize in the production of goods that use the most of their cheapest inputs.*

By exporting guns and importing butter, therefore, this society is able to escape its domestic production possibilities curve PP. If the society exports the difference between the amount of guns it produces and the amount it consumes, and imports

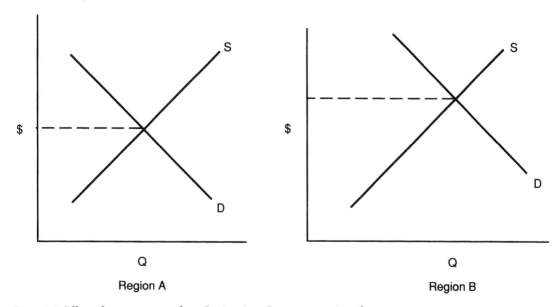

Figure 5.2 Effect of transport costs from Region A to B on comparative advantage

the difference between the amount of butter it produces and the amount it consumes, it will be better off (reach a higher utility level, U_2) than if it does not engage in trade. Thus, *a society realizes some gains from trade (gains from exchange) even without specializing in production.* However, if the society specializes in the production of guns – at the point where the international price line II is just tangent to the domestic production possibilities curve DD – then it will produce G_2 amount of guns (consuming G_1, exporting the difference) and consume B_2 amount of butter, which greatly exceeds its production, B_1, allowing it to import the difference. By buying and selling at international prices, the society is able to consume yet higher levels of both commodities, enjoy a higher standard of living, and reach a yet higher utility curve (U_3) than if it did not engage in trade and specialization. Thus, the gains from trade can be broken into gains from exchange and gains from specialization.

There is an important corollary to this analysis. Note that even if the society specializes in the production of guns for the international market, it still produces a limited amount of butter for its

own consumption. Because the PP frontier is curved, it reflects diminishing returns; at very high levels of gun production, firms require large amounts of capital, making it scarce and expensive. The shift into an increasingly gun-specialized economy, therefore, releases large quantities of labor, making it relatively inexpensive, allowing some domestic producers to compete with imports. Thus, unlike the Ricardian model, the neoclassical model explains incomplete specialization. It lacks, however, one vital aspect necessary to understanding economic landscapes, transportation costs.

Comparative Advantage and Transportation Costs

In reality, trade occurs across vast distances, forcing producers to transport their goods to market as well as produce them. Transport costs can exert a significant influence on the prices of imports and hence on the realization of comparative advantage. Figure 5.2 illustrates two societies with markets for the same good. Because Region A produces the good more cheaply than B, it can export its output to B.

If there were no transport costs, the good would sell for price P_1; however, the addition of transport costs to the production price forces the import price up to P_2, discouraging some consumers in B from purchasing these goods. Without as large a market in B, producers in A will cut back on their production and exports of this good. Thus, *by making imports more expensive, transport costs inhibit the formation of comparative advantage*. The same holds true for all barriers to trade such as quotas, tariffs, and nontariff barriers. (The same analysis can be incorporated into Figure 5.1: when transport costs are added to the costs of imported butter, the international cost of butter relative to guns rises, or, conversely, the price of guns falls, discouraging the society from specializing as completely in the production of guns.) When transport costs decline, conversely, new comparative advantages can be created. The introduction of refrigerated shipping in the 1890s, for example, allowed New Zealand to become a leading producer of mutton for European and U.S. markets. The completion of the transcontinental railroads in the same decade allowed the Pacific Northwest to become an important lumber producer.

COMPARATIVE ADVANTAGE, ECONOMIC BASE ANALYSIS, AND LOCAL ECONOMIES

We have seen how capitalism produces regionally differentiated landscapes through the formation of comparative advantages, which vary broadly across space and time. Firms and sectors in which a region or city possesses a comparative advantage are vitally important to local economies; when they sell their output to clients and customers outside of the local area, export-related firms bring in nonlocal revenues for a region, linking a particular locality to the wider national, or international, division of labor. *By producing and selling what a place produces more efficiently than other places, such firms form the "backbone" or "motor" of local economies*, generating much of the income that allows regions to purchase imports that they cannot produce efficiently for themselves. Note that "exports" in this context does not necessarily refer to foreign or international exports: from the perspective of a local company, it matters very little whether the clients purchasing its products are located in the neighboring city, state, or halfway around the world: all of them generate revenues. For this reason, export-related firms are an ideal conceptual starting point for analyzing the economic structure of local areas.

A common misconception in the analysis of regional economies is that only manufacturing firms export their output nonlocally, and that service firms serve only the local market, making them totally dependent upon local clients. *The notion that only manufacturing firms export their output nonlocally, and that services serve only local clients, is absolutely untrue:* **both** manufacturing **and** service firms are capable of serving local as well as nonlocal markets. There is, therefore, no simple correspondence between manufacturing and export-related firms or services and locally related firms; reality is much more complex than this. For example, consider cities that "export" services (i.e., realize a comparative advantage by selling to nonlocal clients). The most obvious examples include cities such as New York, with a large concentration of financial and business services: New York banks, securities firms, insurance companies, legal services firms, and advertisers serve clients throughout the U.S. and the world. Los Angeles is famous for its production of films and movies. Other examples abound. Hartford, Connecticut, is a significant center of the insurance industry. Tourism represents an export of services as well: when tourists spend funds on hotels, meals, and entertainment, they are pumping money into local economies from elsewhere. Cities such as Las Vegas, or Atlantic City, New Jersey, or New York, or Los Angeles, and even the economy of entire states such as Florida or entire nations, such as island countries of the Caribbean, rely heavily on tourism.

Many cities export government services as well, paid for by taxpayers distributed over broad areas. Washington, D.C., for example, is largely funded through tax payments from the rest of the nation – a service export; similarly, many state capitals derive substantial revenues from taxes from the rest of the state in which they are located. Unsurprisingly, therefore, cities such as Columbus, Ohio, Sacramento, California, or Albany, New York, often thrive while cities with comparative advantages in private sector activities tend to feel the impact of business cycles more severely. Less well-known examples of service exports abound: every university town draws students from a wide area, whose tuition payments therefore constitute a service export from that place; the software firm in, say, Seattle, that contracts to a client in Alaska; the engineer in Chicago working on a hydroelectric dam in Arizona; the television station in Atlanta that sells advertising time to a firm in Pennsylvania; the law firm in Cleveland that handles probate in Florida; the management consultancy in San Francisco that assists a firm in Oklahoma; medical clinics (e.g., the Mayo Clinic in Rochester, Minnesota) that export specialized medical services to people from all over the nation or the world. Unfortunately, little data are collected, and little is known, about the empirical patterns of inter-regional trade in services.

While the conceptual difference between firms that export and those that sell their output locally is fairly easy to grasp, in fact the boundaries between export and local firms are often hazy; many firms serve both markets, usually relying upon one more than the other. A typewriter manufacturer, for example, may export 95 percent of its output to clients throughout the nation or overseas, but also serve local consumers. An attorney who specializes in, say, medical malpractice suits, may spend most of his/her time serving clients in a distant city and yet also serve some local residents of his/her community. Thus the division between export and locally dependent firms and sectors is somewhat artificial and, at times, even arbitrary.

To complicate matters further, the specialization of local economies may occur not only *among* different economic sectors (e.g., wheat, steel, autos, or banking), but *within* sectors as well. At different stages of the production process, the locational requirements of firms change; very labor-intensive processes such as management or research and development, for example, may operate best in large cities, where they have rapid access to large pools of skilled labor, ancillary services (e.g., law or advertising firms), and specialized circuits of information, while relatively routinized, capital-intensive processes, such as assembly operations or data entry, may operate more profitably in small towns or rural areas, where labor and land costs are generally much cheaper. (Indeed, as will be seen later, central cities often enjoy a comparative advantage in labor-intensive forms of production while peripheral or suburban regions enjoy a comparative advantage in capital-intensive ones.) Many firms maximize their profits by spatially separating different parts among different places. Thus, large cities (or their adjacent suburban areas) often exhibit complexes of headquarters (of both manufacturing and service firms), with large numbers of managerial, administrative, and clerical workers. In this sense, a city may specialize (realize a comparative advantage) in one part of a production process that cuts across several different economic sectors.

Location Quotients and Regional Specialization

One very common way of measuring comparative advantage empirically is through the use of location quotients. Location quotients measure how specialized a town or region is with respect to some larger areal unit such as a state or nation. For example, the degree to which Michigan specializes in automobile production, i.e., its comparative advantage, can be measured by examining the proportion of employment (or sales) in automobiles in Michigan compared to the United States. If the proportion is higher, we can assume Michigan is specialized in this sector; if a region's proportion of employment is lower than that for the U.S., we can assume it is less specialized than the nation in that sector and must import those goods and services from the rest of the country to satisfy local demand; finally, if the location quotient is the same as that for the U.S., we assume Michigan's automobile employment serves only the local demand, i.e., it does not export or import automobiles.

In mathematical terms, we can define the location quotient (LQ) as

$$LQ_{ir} = \%E_{ir}/\%E_{iR} = E_{ir}/E_r \ / \ E_{iR}/E_R$$

where E refers to employment,
    i refers to a given industrial sector,
    r refers to a region,
    R refers to the comparison region.

Thus $\%E_{ir}$ refers to the percentage of employment in industry i in region r, while $\%E_{iR}$ refers to the share of employment in the same industry in the comparison region, e.g., the U.S. (Location quotients can use any measure of regional specialization, such as sales or output, but because employment data are usually the most specific and easily obtainable, they are the most commonly used.) As an example, let us use the case of a mythical city called Bankville (r) which has a comparative advantage, and thus specializes in, the banking industry (i). Say total employment in Bankville is 100,000, of which 12,000 are employed in banking. Say the region to which we compare Bankville is the state of Columbia, in which 1,000,000 people work, including 100,000 in banking. Then using the notation above, for Bankville $\%E_{ir}$ = 12,000/100,000 = 12%, and for Columbia $\%E_{iR}$ = 100,000/1,000,000 = 10%, so the location quotient for banking in Bankville is $\%E_{ir}/\%E_{iR}$ = 12%/10% = 1.2. If the location quotient exceeds one, we assume Bankville is specialized in banking; if its location quotient is lower than one, we assume it is less specialized than the state of Columbia in that sector and must import those goods and services; finally, if the location quotient is one, Bankville is as specialized in banking as Columbia, and we assume it does not export or import banking services. Since Bankville's LQ exceeds one, we conclude it is more specialized than Columbia as a whole and therefore enjoys a comparative advantage in that sector. Note that a region cannot be specialized in everything: if it is specialized in some sectors (its LQ is above one), it must be less specialized than average in others (its LQ will be below one). Also note that this technique assumes that regional levels

of productivity equal those of the comparison region, which may not in fact be true if firms are attracted to a place because it is more efficient than other places. Thus the location quotient technique has its limitations, but as a quick and easy way to study regional specialization it can be very handy.

A more realistic usage of location quotients is found in Figures 5.3 through 5.19, which depict location quotients of employment in several selected industries in the U.S. These maps depict the degree to which regional economies in the U.S. are specialized in some industries and not others, i.e., the way in which the spatial division of labor in the U.S. has created an uneven topography of production. The reader should be cautious not to confuse the degree of specialization of a state with the size of its economy; location quotients only depict the degree to which a region's economy is concentrated in a sector relative to the nation as a whole, not the absolute size. Thus, a large, relatively unspecialized state may in fact employ more people in a certain sector than a small, highly specialized one. In wood products (Figure 5.3) the most specialized states are located in the Pacific Northwest (with location quotients of 5 to 9, indicating that their economies are five to nine times more specialized in that sector than the nation as a whole). The mining sector (Figure 5.4) is most important to states such as West Virginia, with its large coal industry, as well as Wyoming and Oklahoma. States most specialized in food products (Figure 5.5) are located mostly in the agricultural Midwest, such as Iowa and Arkansas, as well as Idaho. Petroleum (Figure 5.6) is disproportionately important to states such as Louisiana, where it comprises seven times the proportion of employment as it does nationally. Although Texas and Oklahoma are also specialized in this sector, as indicated by their location quotients between two and four, and they may produce more than Louisiana in absolute terms, their economies tend to be more diversified. States that rely heavily upon chemicals (Figure 5.7) include West Virginia (where Dow Chemical is a critical corporation) and South Carolina. The textile industry (Figure 5.8) plays the greatest role in the economies of North and South

Carolina, while apparel production (Figure 5.9) is also most important in the South, including South Carolina, Tennessee, and Mississippi. Employment in primary metals production (Figure 5.10) such as iron and steel is a principal industry in Indiana and West Virginia as well as some other traditional Rustbelt states such as Ohio and Pennsylvania. Specialization in fabricated metal products (Figure 5.11) is likewise greatest in the industrial Midwestern states of Ohio, Indiana, Michigan, and Wisconsin. The rubber and plastic industries (Figure 5.12) employ the greatest shares of workers in Ohio, Indiana, and South Carolina. States with electronic products (Figure 5.13) are Indiana and Michigan (although as noted above, total employment or output in this sector may be larger in less specialized states such as California). Transportation equipment (Figure 5.14) includes the automobile and aerospace industries, and is most pronounced in states such as Washington (home to the Boeing Corporation) and Michigan (including the traditional automobile complex around Detroit). Note that states often specialize in several sectors that are closely related to each other in the production process, such as textiles and apparel, indicating that functional connections among sectors often give rise to geographic proximity, an issue we shall explore in more detail later.

Location quotients for various types of services reveal that they too play uneven roles in regional economies, although generally they tend to be more broadly distributed among states than manufacturing, as revealed by their lower location quotients. The finance, insurance, and real estate sector (F.I.R.E.; Figure 5.15) is above all most significant to the state of New York, including New York City, and Delaware, which has attracted numerous banks in recent years. Employment in various types of business services (Figure 5.16), including advertising and public relations, is significant to the economies of Maryland (many of which serve the Washington, D.C. area) and, surprisingly, Utah. The map of location quotients for legal services (Figure 5.17) is unusual, indicating that the District of Columbia is by far the most specialized region in the country in this sector;

with a location quotient of 8.4, it is more than eight times as specialized in law firms as the nation. In engineering services (Figure 5.18), states in the Southwestern and Northeastern U.S. tend to be more specialized than the national average. Finally, the geography of specialization in the motion picture industry (Figure 5.19) indicates that employment in California, led by the famous complex of firms in Hollywood, is more than three and a half times more likely to be found in this sector than nationally.

Export Base Analysis

A common approach to studying the impacts of export-related firms or sectors on local economies is economic (export) base analysis, which conceptually and empirically relates nonlocal exports to the structure of local labor markets. The analysis begins by differentiating between basic sectors, or export-driven firms or industries on the one hand, which may include services (i.e., those sectors that primarily cater to nonlocal clients), and nonbasic sectors on the other hand, i.e., firms and sectors that primarily cater to local demand. Examples of nonbasic sectors include retailing (except in cases such as outlet centers or tourist shopping centers), repair and maintenance, tax consultants, personal services (e.g., barbers), and restaurants. Most employment in most regions is in nonbasic sectors. However, remember that the dividing line between basic (export-driven) and nonbasic (locally dependent) sectors or firms is often artificial: most firms cater to both local and nonlocal clients, specializing more as a matter of degree.

Once we establish a division between basic and nonbasic sectors, we can explore their interrelations. Total employment in a city or region, T, is equal to basic employment, B, plus nonbasic employment, NB, or

$$T = B + NB \qquad (5.1)$$

To assess the impacts of basic employment on a local economy, we are interested in the relations between total employment and basic employment, often known as a multiplier, m.

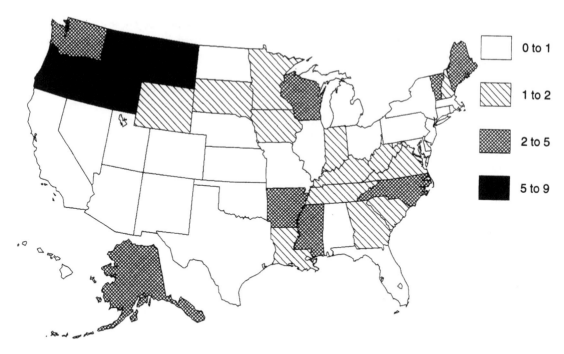

Figure 5.3 Location quotient for employment in wood products, 1990

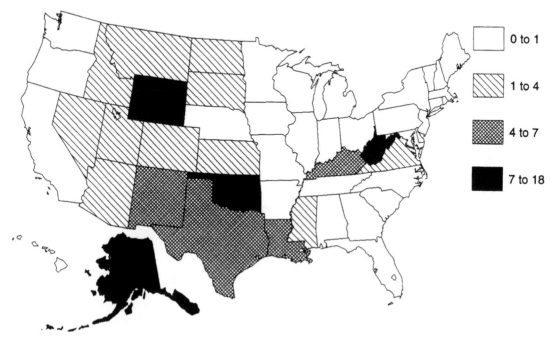

Figure 5.4 Location quotient for employment in mining, 1990

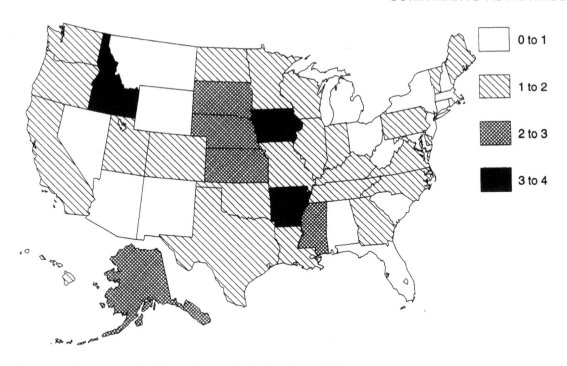

Figure 5.5 Location quotient for employment in food products, 1990

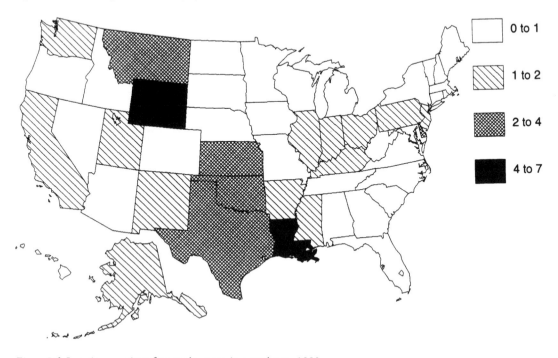

Figure 5.6 Location quotient for employment in petroleum, 1990

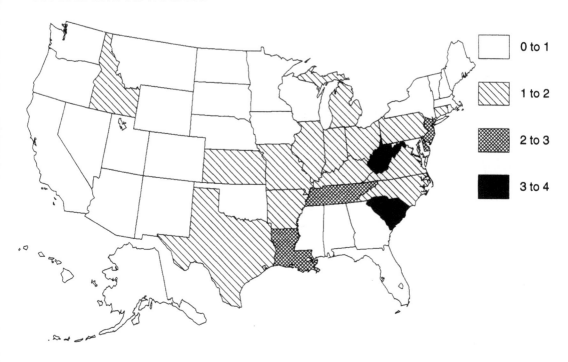

Figure 5.7 Location quotient for employment in chemicals, 1990

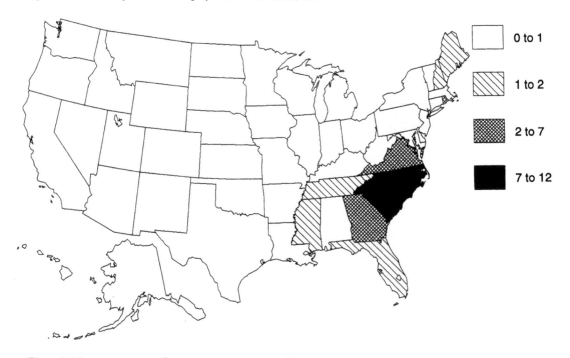

Figure 5.8 Location quotient for employment in textiles, 1990

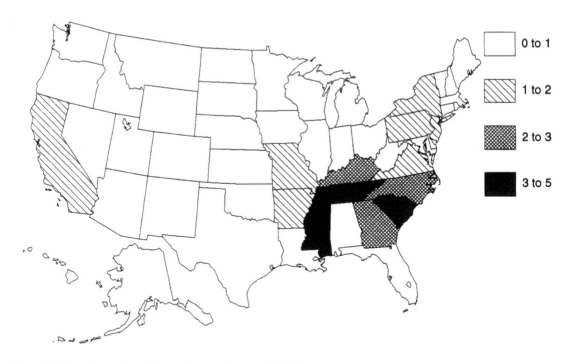

Figure 5.9 Location quotient for employment in apparel, 1990

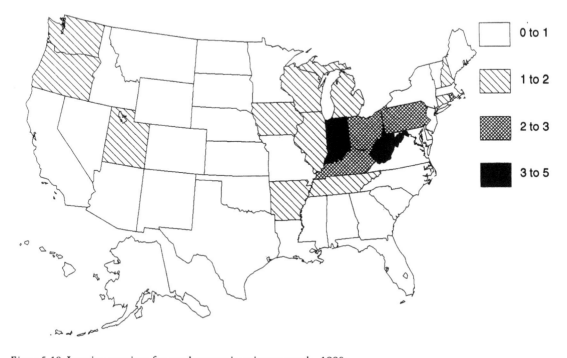

Figure 5.10 Location quotient for employment in primary metals, 1990

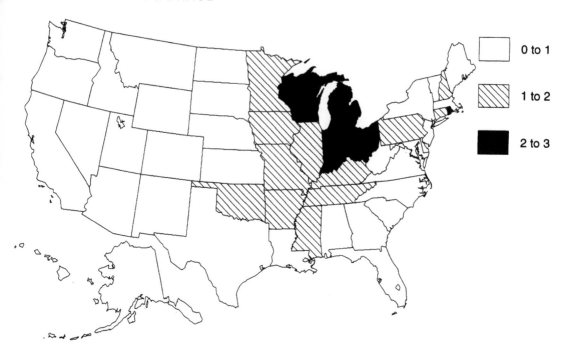

Figure 5.11 Location quotient for employment in fabricated metal products, 1990

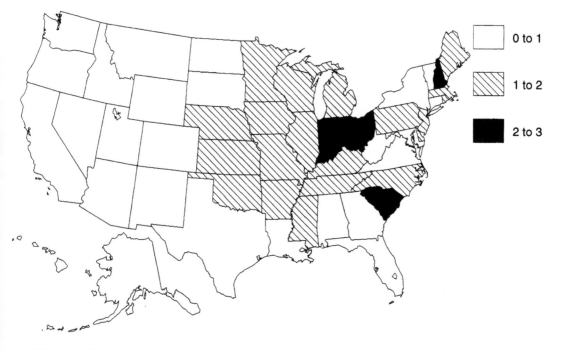

Figure 5.12 Location quotient for employment in rubber and plastics, 1990

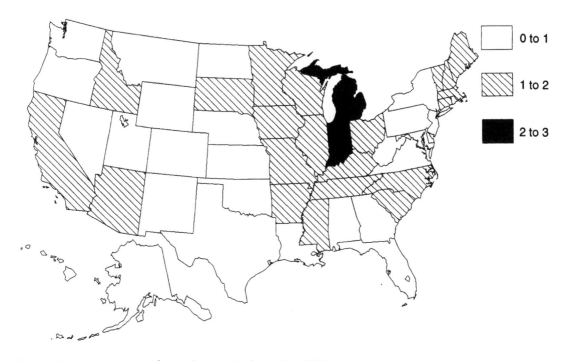

Figure 5.13 Location quotient for employment in electronics, 1990

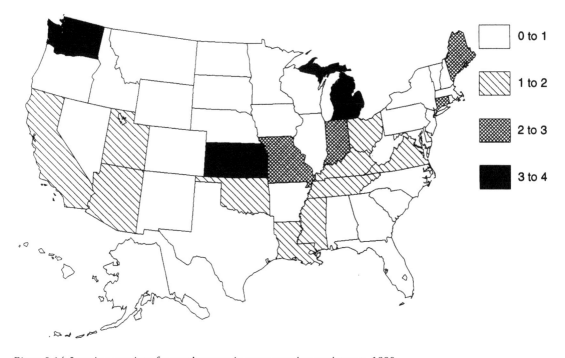

Figure 5.14 Location quotient for employment in transportation equipment, 1990

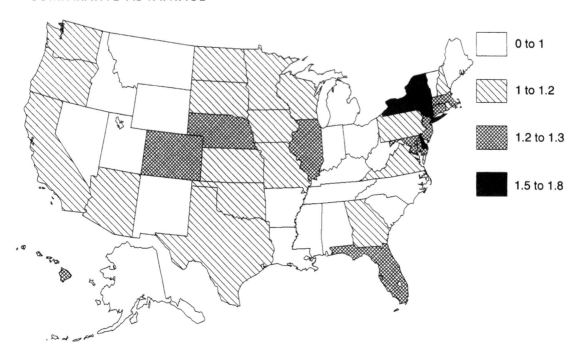

Figure 5.15 Location quotient for employment in finance, insurance, and real estate, 1990

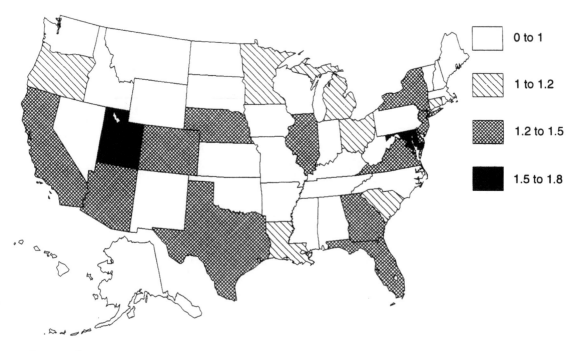

Figure 5.16 Location quotient for employment in business services, 1990

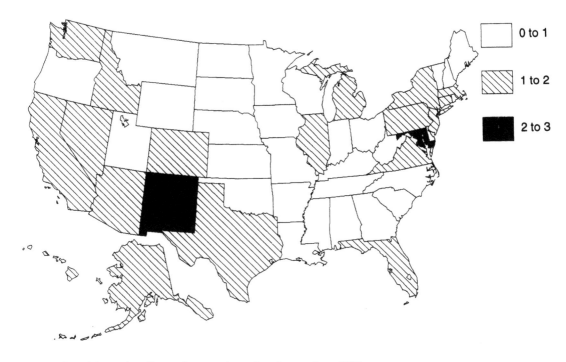

Figure 5.17 Location quotient for employment in engineering services, 1990

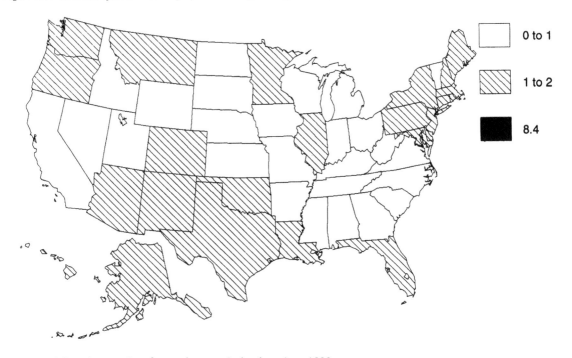

Figure 5.18 Location quotient for employment in legal services, 1990

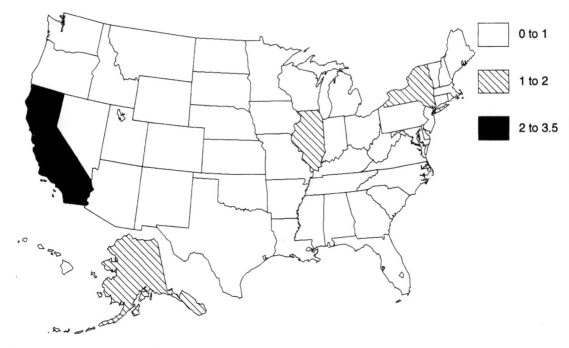

Figure 5.19 Location quotient for employment in motion pictures, 1990

Multipliers express the "ripple effects" of economic transactions as they reverberate through a local economy, and will be discussed in more detail later. Let us define m to be the ratio between total and basic employment, or

$$m = T/B \qquad (5.2)$$

Now, since $T = B + NB$, it must be that

$$m = (T + NB)/B = 1 + NB/B \qquad (5.3)$$

In other words, the multiplier can be approximated as one plus the ratio of nonbasic to basic employment. Once we know this, we can understand how changes in basic employment over time affect total employment in an area, that is,

$$\Delta T = m \, \Delta B \qquad (5.4)$$

where Δ (delta) is the mathematical sign denoting a change.

For example, suppose that a town called Jonesville has 1,000 people in its labor force (which is not to say that its population is equal to 1,000; the relations between total population and labor force size reflect local labor force participation rates, the age composition of the local populace, and so forth). The chief employer in Jonesville is a lumber plant that employs 400 people. If the lumber company adds 100 people to its labor force, what will be the effect on the local economy? We can answer this problem straightforwardly. Unless the people of Jonesville have an insatiable appetite for lumber and consume all of the company's output locally, most of the lumber company's output will be sold nonlocally, i.e., it is a basic sector. We know that $T = 1,000$. Therefore,

$$m = (B + NB)/B = 1 + NB/B = 1 + 600/400 = 1 + 1.5 = 2.5$$

In other words, every addition to basic sector employment will create 2.5 jobs total (including the one generated in the basic sector). Now we know that $B = 100$ jobs; therefore, the total change in Jonesville's employment is

$$T = m \, \Delta B = 2.5 \,(100) = 250$$

Thus, the change in total employment in a local area is much greater than the changes in basic sector employment. Why is this so?

Multipliers express a fundamental facet about all economies, that industrial sectors (and, ultimately, individual firms) are deeply interconnected to one another, a phenomenon generally conceived as a series of **linkages**. Linkages are the purchases and sales of goods and services that tie sectors and firms together. When firm A purchases something it needs to produce its output from firm B, for example, it is spending money on a backward linkage; but what constitutes a cost to firm A is a sale to firm B, i.e., a **forward linkage** to a client. *In other words, one firm's backward linkage is another's forward linkage, and vice versa*. When firm A purchases an input from firm B, its costs become part of firm B's gross revenues. Firm B will, in turn, spend money on inputs it needs, i.e., on its own backward linkages, which form revenues for firm C, which has its own backward linkages, and so forth. Linkages bind firms together into integrated production complexes, and may take many forms. A truck or train or airplane hauling parts from one firm to another is a concrete, empirical form of linkage; so is a meeting between a food producing company and an advertising firm, or a telephone call between an engineer and a client. Linkages therefore express both the economic and geographic structure of economies. When the revenues accrued to firms A, B, C, D, and so forth – and the jobs generated by these revenues at each step – are added up, they are much greater than the first transaction, from A to B. Depending upon the size, number, and distribution of linkages, multipliers reveal the ways in which changes in total output or employment reverberate far beyond changes in basic sector output or employment.

To illustrate the concepts of multipliers and linkages further, let us examine a more realistic example. In the 1980s, General Motors announced that it would build a new automobile plant to manufacture its Saturn cars. Numerous local governments, including states, counties, cities, and towns, solicited bids to attract GM, often promising tax rebates, labor training programs, zoning changes, and other advantages in the hopes of luring jobs. After an extended search, GM decided upon Spring Hill, Tennessee. Clearly, because the output of the new plant will not be consumed locally, this plant comprises a basic sector of the local economy, with forward linkages to consumers (including other firms) all around the nation and in other countries. The creation of jobs and incomes in Spring Hill clearly depends upon the "export" of automobiles and the revenues that they generate; should sales drop off, the impacts will be felt in local declines in employment.

CASE STUDY

General Motors' Saturn complex

In January 1985, General Motors announced plans to invest $3.5 billion in its new Saturn subsidiary, eventually to produce 500,000 compact cars per year. The new subsidiary and its new facility would use new or imported principles of supplier organization, worker organization and pay, automation of mechanical and information processes, plant layout, and logistical flow. Projections called for 20,000 auto-related jobs: 6,000 directly employed by Saturn at the project's completion, and the rest expected in supplier establishments that would be drawn to the area. *What* area? Upon making the announcement, GM said only that the facility would be within the U.S. GM established a task force, headed by its real estate subsidiary and a New York City based consulting firm, to conduct a massive industrial location search. The size and publicity of the announcement motivated 36 states and many more

Figure 5.20 Location of General Motors' Saturn complex

municipalities to propose over 1,000 sites and incentive packages, some of which were worth hundreds of millions of dollars. Saturn and GM announced their decision in July 1985: 2,400 acres in Spring Hill, Tennessee, a settlement of 1,000 people just beyond the suburbs of Nashville – only 35 miles from the 1983 Nissan assembly plant at Smyrna, Tennessee (Figure 5.20).

What motivated this important decision? The mid-section of the U.S. was heavily favored for **transport cost** reasons: many of the industry's suppliers are in the industrial upper Midwest, a reflection of the industry's history. More important for Saturn, and explaining the south-central sites of Japanese auto-assembly plants, is the southern and western movement of U.S. population and buying power. Southern attitudes toward **unions** and workers' rights are also attractive to large companies.

However, several states and untold potential sites are in this southern mid-section of the country. Why Tennessee, and why the Nashville area? Transport considerations, cited as key by Saturn officials, are favorable. Three-quarters of the U.S. auto market is within one day's transport of Nashville, via railroad and three Interstate highways. The state's voters agreed in the early 1980s to an increase in the state sales tax to augment local property taxes' support of public **schools**, specifically targeting math, science, and computer skills; Tennessee has no personal **income tax**. **Workers' attitudes** toward their work were also cited by Saturn and Tennessee officials, as well as by other manufacturers new to the state. **Electric power** (the facility requires 80 megawatts per day) and **fresh water** (the facility requires 4 million gallons per day) were plentiful and cheap. Using these quantifiable, cost-driven criteria, Spring Hill was a leading contender. However, these factors are replicable (or nearly so) 15–35 miles from a number of medium-sized cities in the middle of the U.S. Among the finalists reported in the press (not confirmed by the company) were Shelbyville, Kentucky (30 miles east of Louisville) and Schoolcraft, Michigan (12 miles south of Kalamazoo). However, only the Nashville area contained the two-year-old Nissan facility, which had induced a range of **agglomeration economies**: (1) supplier locations; (2) public sector education and training centers with programs relevant to advanced production methods; and (3) local labor market familiarity with new industrial work practices. While the region was becoming familiar with auto manufacturing, it was not familiar with the United Auto Workers union or with management–labor history at any **other GM facilities**. Distance from other UAW facilities reduced the interaction between Saturn workers and auto workers operating under more traditional contracts. On the other hand, GM and the UAW may have hoped that bringing a unionized facility so close to the Nissan plant might assist UAW efforts to represent Nissan workers, benefitting the union directly and benefitting GM by bringing a Japanese-owned competitor into a similar management–labor relationship.

Tennessee's constitution prohibits direct payments or credits to private entities, so no cash incentives were offered. However, the state agreed to spend $72 million on roads, worker training and education. The massive Saturn complex was financed through industrial revenue bonds sold by the county, to be reimbursed by the corporation, which thereby gained financing in the lower cost municipal bond market. Property taxes *per se* are avoided by having the county's industrial development agency hold title to the land; GM will pay fees to the county in lieu of, and less than the otherwise applicable, property taxes.

A year after the announcement, Spring Hill and its adjoining farmland had seen real estate appreciation up to 2,000 percent. However, GM's commitment to hire laid-off and mobile GM employees from elsewhere made it questionable how many local people would be hired when the plant was completed. Unlike the Japanese transplants, Saturn hired experienced and unionized auto workers, resulting in a sizable migration of workers from Michigan and other GM locations to Tennessee. For the corporation, the willingness to hire unionized labor, along with the union's participation in planning the new corporation and facility, bought close cooperation, wage concessions, and work-rule changes from the union.

Sources

Alexander, C.P. (1985) "GM picks the winner," *Time* 5 August: 42–3.
Borden, A. (1986) "G.M. comes to Spring Hill," *The Nation* 21 June: 852–4.
The Economist (1985) "Rings of Saturn," 3 August.
Engardio, P. and Edid, M. (1985) "Why a 'little Detroit' could rise in Tennessee," *Business Week* 12 August: 21.
Fisher, A.B. (1985) "Behind the hype at GM's Saturn," *Fortune* 11 November: 34–46.
Production (1986) "The General Motors 'Saturn' experience," 97 (3): 74–6.
Seamonds, J.A. (1985) "Will Nashville become the Detroit of Dixie?" *U.S. News & World Report* 12 August: 53–4.

How can we grasp the effects of the Saturn plant upon the economy of Spring Hill? To do so, we must address its backward – as opposed to forward – linkages. Because automobile production essentially involves the assembly of thousands of parts, the new plant will require numerous inputs from various economic sectors (backward linkages), including labor, steel, glass, rubber, plastics, textiles, and many other materials (most of which will already be processed in the form of prefabricated parts) as well as services such as engineering, legal services, advertising, accounting, and, if the plant is not internally financed by GM, banking. The Saturn plant's purchase of these goods and services – its backward linkages – are forward linkages for the firms that provide them.

What will be the impact of such a plant on the local economy? Rather than a simple basic–nonbasic bifurcation, a realistic analysis explores the types and nature of the backward linkages. Clearly most of the labor necessary to assemble the automobiles will be located near the plant, generating jobs and incomes for the people of the town. Many of the parts and supplies, however, will not be located in the immediate area; while some will be located (or induced to locate) within a fairly close distance, others will remain situated in other areas of the country (for example, near the traditional automobile producing centers in Michigan). To the extent that the Saturn plant relies upon parts "imported" into the area from other parts of the nation, the creation of local jobs and incomes is minimized. In other words, *backward linkages that extend into other regions constitute an "import" from one area and an "export" for other areas.* When GM subcontracts with local suppliers (e.g., a tool and die manufacturer, or janitorial services company, or printing company to print instruction manuals), creating revenues and profits for them, these firms in turn hire new employees. Thus, in addition to the **direct** effects of the plant at the factory itself, there are numerous **indirect** effects generated through backward linkages. Finally, the employees of the GM plant and of the local suppliers will spend their wages (and save some), in turn creating more jobs and incomes, or **induced** effects, in the establishments where they spend their incomes (e.g., local banks, movie theaters, retail establishments, and other nonbasic companies). Thus the impacts of the Saturn plant are complex, and depend upon the size, number, type, and distribution of its backward linkages and the resulting multiplier effects.

As a further example, consider the likely impacts of your college or university upon the area in which it is located. What are a university's forward linkages? Many students are quick to point out tuition, which is a principal source of revenue. However, depending on the size and nature of the university (i.e., whether it is privately or publicly owned), it may receive additional funding from state tax revenues, endowments from alumni, and grants awarded to its faculty. What are the impacts of a university? Unlike a steel plant, we must consider the impacts of the university itself and the impacts of the students it attracts. Clearly the student body will have an impact on the local economy, including the local housing market, laundry facilities, bars, movie theaters, and so forth, which may be considerable, especially in small "college towns" where the university is the largest employer. The university itself, however, will have a different set of impacts. Because education is a very labor-intensive process, most of these impacts will occur directly

through expenditures on salaries, including administration, faculty, and staff, and the resulting induced effects will be large. However, a university will also have other backward linkages, including some that will extend to other cities and communities and some that will generate revenues and jobs locally (including, for example, water and electricity utilities, janitorial services, maintenance and repair, landscaping, and numerous other services that are not necessarily performed "in house"). The number of jobs generated by a university will, therefore, differ markedly in number and type from those generated by an automobile plant. Different industries, employing differing quantities and types of labor and relying upon different mixes of inputs (including nonlocal subcontracting), can have widely varying impacts upon communities.

From these examples, some conclusions about the determinants of local multipliers – the relation between basic and total employment – may be offered. In particular, five factors influence the size of multipliers (and, hence, the relations between basic sectors and the regional economy as a whole):

1 The *number of backward linkages* is an important determinant of how many jobs (incomes, taxes, etc.) a particular industry or plant will generate. Obviously, all else being constant, an industry with more numerous linkages to suppliers and subcontractors (e.g., an automobile plant) will generate more jobs per unit output than one with relatively few linkages (e.g., lumber mills).

2 *Technological relations and factor input prices* exert a significant effect on the form and size of backward linkages, and, therefore, on multiplier effects. Technological relations express the amount of input necessary to produce a given quantity of output (i.e., efficiency). The more efficient an industry is (i.e., the fewer inputs per unit output it requires, including labor), the smaller its impacts upon suppliers will be. This notion is useful both in studying one industry over time and for comparisons of different industries. Similarly, factor input prices affect the choice of inputs that firms make in order to maximize

profits. As the cost of one input rises relative to others, firms tend to substitute into less expensive ones (as in the Heckscher–Ohlin model of comparative advantage). For example, when petroleum prices rose dramatically in the 1970s, automobile producers began to manufacture more fuel-efficient cars, partly by making them lighter by using more plastic and less steel, thus changing their patterns of backward linkages and associated indirect employment effects.

3 *Leakages* from a local economy constitute a form of regional "import," and hence forgone jobs and incomes, and an "export" for another region, generating jobs and incomes elsewhere. The spatial distribution of backward linkages (subcontracting) is thus vitally important for understanding the impacts of an industry on a region. Industries with large leakages will have relatively few impacts; conversely, those that rely extensively upon local suppliers will generate correspondingly more jobs. Often, local governments require firms to purchase inputs from local manufacturers when bidding on government contracts as a way of minimizing job losses in their region.

4 Expenditures by workers employed in a basic industry and its suppliers create *induced effects* that are an important part of the total impact of an industry on a place. **Induced effects** reflect several factors pertaining to the levels and nature of consumption, including wage and salary levels (which in turn reflect the occupational composition of employment, levels of local unemployment, and so forth); the extent to which workers save or spend their incomes; local taxation levels; the demographic composition of the labor force (i.e., its age and sex makeup; whether it has large numbers of young workers in their family-forming years, etc.); its ethnic composition; and preferences for particular types of commodities and services, all of which vary over time and space.

5 Lastly, *how one defines "local"* is critical to estimating multiplier effects. This is much more than an abstract semantic question; the boundaries of "locality" influence one's understanding of regional imports and exports, subcontracting, and so forth.

For example, if one's definition of "locality" includes a relatively large region (say, a state), there is likely to be a higher probability that basic industries will use "local" suppliers than if one adopts a very limited, constrained definition. In general, the larger the area encompassed by one's study area, the a higher the multiplier effects will be, in large part because broader areas will, by definition, have fewer leakages. As Figure 5.21 illustrates, varying definitions of "local" will capture different backward linkages and yield different estimates of multiplier effects, and, thus, total impacts. In this conception, the object to be analyzed (e.g., a new factory or building) is located in the center and firms supplying goods and services to it through backward linkages are indicated by the dots connected by lines to the center. A definition of local using the innermost circle, labelled I, would capture zero backward linkages and hence yield an indirect impact of zero; only the direct effects would be estimated. A definition based on circle II would include three backward linkages, and include more backward linkages than would one using circle I, while a definition using circle III would include six suppliers (three within circle II and three more within circle III). Note, however, even the largest circle does not include all the impacts of the facility in the center, such as international subcontracts to firms in other countries. In economic impact analyses, the choice of scale to be used in defining "local" is often arbitrary, and may lead to varying estimates of the impacts of a given event.

It is also important to remember that multipliers reflect the relation between *changes* – not only *growth* – in basic and total regional employment. (Employment multipliers are one form of this relation; so are income multipliers, output multipliers, and so forth.) Changes in basic employment, however, can be negative as well as positive, that is, multipliers can work as a "double-edged razor." The closure of a steel or auto plant, for example, generates multiplier effects in exactly the same way as the opening or expansion of a plant: in addition

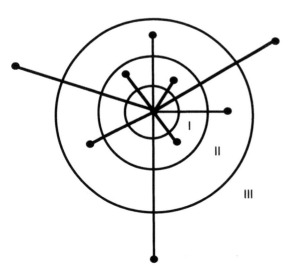

Figure 5.21 Linkage structures under varying definitions of "local"

to the direct loss of jobs, there are declines in sales to subcontractors (both local and nonlocal) as the basic firm cuts back on its purchases of inputs from them, and as workers are laid off, suffering declines in their incomes, the induced effects generated by their wages (from expenditures on housing, retail goods and services, and so forth) will create further rounds of unemployment. Thus, if a local multiplier for a glass producing factory is 2.4, the addition of 100 basic glass-producing jobs will generate a total of 240 jobs in the local economy; conversely, however, layoffs of 100 glass workers will create a loss of an additional 140 jobs (plus the 100 in the glass plant) in the local area. Local governments are also deeply tied to the health of local economies through tax collections: when employment and total regional incomes rise, so too do local property, income, and sales taxes; when local basic industries close down or relocate (as many did in the U.S. Rustbelt in the 1960s through the 1980s), local governments find their tax revenues dropping, inducing difficulties in their ability to provide public services (e.g., schools, police, fire, snow removal, etc.).

What does our examination of comparative advantage, export base, linkages, and multipliers tell

us about the health of regional economies? It is clear that regional economic specialization around a few industries in which an area has a comparative advantage brings with it both benefits and costs. Specialization is a useful way of stimulating export-driven job growth, an important mechanism for local economic development. Cities, states, and nations that specialize around growing industries generally enjoy steady, even rapid, increases in employment, incomes, and tax revenues. Texas, Louisiana, and Alaska, for example, whose economies rely heavily upon petroleum refining, saw explosive growth in their economies during the 1970s, when oil prices rose steadily. New York City saw a prolonged period of growth in the 1980s as its financial services sector grew rapidly, generating numerous well-paying jobs.

But the flip side to rapid growth, however, is rapid decline. Concentration around a few industries has severe costs when those industries suffer declines in their sales or prices of their output (such as farmers saw the price of grain fall in the 1980s). Thus, when petroleum prices plummeted in the late 1980s, Texas, Louisiana, and Alaska all suffered disproportionately, with rising unemployment and declining property values and tax revenues (while other regions enjoyed the benefits of cheap oil). When the national and global financial markets began a downturn in the late 1980s, New York City saw its specialization around banking work to its disadvantage.

In short, specialization can be dangerous. Because export industries link local areas to broader markets (the national and international division of labor), the degree of specialization can make a local economy vulnerable to fluctuations in the demand for its output (i.e., through business cycles). In general, therefore, the more diverse a local economy is, the less sensitive it will be to rises and falls in the fortunes of a few industries. Regional economies with several different basic sectors – a broad comparative advantage – tend to exhibit more stable employment patterns over time, as the growth of one industry allow firms and workers to adjust to declines in another. The southern California region, for example, has several different basic sectors – including aerospace, electronics, banking, tourism, motion pictures, and the port of Los Angeles, the nation's largest – and is thus well poised to ride fluctuations in the national and global economies. Conversely, many rural areas, with economies centered around the export of raw materials (minerals, lumber, agricul- tural goods, oil) tend to suffer more dramatic "boom and bust" cycles over time. Regional policy makers, therefore, frequently attempt to attract a diverse mix of industries to their local area in an attempt to min- imize disruptions over time.

SUMMARY

This chapter has explored the essentials of comparative advantage, a critical component in understanding industrial location, regional devel- opment, and international economic relations. To reiterate, capitalism creates uneven development of production and of markets over time and space – it is fundamental to the structure of market economies. For individual regions, this process entails specialization around one or a small handful of industries, or export-led "basic" sectors. In the absence of internal subsidies or external barriers to trade, these sectors include the activities for which the region is a good, rela- tively productive location. Geographic factors (climate, materials, topography), economic factors (local markets, labor, expertise), and institutional factors (the structure of government, labor rela- tions, support for entrepreneurship) all affect the comparative advantage of a region. The relations between a basic sector and the remainder of the local economy, the multiplier effect, are related to the number, type, and distribution of back- ward linkages from that industry. Thus, changes in the fortunes of a basic sector – for better or worse – will have disproportionately greater effects on the local economy. For this reason, regional economic specialization brings costs as well as benefits, and diversity allows regions to ride out local, national, and international busi- ness cycles.

SUGGESTED READING

Ellsworth, P. and Clark Leith, J. (1984) *The International Economy*. New York: Macmillan.

Porter, M. (1990) *The Competitive Advantage of Nations*. New York: Free Press.

Samuelson, P. (1976) *Economics*. New York: McGraw-Hill.

Walsh, V. and Gram, H. (1980) *Classical and Neoclassical Theories of General Equilibrium*. New York and Oxford: Oxford University Press.

TECHNOLOGY AND LOCATIONAL CHANGE

is chapter explores the nature and causes of iations in technology across establishments, npanies, locations, and times. We'll present some the ways in which technological change affects ility and industry location, and then some of the ys in which the local setting of an activity affects rate and direction of technological change.

ARIATIONS IN TECHNOLOGY

chnology is merely a systematic way of omplishing a particular task. Each of you has a hnology for reading this text and using it as a t of your study in industrial location. Some of a read hurriedly to skim ideas, and then may iew the text after hearing a lecture or partici- ing in a class discussion. Others of you read very ensely, highlighting and noting nearly every inition and point made and then not reviewing se notes at all until a final examination. Your rpose or your product is the same – to learn terial to be covered in some examination or future dy. However, based on your own abilities, engths, time constraints and past experiences, you ve devised different technologies for the same sired output. Similarly, different companies may velop different technologies for production of very nilar outputs. Is one technology "better" than ther? Yes, if one method of production yields tter or greater output with the same inputs of time d capital. Given the range of inputs available to u and your classmates, or to different companies,

or in different regions of the world, there is little surprise at the range of technologies in use simul- taneously to produce similar outcomes.

A given technology includes some possibility of **substitution** or modification in response to circum- stances. Perhaps your instructor suddenly changes the date of an examination, giving you a week's less time to prepare. Over the years, you've developed ways of coping with such problems, by changing your usual study technology. Most organizations can use their particular mix of technology in slightly different ways.

People have always been attuned to better ways of doing things. We may not always want to change the way we study, dress, dance, travel, cook, or work, but we are very sensitive to these **innovations** – new products or new ways of doing things. In a market-driven setting, organizations are also eager to become aware of innovations, which can greatly affect their own technologies or determine whether there will be a continued market for their products or services.

Technological Change

In addition, time brings *changes* in technology. Turning again to your personal study technology, your own methods have changed and will continue to change. The changes reflect your increased abilities to read and comprehend, as well as your increased knowl- edge of particular fields of study. Next year, more conversant with geographic and economic thinking, your reading of basic texts such as this will be much quicker, without as much re-reading or careful reviewing. In addition to these personal changes based on individual learning, the general technology of study will change. Twenty years from now, your children may find reading paper texts an amusingly time-intensive way of learning, compared to the communications and learning technologies available to them.

In analogous fashion, the experience of a company's managers and workers generally leads to changes in the technology used. The development of improved or totally new technologies is a constant part of contemporary business life. These changes, their geographic manifestations, and their regional sources, form the topic of much of this chapter.

Based on the previous chapters, it is easy to see some of ways in which technology and technological differences affect location. Cost-minimizing approaches to facility location are very sensitive to the input-factor proportions of production operations. An operation's process technology determines its mix of material and labor inputs, and determines how its input mix responds to variations in factor costs (including costs arising from transportation). The location analyst needs to know the degree of substitutability that is possible within a given technology, so that the assessment of potential sites can reflect the mix of factors that would be used at that site. Technologies differ in their scale economies, thereby affecting operations' requirements of a central versus peripheral market location. A basic way of distinguishing technologies refers to their **factor intensity**, the relative values of capital inputs, basic labor inputs, and highly skilled labor inputs per unit of production. Clearly, factor intensity influences the suitability of production locations, in terms of cost and local availability of suitable workers. Technologies and the resultant locational needs differ across companies and facilities in the same industry, because of their internal development of proprietary technology and because their adoption of particular non-proprietary technologies reflects their particular resources and strengths.

In most industrial settings, technological change has led to a reduction of the labor used per unit of output, especially in manufacturing industries

since the Industrial Revolution. However, just as technology can be simply defined as an organized means of accomplishing some task, technological change can be defined as any change in that organization or those means. Thus, technological change may be **factor neutral**, in which factors are used in the same proportions, but hopefully less of all factors are used to produce a given level of output. Factor bias in technological change can take many forms, from a reduced labor proportion or processing time, to reduced raw material input based on more efficient use of material input.

In the following sections of this chapter, we will investigate technological change and its influences on location of productive facilities at two very different temporal scales. We will first inspect forms of incremental innovation and technical change that are very important in the short term. This is roughly the same temporal scale as the period over which the production process for a specific product changes from design intensive to routine. Then, we will take a long, historical view of changes in manufacturing, and changes that may be expected in service provision. We will investigate the dramatic reorientation in manufacturing and service production in the 1980s and 1990s. This reorganization takes many forms and many names, but we will refer to them here as increased flexibility of production processes. Finally, we will ask what principles of industrial location, formulated in a time of expensive transport of heavy materials, are relevant in an age when production and competition relies heavily on information. For some companies and in some industries, the development of new proprietary technologies or the rapid adoption of new, linked technologies is of critical importance. At the end of this chapter, we will look at technological development itself as a locational factor.

Technology Diffusion

Not all technology is proprietary, or company-specific. Just as you have gained some awareness of your classmates' study styles and you may have been taught study styles at some point in your schooling, companies share information about technologies. This sharing is part of the process of **innovation diffusion**, the circulation of information about a new technology and the subsequent adoption of the technology by people or organizations. Like any process of social diffusion, the pattern of diffusion of a specific innovation can be studied across time, distance, and organizations. Innovation diffusion occurs through many different media. One very important medium for diffusing new or improved technologies is the *technological linkage* across industries. The new products of one industry are often the bases for improved production technology in another industry. Among the most important examples of technological linkages are the effects that changes in information handling and communication have had on industrial production. Rapid, regular increases in the power and speed of electronic integrated circuits provide the basis for regular increases in the power and speed of microcomputers and computer work stations. Combined with regular improvement in computer software to perform functions and to link computers, these new products have allowed a vast range of companies to modify their technologies for product design, manufacture, production logistics, marketing, and general management. The companies that purchase the new equipment, software, or services are adopting new products which become part of the companies' changed process technologies. Innovation diffusion by way of inter-industry linkages is pushed by two powerful motivations: the desire of the new product developer to produce and sell goods and services to other companies, and the desire of the purchasing companies to maintain a production technology as good as their competitors, who may also purchase the new goods and services.

There are other media of innovation diffusion. One general, open, and relatively inexpensive medium is the *trade journal or trade meeting*, through which new ideas and techniques are shared. The limits to this inexpensive medium include the necessity for each company to maintain staff with sufficient time and expertise to gain useful information this way, and the limited, general (but still potentially useful) kind of technological information

that will be disseminated in such a medium. *Employees* are a third medium of diffusion. Skilled employees come to their respective organizations with technological information – whether it is a technology for organizing market information, for evening production flow, or for improving product quality. The sources of their information include their formal training, their on-the-job experience and their former employers. While some companies make attempts to prevent departing employees from using extremely sensitive technological information in another company, this medium of technology diffusion remains very important in industries whose technologies are largely embodied in skilled people rather than in equipment, software, or off-the-shelf process design. Finally, there are more *company-specific channels* for sharing technological information. Companies intent on technological improvement often form strategic alliances or joint ventures to develop and disseminate technological changes. These new, sometimes short-lived organizations rely on their parent companies' complementary strengths in the complex process of identifying needs, developing technologies, and marketing new products or processes. After developing a useful innovation, a company or group of companies may sell full specification of the technology and the rights to use it, via a licensing arrangement. This often occurs when the innovating company does not have the resources to produce the new product or to use the new process in sufficient quantity to satisfy worldwide demand. Less sanctioned media for company-to-company diffusion include conversations between employees of different companies, or industrial espionage in which a company may attempt to uncover a competitor's technology via reverse engineering of products or surreptitious observation.

INCREMENTAL TECHNOLOGICAL CHANGE

Learning and Experience Curves

In most complex activities, productivity increases as the same person or group performs the activity repeatedly. Many terms have been developed to describe this process: learning by doing (especially relevant for the human learning component of the productivity increase), experience-based learning, the **learning curve** (a measure of the proportion by which unit labor costs fall as cumulative production of the same product doubles), and the **experience curve** (a measure of the proportion by which total inputs fall per unit of output, as an organization doubles its cumulative output of a product or product line). Among the causes of this unit-cost reduction with greater cumulative output are:

- *labor training* on the job as workers routinize their tasks and gain trouble-shooting expertise;
- *improved logistics* flow within the service or production process, with the removal of bottlenecks observed after the system is started;
- *reduction in waste or inefficiency* resulting from analysis of the production process after the system is started; and
- *improved communication* among production, design, and marketing within the organization, and between the organization and its suppliers and markets, as the participants learn what questions to ask in what format. This learning comes from exposure to the particular production and inventory process, and entails a systematic step beyond management learning about specific steps or processes or control to organization-wide learning about the entire process.

Note that experience-based learning is not automatic, but requires organizational attention and benefits from skilled people doing the work. Without a basic understanding of why something works, making it work better is purely hit and miss: good enough 50 years ago, but not today. This implies the following:

- Resources should be deployed to increase learning.
- The deployment of these resources should be timed in recognition that whole new technologies and systems will come along eventually (see the next form of technological change), which will reduce the utility of the earlier experience-based learning.

- What has been learned needs to be spread throughout the organization, so that no one person or group is indispensable. This spreading is not always possible.

Technological Change over the Product Life Cycle

The origins of the product life cycle concept are complex. This complexity is reflected in the many uses to which the concept has been put. The first use of the concept was in business marketing. The regularity of product introduction, growth in demand for a successful product, and the eventual decline of demand for a product have been noted since the dawn of consumer-oriented manufacturing. The interest of researchers in this regularity has varied from concern over the sources of innovation, the role that marketing can play in extending the growth of a product, to the determinants of eventual market decline. The locational implications of a product cycle were first explored by Edgar Hoover. In his books *Location Theory and the Shoe and Leather Industries* and *The Location of Economic Activity*, Hoover noted the role that urban centers play in the "incubation" of new products and processes. The dense interconnections of suppliers, the availability of highly skilled labor, and the communication with potential markets and sources of demand all encourage innovativeness – new products and new processes – in core urban areas. Hoover and Raymond Vernon, in *Anatomy of a Metropolis*, a study of the New York metropolitan region in 1959, revisited this concept of the urban incubator. They recognized that the land and labor expense of central New York City precluded the growth of most manufacturing industry there. Indeed, one of the principal findings of the New York regional planning study was the provisions for manufacturing activity in suburban locations of the region. However, there was a continued innovative or incubator function for central New York. The resultant manufacturing was generally of small scale, and the products were reasonably expensive. If and as demand for products took hold, there was a need for physical expansion which was most often met in more outlying areas of the region.

International Product Cycle

Vernon recalled this domestic manifestation of a product cycle and location years later. In the mid-1960s, he was concerned with economists' finding that United States exports had a higher labor/capital ratio in their production than the group of industries in the U.S. which faced extreme import competition. This finding contradicted the usual understanding that the United States has much capital available at relatively low interest rates, while American labor was relatively scarce and expensive. Traditional models of international trade suggest that a country with such characteristics should specialize in and export products that require much capital and little labor in their production. Vernon suggested that the process and the input mix for the production of products varies over those products' life cycles: labor intensive shortly after their development, becoming progressively more capital intensive. This variation in input mix can result in changes in the preferred location of production over the product cycle.

Specifically, Vernon noted that his model of dynamic trade and investment locations pertained to particular products and relied on particular assumptions. He was concerned only with manufactured products, and specifically those manufactured products that faced income-elastic demand. This means that as per capita income in a region increases, demand for these income-elastic products increases more rapidly than income. We can suggest two reasons for this income elasticity of demand. For consumer goods, households' demand for such products increases more rapidly than the households' incomes. These "superior goods" (automatic dishwasher machines, vacuum cleaners, home freezers) are demanded at an increasing rate by increasingly wealthy households because they save time within the household or satisfy needs that become more salient with increased income. Capital goods (automated equipment, specialty steel, more powerful motors, jet aircraft) may be income elastic if their use in a productive activity reduces the amount of labor required for that activity. Thus, as per capita incomes and wages increase in a region, the demand for such labor-saving capital inputs increases.

In addition, Vernon was thinking about international trade in a particular temporal context. In the mid-1960s there was a very recognizable hierarchy of nations when ranked by per capita income. The United States had the highest per capita income, followed by certain West European countries, then Japan. Other regions of the world faced rapid industrialization and rising per capita incomes. Per capita incomes, wages, and level of industrialization matched each other within this hierarchy.

Vernon's product life cycle model begins with the invention of a *new product*. This invention is assumed to occur where the first demand for this product is recognized. The coincidence of invention and demand suggest a market-led conception of innovation. (Alternatively, innovations could be based on improvements in materials or processes.) The first production of this new product is assumed to occur at or near the location of the invention. Where is this likely to be? Given our focus on income-elastic goods, and given our assumption of national rankings by per capita income, we would expect this first production to occur within the United States. (Note that the strict ranking of countries by per capita income is impossible in the late twentieth century. However, the general concept of market-led innovation and location of production is still one worth studying.)

What are the characteristics of this first production of a new product? Because the demand is small and uncertain, the scale of production is small. The labor inputs to the production process are large and entail skilled and technical labor. This reflects both the need to modify the new product in response to refinements in technology and demand, and the inability to rely on capital-intensive, mass-production techniques for a product that still faces an uncertain future.

During this new product stage demand is local, highly income elastic, and quite cost inelastic. The first users or purchasers of the new product are households with high incomes or great desire for the new product, or those productive activities that can gain the most profit from early use of the product. Thus, the price of the product is of secondary importance. As a result, the cost pressure on the first producer (who, by definition, is a monopolist for this unique new product) is not intense. This combination of process requirements for skilled and technical labor, the need for interaction with a rapidly changing market, and relatively low-cost pressure on a producer, all suggest that the production of the new product will occur at the first market, despite its high wages and costs. In the mid-1960s this meant in the United States. Thus do we explain the paradox of labor-intensive production in a capital-rich, high-wage setting.

During the *growth phase* of a successful product, demand rises to high absolute levels in the home market. In addition, demand first appears in countries or regions with slightly lower per capita incomes. This happens for two reasons. First, the postwar experience of the industrialized countries suggested to Vernon a steady increase in per capita incomes across nations, maintaining the rank hierarchy of nations. Thus, as time passes, nations with slightly lower per capita incomes than our innovating country will contain a small amount of demand for the

product. In addition, increases in scale of production and the possible advent of competition among producers of this growing product will lead to falling prices. As prices fall, the product sees effective demand in more locations. At first, this foreign demand is met by exports from the first or innovating country. Therefore, first-country production may rise even faster than first-country demand (see Figure 6.1).

In the third or *maturing stage* of the product, the product and production process lose some of the characteristics of newness. Local production is likely to commence in second and third markets, as demand in these markets justifies the establishment of reasonably-scaled facilities. In addition, local producers in these markets may put pressure on their governments for trade protection against imports from the first country, encouraging local production for the local market. The local producers can include foreign subsidiaries or licensees of companies in the first country, who gain access to those companies' production and marketing expertise. Production is likely to follow the expansion of demand in second and third markets.

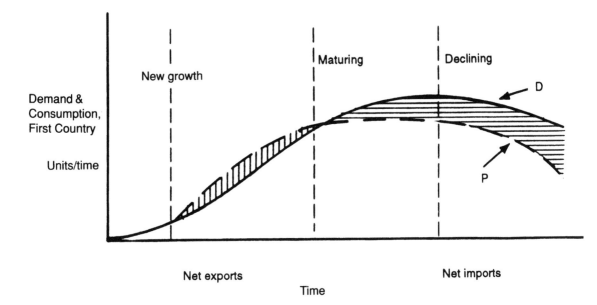

Figure 6.1 Demand, production, and trade over a four-period product life cycle

By the final, *declining stage* of a product cycle, cost cutting becomes the most important means of producer survival in this declining (but still quite large) industry. Standardization of the product's characteristics allows the use of large-scale, capital-intensive production techniques. In many cases, the labor content is both reduced in quantity and changed in quality: relatively little skilled or technical labor is necessary. Production will occur however and wherever cost can be minimized. Given the greater mobility of financial capital than of most workers, financial capital is invested where people can be hired at low wages. Thus do we explain the seeming paradox of capital-intensive production in labor-rich, low-wage locations.

Financial capital is especially mobile within a single, large corporation. In addition, the production technology for low-cost, mass production has most often been developed within large corporations (at least until the last quarter of the twentieth century). Finally, it is the large corporations that maintained the trademarks and distribution networks that allowed for national and international markets for mass-produced items. Thus, the product life cycle helped explain the tremendous advantages of multinational corporations in organizing world-wide production and distribution. In the next section, we will see the limitations of this model of production and investment.

The product life cycle resolves the paradox (observed in the 1940s and 1950s) by explaining that the United States exported manufactured goods when they were labor intensive and imported those same goods when they were capital intensive. Note that the product life cycle is a partial explanation, pertaining only to manufactured goods that face income-elastic demand. Some of its empirical assumptions bear little relationship to the world at the end of the twentieth century, but the ideas of production modifications and of mobile capital remain valuable.

Fundamental to the international functioning of the product life cycle is the mobility of capital and technology. As has been noted before, financial capital is extremely mobile. Technology, however, is most mobile within the context of an organization. An organization – a government, or a private enterprise – can allocate the flow of technology among its parts without negotiating the price to be paid for technology. The multinational corporation is the conduit for both the capital and technology flows in the international product life cycle. Note that the productive factor that is *not* internalized within the multinational corporation – labor – remains a characteristic of places. This relative immobility of labor forms a fundamental motivation for the locational change postulated during the product life cycle.

So we have a model that includes some interesting features. Certain countries or regions export newer, labor-intensive products because those regions are adept at product innovation. That deftness lies in their institutions and their highly skilled labor, each of which is difficult to transplant or relocate. Capital-intensive production processes can be located in labor-rich settings, because the cost of capital for a large corporation does not depend substantially on where the capital is invested in production. The availability, cost, and conditions of labor do vary geographically, and especially internationally. This reflects the sovereignty of nations to establish their own labor regulations and to restrict immigration (labor mobility). Finally, there is a rapid international movement of technology, within the large corporation.

Industrial Filtering

At the sub-national, inter-regional scale, product cycles (and broader industry cycles) have been used to present general models of **industrial filtering**. The assumptions are similar to Vernon's international product cycle, with standardization and scale allowing the development of specialized manufacturing equipment that embodies technology, thereby substituting rare, highly skilled workers with much less highly skilled workers. The wage rates for relatively unskilled workers are assumed to be lower in more peripheral, less market-oriented regions than in central regions with relatively many highly skilled workers. When the industry's scale and technology have advanced to the point that much of the technology is designed into specialized, fixed capital, the greatest locational pull is that of plentiful, low-cost labor. Again, the model pertains to industries whose inputs and products are not difficult or expensive to transport, so that non-transport considerations loom large. For large companies in mature industries, financial capital is not a locational constraint. Therefore, the optimum production location becomes a low-wage (presumably low-skill-endowment) location. Where are these low-wage and low-skill-endowment areas? They include small urban centers, rural areas, and less industrially developed regions.

This description of industrial filtering differs from the international product cycle in the lack of a threat of trade barriers and in the generally smaller wage differentials across the regions of a country than across the countries of the world. Like the international product cycle, it explains certain historical periods and certain types of manufacturing better than others. The decline in textile production in the Northeast United States and the industry's concomitant investment in the Southeast during the early twentieth century have been explained in these terms. Mid-twentieth-century trends toward the industrialization of non-metropolitan areas of high-income countries have been explained with this logic, along with the *general* technological changes of this period (especially the automobile, truck, highway, and potential for smaller scale plants).

CASE STUDY

U.S. semiconductor components companies

Two very different kinds of models have been used to describe the dynamics of manufacturing investment in the semiconductor industry, responsible for the transistors and later integrated circuits and microprocessors that power the information revolution. Models of product life cycles or industrial filtering emphasize the industry's movement from its World War II origins in metropolitan New York, to its postwar, military-related development there and in/around Boston, to its rapid expansion and commercialization in and near Palo Alto, California. From these centers of innovation (which include more single company focused centers in Illinois, Texas, and Arizona), the industry has spread to nearly every state in the U.S., to strongholds in Japan and western Europe, and then to rapidly industrializing countries, especially in East and South Asia. On the other hand, models of spatial division of labor emphasize that these different kinds of locations are home to very different kinds of operations. Most of the product innovation, and much of the process innovation, originates in facilities at or near corporate headquarters in California, Japan, Germany, and the Netherlands. Processes that require close tolerances and continual engineering input have spread to lower cost locations, often within the home country of the company. Companies quickly located labor-intensive assembly operations in low-wage countries, though this pattern became more complex in the late 1980s, as assembly operations became more automated and as low-wage countries became important markets for electronic components of final products assembled in those countries.

However, both types of models treat the industry as the basic unit, when in fact it is the individual company that makes these fixed investments. Companies make these investments in light of their individual corporate strategy or circumstances: capital availability, product mix, market segment and location, and vertical integration. Along all these dimensions, companies such as IBM, Intel, and a small manufacturer of semi-custom integrated circuits are very different. Their distribution of semiconductor manufacturing facilities are very different, as well. IBM has invested heavily in a few, massive facilities in a few countries where the company needs the most components: all of IBM's electronic components are sold to IBM computer and systems divisions. Intel has a worldwide network of design, fabrication, assembly, and testing operations, following both market and cost-minimizing orientations. Small manufacturers, especially those producing small numbers of relatively customized products, appear in large numbers in electronics agglomerations such as Santa Clara County ("Silicon Valley") and metropolitan New York, but are very rare outside of such agglomerations. Thus, individual companies in the industry may combine elements of industrial filtering and spatial division of labor (especially in an industry such as this, in which rapid innovation occurs simultaneously with unskilled-labor-intensive, non-market-oriented processes), or may have a locational configuration that differs from either model. Understanding the needs of the individual company is required.

A final complication in the investment decision of semiconductor companies is the prevalence of **subcontracted** work. Worldwide, thousands of companies have as their principal activity the design, fabrication, testing, or assembly of electronic components to the exact specifications of individual clients. Testing or assembly subcontractors receive the components from the client, and return the components to the client. The location decision for subcontractors reflects the location of clients' operations, the location of potential clients' operations, and the labor requirements of the particular operation. It is not surprising that companies whose business is mainly subcontract work are concentrated in electronics agglomerations. For any electronics component manufacturer, the presence of subcontractors influences both location of owned facilities and decisions to invest in a facility versus subcontract. Small companies are more likely to subcontract than are large companies, to take advantage of subcontractors' scale economies in a specific operation. Products for which strict quality control is required (for example, products for military use) are more likely to remain in-house, under the full control of the principal company. Components that require less sophisticated equipment (for example, older, less intricate components) are more likely to be produced totally in-house.

The complexity, value, and transportability of this industry's products makes the decision to invest in a company-owned facility, as well as the location of the facility, contingent upon process-specific and company-specific characteristics. The widespread notion that these facilities are "footloose" is not very accurate. Rather, the investment needs are very complex, and sometimes very subtle, not dependent on a single concern such as cost, materials, markets, or even engineers.

Sources

Harrington, J.W. (1985) "Corporate strategy, business strategy and activity location," *Geoforum* 16(4): 349–56.
Malecki, E.J. (1985) "Industrial location and corporate organization in high technology industries," *Economic Geography* 61: 345–69.
Scott, A.J. and Angel, D.P. (1988) "The global assembly-operations of U.S. semiconductor firms: a geographical analysis," *Environment and Planning A* 20: 1047–67.

TECHNOLOGICAL CHANGE OVER THE LONG RUN

General Trends in Industrial Technology

If we go back to our cost-minimizing model of industrial location, we can see how changes in industrial production during the past one hundred years have changed the basic nature of industrial location. One of the most dramatic changes in industrial production has been the *improvement in transportation* capabilities and infrastructures. In all regions of the world, governments and corporations have invested in roads, railroads, canals, ports, airports, and freight handling facilities of all sorts. Along with technological improvements in transport carriers, these improvements have dramatically reduced the cost of transporting a ton of any given item over any given distance. What does this do to industrial location, given our cost-minimizing model? Clearly, the expense of transporting inputs to some central production point, and of distributing products to market, declines. This means that the relevant market area for which a production facility is contemplated is likely to increase. Indeed, the period of railroad building in the United States during the second half of the nineteenth century allowed the development of nationwide markets. This, in turn, had a dramatic effect on the corporate distribution of activity in the U.S. Companies found it possible to invest in a trademark or patentable innovation, to market the resultant product nationwide, and to charge a premium nationwide for the assurance consumers felt because of the trademark or the utility derived from the innovation. In many industries, there is nationwide monopolistic competition based on product differentiation. The development of nationwide markets and nationwide trademarks created the twentieth-century industrial landscape and industrial competition. Large companies could develop technologies that exploited and created great economies of scale and could invest in the technology and the marketing, including advertising, of proprietary trade names and proprietary product standards.

As the importance of transport costs decreased, the relative importance of immobile factor differences increased. This means that regional differences in labor wages across the U.S. – and increasingly around the world – could pull cost-minimizing manufacturing investment far away from some absolute minimum of transport costs. This allowed the industrialization of areas around the United States and around the world, including the southeastern United States, and more recently Southeast Asia and Latin America. However, the fact that not all low-wage locations have seen industrialization suggests the concomitant increase in importance of other immobile characteristics of places. These include government policies toward external investment, taxes, political

stability, and investors' willingness to invest in particular areas or regions or countries.

At the same time that transport improvements reduced the importance of logistical considerations relative to other regional characteristics, *manufacturing has become much less oriented toward heavy, bulky, weight-losing raw materials*. For one, the utilization of raw materials has become more efficient with less waste or unused by-products. More importantly, the world's economy has become more complex as increased market size allows facilities and whole industries to break manufacturing and distribution into fine distinctions. The increased number and complexity of industries means that a much higher proportion of manufacturing operations start with finished components and add value by assembly or modification, or mere breaking of bulk. From our cost-minimizing model, reduction in the amount and number of bulky, weight-losing input increases the prevalence of market locations for manufacturing facilities.

Thus we see that merely these two large changes in industrialization over the past one hundred years – improvements in transportation technology and infrastructure, and increases in the complexity of the manufacturing–distribution–marketing sequence – can lead to substantial differences in the basic nature of industrial location. Regions or countries that are the sources of bulky raw materials have seen their manufacturing base wane relative to regions that have large market areas or regions that have other important immobile characteristics, such as inexpensive labor, low taxes, or other characteristics that encourage production investment.

Temporal and Spatial Clusters of Technological Change

The changes presented above did not occur all at once, of course. Nor did they occur in a smooth sequence over the past 225 years. The series of improvements in transportation and the changes in production structures and methods were not unrelated. Indeed, many economic historians have marveled at the temporal and spatial connections among major innovations. Economic history in western Europe and the United States is widely interpreted as a series of periods of intense, interrelated technological and economic changes, each followed by a period during which these changes were digested by the economic, social, and political systems. These long cycles of change have been named Kondratieff cycles, after the Russian economist who proposed them in the early twentieth century.

The first such cycle began with the Industrial Revolution in Great Britain and the Northeast United States, toward the *end of the eighteenth century* (see the discussion of the Industrial Revolution in Chapter 7). This first great spurt of mechanized manufacturing was built upon *steam power*, which allowed operation of large-scale factories with heavy equipment. Large, centralized plants affected the way workers lived, demanding physical separation of home and work and temporal separation of home-day and work-day. Each plant and each local area had a functional and industrial specialty, with textiles being one of the first sets of products to yield to such mechanization. However, because of limited transport (horse-drawn carts and longer distance canal and river barges) market areas were relatively small.

In the *second half of the nineteenth century*, the *railroad* allowed much larger market areas and much greater regional specialization, in coal, steel, textiles, and heavy machinery. Businesses began to organize to take advantage of budding national markets and plant specialization: this process took the entire period, culminating in giant companies that gained national monopolies. It also culminated in the vast manufacturing and transportation centers of the British Midlands and the American Northeast, within which factory life became the norm. Within the plant, work tasks were even more finely defined, including the beginning of geographic separation of headquarters and marketing functions from the production facilities.

In the early twentieth century, manufacturing, urban form, and urban life were transfigured by widespread *electric power and the automobile* and truck. Electricity allowed smaller scale factories (and the concomitant decrease in the sizes of markets that could be served by a factory; now a range of factory sizes and market areas was feasible). Automobiles (and trucks or lorries) gave people and companies private, flexible, fast transport with low fixed and low terminal costs. This encouraged suburbanization of industry and of residence, and a slow movement toward industrialization in formerly remote areas (most especially the Southeast United States). Automobile parts, assembly, fuel, and repair activities dominated the mid-twentieth-century industrial landscape in the industrialized countries.

After World War II, several key innovations grew out of the massive technological efforts in support of the war. *Synthetic materials* reduced industrial dependence on extracted input linkages (iron ore, basic metals, textile fibers), increasing the number of manufacturing steps in most production. *Electronic communication* and *air transport* have allowed increased separation of corporate control and manufacturing operations, and have helped create worldwide markets and worldwide production capability.

In each period, key, linked industries grew most rapidly in a particular set of regions, which took on industrial, labor, political, and educational characteristics suited to the region's dominant and (for a while) rapidly growing industry base: New England mill towns, Pittsburgh steel plants and Buffalo machinery works, Detroit automobiles, Silicon Valley electronics and southern California aerospace (Figure 6.2). Note the historical rhythm of regional dominance and decline. We will make more of these regularities in Chapter 7.

Technological Change away from Mass Production

The historical interpretation of long waves or cycles of economic activity, based on significant, general changes in technology, can be augmented by analysis of social and institutional changes. A set of powerful interpretations result, which focus on fundamental changes in the organization of production, consumption, and social institutions during the last twenty years of the twentieth century.

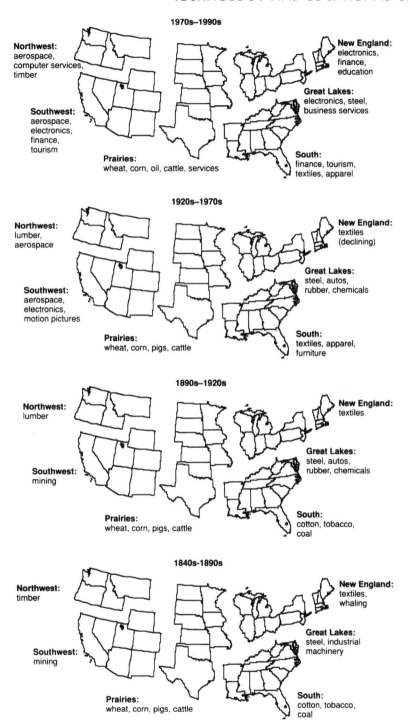

Figure 6.2 Regional industrial structures of the U.S. in four Kondratieff cycles

Mutually supportive systems of production technology, development of markets, organization of labor and of enterprises, and governmental policy have been called "**regimes of accumulation**." Recent interpretations have focused on a perceived shift in these regimes, a gradual change in the organization of industrial societies. The system of institutions and technology that has dominated the U.S., Canada, western Europe, and Japan has been given the label Fordist. The **Fordist system** entailed implicit agreement among business, workers, and government to support the development of mass markets to consume the rapidly increasing output of large, efficient plants and organizations. The institutions that were formed, repeatedly clashed, and reached working agreements included large corporations that claimed competitive advantage based on mass production and scale economies, large labor unions, and (essentially capitalist-supporting, but worker-supported) labor parties in government. The instruments of the agreement included progressive wage settlements and tax policies, mass-production technology, public infrastructure (to support geographically extensive growth and tremendous suburbanization), credit availability for investment and consumption, and social security. Continual capital intensification of production provided regular increases in labor productivity, which supported regular increases in labor wages. The increased wages supported consumption of new products and services, employing both new workers and those workers displaced from capital-intensive industries. Rapid rates of corporate and residential investment accelerated economic demand.

The successful operation of the Fordist system over three or four generations created several problems for the system's continued operation. First, the expectation of steadily rising wages slowly but surely surpassed actual rates of productivity increase, causing rising inflation and smaller profits. Second, more and more finely divided tasks for individual workers and for individual facilities reduced the interest of workers in their jobs and increased the reliance of the system on massive, sustained demand. Third, the steady expansion of markets, required for

ever-increasing production scale, led to corporate and national searches for export markets, ultimately increasing international economic competition to acrimonious levels. Fourth and related to this is the increased logistical and managerial difficulty of corporate organization of far-flung, interdependent establishments and supply linkages, and the increased sensitivity of the system to supply shocks (such as oil availability and price) anywhere in the world. Fifth, the operation of the international product cycle model, aided and abetted by indigenous industrialization in low-wage, labor-rich countries, created export platforms where (for a generation or so) production occurs at wage costs that do not support consumption by local workers. Thus, all that production is exported to wealthy countries, whose basic manufacturing workers, suddenly in competition with global labor markets, are squeezed financially. Global production capacity increased more rapidly than global consumption of mass-produced business and consumer goods.

Some observers have suggested that new technologies — computer software development, ever-higher quality communication systems, biotechnology, advanced materials science — will reduce these difficulties, by increasing productivity, reducing material requirements, improving international logistics, and providing new products for design, production, and marketing within the advanced countries. Others argue that the current period (the last 20 years of the twentieth century) is a confusing period during which such solutions are being attempted by Fordist institutions (e.g., large corporations and large labor unions), even as a post-Fordist system is evolving.

A **post-Fordist** system or regime of accumulation uses new technologies and the global spread of industrialization to supplant large, centrally controlled corporate, labor, and even social institutions with complex, interconnected webs of smaller institutions. Signs of this shift surround us: the increase in entrepreneurship in many countries since the 1970s, halting attempts by labor and management organizations to coordinate goals and reduce restrictive occupational classifications, the increase in materials and services supplied to large corporations from separate but closely coordinated companies. More generally, the post-Fordist scenario entails increased use of more flexible employment relations (like subcontracting, part-time work, and unofficial employment); increased reliance on temporary networks of companies that join forces to develop, produce, and market particular goods or services; increased reliance on the skills and information of individuals who can broker these networks and devise creative means of understanding market needs, producing quality goods or services, and continually monitor individual clients' needs. Goods and services should become more specialized, adding greater value to their users than standardized products, and therefore commanding a price that reflects their closeness to business or consumer needs.

The stable, large corporation and the labor union become less central actors, and less critical providers of employment, health care, pension benefits, or charitable contributions. New forms of social control and social assistance must be devised. This implies new, yet uncertain, roles for governments. In addition, communication, transport, and familiarity become critical requirements for the forming and re-forming of production networks. The implications are unclear, and are probably heterogeneous. On the one hand, advanced means of communication and control allow far-flung networks to form. On the other hand, there is an undeniable benefit of proximity of key members of a network – not in single-industry agglomerations, but in localized regions of related production, service, and information activities. This chapter's final section explores this regionalization of technology and production. Additional heterogeneity will result from the steady entry of more of the world's population into a mode of consumption. Some of this consumption may be satisfied by (and financially supported by) mass production, likely within the emerging markets.

The concept of "flexibility" appears in different manifestations within discussions of these issues. We can present four groups of changeable, transient, adjustable systems that are present in current and evolving production, consumption, and social systems.

The first and smallest system is the **flexible machine**: programmable tools and robots that can replicate a range of actions to contribute to the production of many different products, with little or no "down time" between product types. This allows a single production line to yield a variety of products in a capital-intensive way that modifies the requirements of traditional mass production.

Next comes the **flexible manufacturing system** (FMS), which is an integrated system of design, manufacture, and distribution that relies on continual information from the production lines and the external market, as well as computer software to use this continual information to match supply flow, production, and distribution on a product-for-product basis. Optimally, shifts in demand result instantly in changes in supply orders, production scheduling, and if necessary, product design.

These close linkages come together in a production system called **flexible specialization**, entailing (1) flexibility of supplier relationships; (2) flexibility/customization/specialization of products supplied; (3) flexibility of the uses to which workers are put; and (4) flexibility of the numbers of workers used (through the use of part-time, informal, or subcontracted labor). Flexible specialization is an organizational strategy that allows a given firm (a) to specialize in some product, component, or market niche; (b) to respond quickly to changes in product characteristics or market demand; (c) to rely on other firms for inputs or product finishing; and (d) to change its specialization quickly.

Taken together, these forms of flexibility will affect the size distribution of companies, the nature of corporate interactions, the relationship between companies and employees, the role of governments in sustaining the population and the economy, and a potentially dualistic market structure that describes the over-arching system or regime of accumulation called post-Fordism, or **flexible accumulation**.

In some ways, the evolving system provides more capability for small companies with astute, entrepreneurial managers to enter global competition. Such a company must be linked into networks of other companies. To the extent that these networks benefit from localization, this will be a powerful locational need on the part of such smaller companies.

Industrial Location in an Information Age

The **information economy** can be defined as the set of activities and economic relationships shaped by and reliant upon "**infomatics**," the combination of electronic telecommunications and computers. This powerful combination is occasioned by the conversion of telecomms to digital technology. The information economy includes the production and improvement of the equipment and software that makes the combination possible, but encompasses much more than those sectors. Infomatics is very powerful and affects so much, because it brings the capability to manipulate, combine, use, view, and analyze images, text, instruction, and numbers in different places and at different times, anywhere the communications infrastructure is present, as long as the parties have access to hardware and software.

However, those caveats – where the communications infrastructure is present, and access to hardware and software – are almost as powerful as the new technologies. With respect to certain elements of the information economy, the location of information (or of the generation of information) is not very important, but access to the information network is very important.

Access depends on communications infrastructure, hardware and software, ability to pay for the information, and organizational relationships among information sources and users. Communications infrastructure was not much of a locational issue within wealthy countries during the middle years of the twentieth century. Governments assured the nearly universal availability of basic telecommunications infrastructure, either through government-operated telecommunications networks or by insisting that private companies grant nearly universal access to basic telecommunications services in return for government sanction of the companies' monopoly status within particular regions and for particular services. Basic telecommunications includes the international network of hard-wired lines for voice communication. However, infomatics includes far more than basic telecommunications, and these newer, powerful services are increasingly offered on a competitive basis by private vendors operating without strong governmental mandate for universal access or low prices. These advanced services (for digital transmission of data, images, software, or entire programs) have been offered among sites within large, private companies since the early 1970s (combining satellite and microwave with hard-wire transmission, and adding data-manipulation capabilities as these have become available), providing substantial interconnection and information transfer among these companies' far-flung operations. In these cases, however, information flow is impeded not by distance but by organizational relationships.

Telephone, cable-television, and computer-communication utilities are currently competing for regulatory permission to offer a wide variety of communications technologies on the open market, to whoever is able and willing to pay. The market structure and pricing strategies that result from the current regulatory competition will influence the availability of remote information management and processing to small businesses and households. This, in turn, will influence the degree to which "local" economic operations are dependent upon local labor forces. An even finer spatial division of labor may result from the ability to operate a small company or a set of functions across individuals in home offices or individually rented spaces.

Issues of hardware, software, and ability to pay create new barriers to the flow of technical, marketing, or other information. Again, these barriers are not purely geographic, but rather create a new geography of knowledge, where the frontiers are between small, under-capitalized firms and larger firms, or between poorer and wealthier households.

For those organizational and geographical spaces where the infrastructure and on-site facilities do exist, each software and hardware improvement increases the amount and type of information that can be gathered, manipulated, and used in far-flung locations, with little cost of time or distance. This reinforces the ability of organizations to operate in many different locations. Despite its name, the information economy still relies on labor, still sells to people and businesses around the world, and still faces a hotchpotch of tax rates and structures and trade barriers around the world. Therefore, the motivation to operate in far-flung locations remains: the need for local presence in far-flung markets, or the ability to use relatively immobile factors like labor. The global organization performing tasks wherever trade barriers, taxes, or labor laws encourage those tasks is strengthened.

Examples of such an international division of activities within the global corporation abound. American insurance companies operate large claims-processing operations in Ireland, using trained local workers to analyze claims, make decisions, and conduct the necessary "paperwork," all via computer. The kinds of information linkages required occur almost as easily across the Atlantic as they might across a metropolitan area. Locations in Jamaica and elsewhere in the Caribbean stand to gain substantial information-processing activity as high-volume telecommunications linkages are established there. These activities do not have to be contained within one enterprise. Brokers are available to find contractors that can receive paper forms from companies, hire people to enter data and text into computers, and send the digital information to the originating company via satellite. Improvements in digital imaging and image-reading technology may reduce the attractiveness of international exchanges that involve only data entry.

The availability of so much information transmission capacity will encourage the evolution of worldwide associations of organizations, sharing information, tasks, and profits (or risk) in certain cases but maintaining organizational identities for purposes of flexibility, managerial adaptability, or nationalism. Information networks can allow for selective sharing of information and resources, according to prior arrangement for sharing risk. For the individual organization, formerly clear-cut questions of control, location, and ownership of operations are becoming increasingly complex. The organization may not control all aspects of its information network (e.g., its input procurement for certain projects may be handled by a partner organization). The organization may regularly augment its productive capacity with capacity within other organizations, in other places. These collaborative arrangements may accompany joint ownership of projects or ventures (a new aircraft, a franchise in a new country). The facility location decision loses some of its immutability, as organizations make use of others' facilities, as needed. These collaborative efforts are information-intensive, so that the ability to transmit and manage the information is the limiting factor in the extent and complexity of the collaborations.

Another crucial part of the information economy is in the human ability to create, manage, and act on useful information – useful in creating value by designing, producing, controlling, leasing, or selling goods and services. In a capitalist system, these human abilities cannot be completely owned by the organizations or individuals that rely upon information. Rather, employees and consultants exist in a mutually voluntary relationship with a particular organization. (Most household heads have to work under some sort of wage-labor situation, but they and their employers each have the ability to terminate a particular relationship.) The employee or contract worker wants to maximize the returns to his or her abilities, from any of the organizations with whom (s)he might work. The employer or contracting organization wants to minimize the costs of finding and paying for the skills required.

For the employee, these needs suggest two options, each requiring some flexibility. He or she can develop abilities that are in high demand in a particular location. This assumes the availability of a range of training sources, as well as requiring the worker's knowledge about opportunities and his or her ability to obtain skills at affordable cost. Given the importance of job-related training and of local, social information networks about employment opportunities, the result of this tendency is agglom-erative, insofar as people develop skills useful in their locale because of the mix of local activities. The other option is to locate where his or her abilities are in high demand. This assumes knowledge and mobility on the part of the worker: attributes not available to every worker. This option also reinforces agglom-eration of activities, insofar as people with skills and abilities important (and sometimes scarce) to particular activities locate themselves where their skills are in high demand.

For the operating organization, these needs suggest agglomeration when the required, human-embodied skills are scarce. When few human-embodied skills are required, when the required skills are not scarce, or when the skills can be easily transferred to workers, other considerations become important, such as labor control or labor wages, or governmental taxes and policies.

THE GEOGRAPHY OF TECHNOLOGICAL DEVELOPMENT

The normative and descriptive approaches we have used for the location of production facilities are of some use in our discussion of the location of technological development. The relevant considera-tions within each approach definitely must be changed – from materials logistics and low-cost labor to information exchange, inter-disciplinary cooperation, technical human capital, and adaptable workers. Technological improvement is a critical determinant of rates of productivity increase and thus of returns to labor, land, and investment capital. Technology development also influences the quality and length of human life, as in health-care, food-raising, or food-storage technologies. From a regional or national perspective, we are concerned with the attributes of places that generate the preponderance of techno-logical innovation: how does our region or nation compare with others? Is there something that public policy or private action can do to improve the rate (and to affect the direction) of technological develop-ment? However, recall that innovations diffuse across

companies and locations. Therefore, we are also concerned about the characteristics that affect the rate and direction of diffusion and adoption of innovations. From a corporate perspective, understanding the locational needs of technology-generating activity (from free-standing research and development to on-going improvements of sales or production methods) allows better allocative decisions to be made. Such understanding will also allow corporations to respond in a more informed fashion to the marketing ploys of regions and nations that proclaim themselves as "technology capital of the world"; and to the demands by national governments for assurances of a particular threshold of research and development before the corporation's foreign direct investment will be allowed in the country.

We will proceed with brief location-oriented discussions of general technological development, of the research and development (R&D) operations of large companies, and of technological diffusion. We will end the chapter with a description of technology-oriented or "learning" regions.

Where is Technology Developed?

Based on our simple definition of technology, clearly it can be developed anywhere, by anyone who works at doing something better or differently. Therefore, one important determinant of where a particular technology is improved or created is where the relevant or precursor activity takes place. This straightforward statement is actually tricky to use in practice, because of the complexity of technological linkages among activities, and the rise of entire new industries out of seemingly unrelated activities. A new process for automobile-parts inventory control in a distribution center near Toronto may arise from innovations in computer hardware and software developed in Japan and Washington State. A new synthetic fiber produced in New Jersey may have been developed in a chemical plant near London, or a university in northern England.

One way to approach the connections between local (or national) setting and technology develop-ment is to suggest the key inputs and markets of new or improved processes and products, and to note which of these inputs are immobile characteristics of places and to what extent the technology is bound to particular markets. As noted above, one important starting point for improved technology is the operation of older technology in some similar activity. The current distribution of industrial activity influences the distribution of technological development. Key questions remain concerning which activities in what locations manage to develop a preponderant share of technological innovations.

Much new or improved technology is embodied in or requires new investment in fixed capital, worker training, and marketing activity: investment with an uncertain future return. This investment, itself a series of locational decisions, requires financial capital. Activities with access to financial capital are more likely to develop and to implement new technologies. For the entrepreneur, capital sources include personal and family savings, equity in property (typically residential), and small loans from intermediaries such as banks. Entrepreneurs and regions with high incomes and savings rates, increasing real estate values, and intermediaries well trained to assess technological risks are thus better situated to access the capital required. Successful entrepreneurs may be able to attract "venture capital," financial capital in the form of part ownership in anticipation of the future sale of the entire company for a large profit to the entrepreneur and the venture capitalist. As part owners of the young company, venture-capitalist firms or individuals often take an active role in the management of the company, sometimes limiting their investments to companies in a given industry and in a given region of a country. Therefore, regions with substantial venture capital availability in a particular set of industrial sectors have an important source of capital for relevant companies' technological development. Large corporations with many on-going activities can finance technology development internally, or through a variety of capital markets, with little constraint or connection between the origin of the financial capital and its deployment. Such corporations can also allocate technological development

and implementation across different facilities and regions. Therefore, financial capital availability is less of a locational determinant for large-company technological development or implementation, just as it is not a locational determinant for such companies' production operations.

Just as important to technological development as financial capital is human capital skills at technology development, and general worker ability and willingness to undertake new production. With the broad definition of technology and technological innovation that we are using here, a wide range of skills are relevant: identifying market trends, understanding current processes, university training in engineering, grasping theoretical concepts from several different fields. The locations and establishments where these skills are brought together have the greatest likelihood of developing salable new products or important new processes. Highly skilled marketers, engineers, and scientists are quite mobile, within corporations and across corporations. Their wages and opportunities for career advancement make relocation worthwhile. However, because the best among them are definably rare, their locational preferences have been closely studied by business people and by academics. Drawing conclusions about the preferences of a heterogeneous set of individuals is difficult, but a few characteristics have been observed often and make a priori sense. First, because of the range of opportunities available to the best of these people, the individuals are attracted to locations where they can choose among many employment (and, potentially, entrepreneurial) opportunities without residential relocation. Metropolitan areas and industry-specific agglomerations have these characteristics. This tendency is increased by the tendency of people to marry similar people, and the prevalence of two-career households: the availability of a variety of technical and professional opportunities is important for the complex job- and location-matching that occurs in such households. Second, and more relevant at a very local geographic scale, their opportunities and incomes allow them to manifest general human preferences for good housing, good schools, and physical security. Workers, supervisors, and salespeople all have critical roles in identifying and implementing improvements in products and processes. While these key employees are located wherever a given company or industry has operations, the willingness and ability of employees to make constant improvements are not distributed evenly across companies or locations. These attributes have been found to depend on intra-organizational communication, management and labor rules that allow task-sharing among workers, and a general attitude of cooperation and improvement that comes from a sense that all will share in the proceeds. If these characteristics are not present in a company or facility, they are difficult to implement.

From a marketing or revenue-generating perspective, the best markets for new or improved products or processes are industrial markets that are growing (so that the investment in new products or processes will pay off) or highly competitive (so that such investment is mandatory to remain competitive), and consumer markets where incomes are high and people are willing to try new products. The actual development of technology is drawn toward such geographic markets (wealthy, growing regions and countries) to the limited extent that the development process requires continual interaction with the market. The product cycle model and the post-Fordist regime of accumulation each contain elements of this requirement of continual, proximate interaction between technology development and technology markets. However, the post-Fordist system and actual observation tells us that high rates of innovation within market-led geographic clusters is only one of several possible scenarios.

In summary, technology development cannot be clearly predicted to occur in any particular set of places. However, the *most fertile* ground would have some pre-existing industrial activity, knowledge-generating activities (marketing planners, research institutions, and corporate, government, or free-standing engineering establishments), and a certain amount of wealth and growth to provide capital and markets.

Location of Corporate and Government R&D

Above, we noted that large corporations allocate technology development among their varied establishments. National governments, many of which undertake substantial research programs for military, commercial, or infrastructural development, also allocate these programs to specific locations (or to specific companies).

By and large, these free-standing R&D facilities are subject to the same locational needs as general activities of technological development, in the previous section. R&D facilities have an additional degree of freedom because of the freer flow of capital, information, and personnel within a corporation than among corporations. However, the requisite employees may not already be employed by the corporation, and cannot be relocated unilaterally by the company. Therefore, the influence of market proximity, and of pre-existing concentrations of professional and technical people is reduced slightly, while the influence of external financial capital sources is reduced greatly. However, R&D facilities seem to face an additional tendency toward corporate or divisional headquarters. This in part reflects the internal sources of capital investment to support R&D, and the extent to which these investments are seen as critical to the future of the corporation. Government R&D facilities are influenced by political considerations and by the locations of other government operations. The high incomes and likely professional in-migration brought by such installations make them very attractive plums for politicians who represent or who desire to placate particular regions. These same benefits, along with the hope of local innovation diffusion, make these establishments part of the regional economic development policies of national and sub-national governments.

What about the potential attraction of R&D facilities to leading universities? Given that there are more leading universities in the United States, Great Britain, and Canada than there are large regional concentrations of R&D facilities, this attraction has to be a selective one. Malecki (1991: 222) concluded that the primary sources of this attraction, university graduates and research findings, are quite mobile. It is true that many graduates find their schools' regions attractive and that research findings can disseminate slowly through the usual academic media. However, graduates will go wherever there are attractive jobs, and the university findings that are of great potential commercial or military import are diffused rapidly. Those few regions that contain more than one truly top-calibre research university seem to develop and attract the kinds of technology-producing companies and establishments that create R&D clusters.

The Geography of Technology Diffusion

Most of the media of innovation diffusion, presented earlier in this chapter, operate quite effectively over long distances. Technological linkages occur whenever technologically active companies are free to sell their products or services. These linkages are motivated by the profit from these sales and by the competitive pressures on the purchasers of new equipment, techniques, or services. To the extent that national trade policies restrict the importation of new products or services, or to the extent that national trade, ownership, and antitrust policies reduce the competitive pressures faced by producers, then this important medium for technological diffusion is restricted. Language and cultural barriers to the transmission of technical information reduce diffusion. Regions and countries whose languages are not those of international science and commerce are at a slight disadvantage, overcome by increasing the number of technical and business people who can read and speak key languages. Restrictions on human migration, within and across countries, reduce the effectiveness of the employee medium of diffusion. These restrictions can be official and deliberate, as in the case of most national immigration policies, or inadvertent results of high-priced or scarce housing and language barriers.

Company-specific channels of technological diffusion probably do benefit from proximity of the companies, but this is clearly not required. Each company in the network or alliance must have something to offer the others: some skill, capital equipment, market linkage, or specific technology. Brokers are required to encourage the establishment of networks or alliance. These can take the form of consultants, key employees within certain companies, or perhaps a government intermediary.

Innovative Regions

Each of you has heard the phrase "high tech," and you may have heard the phrase "high-tech regions." These terms, seldom well defined, sprang from the rapid growth of companies, industries, regions, and countries involved in economic activities that undergo rapid technological development. One usable definition of high-technology industries are those industrial sectors in which companies' expenditures on R&D, divided by total company revenues, and companies' employment of graduate scientists and engineers, divided by total company employment, were greater than some high threshold. In the United States, quite a range of industrial sectors meet these criteria, as shown in Table 6.1. Local regions, or entire nations, with higher than average employment proportions in these sectors have been called "high-tech regions." In addition, many local municipalities, regional governments, and chambers of commerce/boards of trade have proclaimed themselves high-tech regions, appending the word "Silicon" or "Biotech" to their local geographic feature (Silicon Glen, Silicon Gulch, Biotech Corridor, etc.).

The source of fascination with technology-related occupations and activities can be summed up in one word: growth. By the 1970s, regions and countries that had formerly achieved and sustained high rates of economic growth found these rates stagnating and falling. Certain sectors of these economies faced rapid employment growth: retail service activities (including restaurants), service activities to businesses, and manufacturing sectors whose competition relied on technology development. The long-held bias against service sector growth is discussed in

Table 6.1 Manufacturing sectors classified as high technology

Guided missiles and spacecraft
Communications equipment and electronic components:
Radio and television receiving equipment
Communications equipment
Electronic components and accessories
Aircraft and parts
Office, computing, and accounting machines
Ordnance and accessories, except vehicles and guided missiles
Drugs and medicines
Industrial inorganic chemicals
Professional and scientific instruments
Engines and turbines
Plastic materials and synthetic resins, synthetic rubber and other artificial fibers, except glass

Source: U.S. Office of Technology Assessment, based on input–output analysis of the R&D embodied directly and indirectly in products from various manufacturing sectors

Chapter 10: it included the generally low wages and unstable employment of some service activities. A race began to generate or to attract high-technology facilities to each local region, in hopes that employment growth and stability would follow. Of course, the results included mutually counteracting location incentives (see Chapter 8) and many ill-considered policies based on an assumption that high technology was a sector rather than an assortment of activities across a wide range of industrial sectors.

More recently, companies and regional development officials have recognized the critical importance of information in the generation of competitiveness and growth. Information is clearly the basic building block of technology – information held by people, by companies, by a field of endeavor. The combination of these information sources with additional information about financing and marketing allow any economic sector to design and produce what some market, somewhere, will buy. The elusiveness of the appropriate information has engendered the inter-corporate alliances discussed earlier in the chapter, as well as the reliance on information-brokering services discussed in Chapter 10. Corporations and regional planners' emphases are turning from sectoral decisions and attraction policies to the formation of information and production networks to produce new products in new ways. Large

firms may create these networks internally, with small firms, or with individual contract workers. Small firms create these networks among themselves, or with large firms. Individuals increasingly find themselves relying upon their own skills and information sources to create livelihoods, decreasingly able to sign on with an organization for life.

These information-filled, innovative regions are not limited to a set of high-technology industries. Any productive activity, even activities within large companies and quasi-public organizations, can add value by improving the flow of information and the diffusion of innovations. To date, certain regions of the world have been studied for their prototypical formation of information and innovation networks, in Italy, Germany, Japan, and the United States. However, the importance of these improvements is being rapidly recognized. From the brief presentation of post-Fordist industrial and social organization, we understand the needs of innovative regions: support and expectation for continual improvement of labor skills; support for the health and financial savings of individuals who are not within companies; information networks about local and distant commercial opportunities; excellent transportation and communication facilities. These characteristics transcend, but do not replace, the industry-specific locational needs, which have been the focus of this book so far.

SUMMARY

Technology itself is ubiquitous, as are its effects. The ways in which we have tried to understand industrial location so far in this book – both the microeconomic perspective of Chapters 2 and 3 and the more macroeconomic perspectives of Chapters 4 and 5 – rely on technological differences across industries, countries, and time to help explain differences in locational needs and tendencies. In this chapter, we have provided overviews of the systematic ways in which technology varies and is developed. Company- or even facility-specific technology variation can reflect context differences, such as market size or access to capital, information, management, or labor: the location of the company or facility is part of this context. In addition, variations in technologies in use within the same industry influence the appropriate location of facilities within the industry.

The dynamic nature of technology has received a great deal of attention in recent decades. Some technological change is incremental, resulting from process or product changes made by astute planners, managers, and workers based on their experience with the market and the production process. These "learning curves" and "product cycles" have locational manifestations, as well, at domestic and international scales. The international product life cycle, conceptualized during the 1960s to explain post-World War II economic trends, depends on the differential international mobility of technology, financial capital, managerial expertise, and production labor within the multinational corporation.

Technological dynamics exist beyond the scope of an individual industry. Improvements in transportation technology, in material productivity, and economic structural change toward less raw-material-oriented production have created a general tendency of industrialization toward greater material mobility. Along with dramatic improvements in communication in the last 50 years, this has created the heralded "global village." So long as the global village contains heterogeneity of wage levels, government regulations, and market preferences, these general technological tendencies should make investment location considerations even more important.

More and more people are involved in the deliberate improvement of technologies. Like any other activity, the generation of technological change has particular locational tendencies. Because some technological change entails high development costs and low dissemination costs, actions of the public sector are very important in the development of new technologies. These actions, again, are not uniform geographically. The information intensity of technology development yields benefits from specialized agglomerations of similar and related producers, even in the face of instant telecommunications. All these considerations affect the wealth and well-being of particular regions and countries, as well as the fortunes of particular companies.

SUGGESTED READING

Aglietta, M. (1979) *A Theory of Capitalist Regulation: The U.S. Experience*. London: New Left Books.

Gertler, Meric S. (1988) "The limits to flexibility: comments on the Post-Fordist vision of production and its geography," *Transactions of the Institute of British Geographers* N.S. 13: 419–32.

Hepworth, Mark E. (1990) *Geography of the Information Economy*. New York: Guilford Press.

Hoover, Edgar M. (1937) *Location Theory and the Shoe and Leather Industries*. Cambridge, Mass.: Harvard University Press.

Malecki, Edward J. (1991) *Technology and Economic Development: The Dynamics of Local, Regional and National Change*. New York: Wiley.

Piore, Michael J. and Sabel, Charles (1984) *The Second Industrial Divide.* New York: Basic Books.

Stopper, M. (1985) "Oligopoly and the product cycle: essentialism in economic geography," *Economic Geography* 61: 260–82.

Vernon, Raymond (1959) *Metropolis 1985.* Garden City, New York: Doubleday Anchor.

Vernon, Raymond (1966) "International investment and international trade in the product cycle," *Quarterly Journal of Economics* 80: 190–207.

INDUSTRIAL LOCATION AND INDUSTRIAL GEOGRAPHY

Any serious study of how firms make location decisions and their consequences must take into account the economy in which they operate. Corporations and decision makers do not exist in a vacuum. They are produced by, and in turn produce, a set of social and economic relations that constrains their options at every turn. Thus, the industrial landscape is not simply the summation of individual location decisions; it is the consequence of historical processes that transcend the ability (and often the perception) of any individual firm. Thus, to appreciate the complexity of industrial location, it is necessary to understand the contextual background of the environment in which firms operate.

This chapter aims to summarize a great deal of such information by linking industrial location and industrial geography, the study of how economic landscapes are produced and transformed. We approach this subject through a brief summary of five sets of issues. First, we offer a broad historical perspective on the emergence of capitalism and industrialization. Second, we explore in some detail the uneven nature of capitalist development over time through an examination of business cycles and their relation to regional change. Third, we make some comments on the nature of the restructuring process of production and the role of technological change. Fourth, we relate these topics to the question of deindustrialization at various spatial scales. Fifth, and finally, we link these topics to the growth of the global economy in the late twentieth century.

CAPITALISM AND THE INDUSTRIAL REVOLUTION

An adequate understanding of industrial geography requires an appreciation of the historical origins and nature of change of contemporary society. This means that we examine not only how contemporary social and spatial structures emerged over time, but the fact that they are also continually in change and that the economic landscapes of the present will change in the future. We begin this examination of the origins of contemporary industrial society with an examination of the birth of capitalism.

Beginning around 1000 A.D., Europe experienced the gradual emergence of a so-called "commercial revolution," which consisted of several innovations introduced from the Arab world, India, and China, including the stirrup, cotton, sugar, rice, silk, paper, printing, the needle, zero, the windmill, and gunpowder. These innovations changed work relations, productivity levels, and traditional ways of feudal life that had persisted for a millennium. In 1347, Europe also imported the bacillus that causes the "Black Death" or "Bubonic Plague": within four years of its introduction, twenty-five million people, or one quarter of Europe's total population, were wiped out. It took Europe more than two hundred years to recover its pre-plague population. By disrupting feudal bonds and obligations, the Bubonic Plague created a shortage of labor, including shortages of serfs to work on feudal manors. Many economic historians speculate that the Bubonic Plague created a shock to a feudal society that had remained in equilibrium for more than a millennium, and suggest that for this reason the Bubonic Plague was an important element in the birth of capitalism.

Throughout the fifteenth, sixteenth, and seventeenth centuries, Europe underwent an enormous, gradual series of economic, political, spatial, and ideological or cultural changes that reflected the gradual emergence of capitalism and its slow replacement of feudalism. This process did not happen everywhere in the same way or at the same time, but gradually, and was largely imperceptible to the people who made it happen. The birthplace of capitalism is frequently argued to be in northern Italy. Particularly in many of the small city states that traded extensively across the Mediterranean, including Venice, Genoa, Milan, and others, a climate existed that was highly conducive to the emergence of market relations (Figure 7.1). The birth of the new society also changed Europe's relations with the rest of the world: in 1400 A.D., Europe was an unimportant backwater in the world, but by 1800, it dominated the world.

Capitalism as a Unique Form of Production

There are a number of specific features of capitalism that differentiate it from all other forms of production (Table 7.1). First, capitalism is, above all, defined by the existence of *private property* and the *domination of markets* as the principal way in which resources are allocated. Markets consist of buyers and sellers of commodities organized for the specific purpose of exchanging goods and services. The presence of markets means that production under capitalism is dictated by *profit*, that is, production for gain and not for use. The values of goods are determined by their relative price in the marketplace. As capitalism gradually expanded, markets grew throughout much of Europe; merchants organized firms and trading companies that were at first highly limited and regulated by the aristocracy, only gradually to replace the pre-existing feudal relations. As trade expanded throughout Europe and between Europe and the rest of the world, new land and sea routes emerged, giving rise to the "Age of Exploration." With the emergence of market society, we also find the appearance of business cycles, unemployment, and rapid technological change, issues that have been addressed elsewhere in this book.

Second, capitalism is characterized by a distinctive set of class relations marked in particular by the *ascendancy of the merchant class*, or traders (in Marxist terminology, the bourgeoisie). Merchants resented the obstacles to making a profit that were placed upon them by the feudal aristocracy, including limitations on what could be traded, when it could be traded, and taxes. However, feudal kings relied upon the merchant class, particularly to raise funds to wage war. Thus, the aristocracy both required and resented the merchant class simultaneously. The growth and power of the merchant class eventually caused the demise of the feudal aristocracy. This change would occur at different rates in different countries. In Britain, transition and power relations from the feudal aristocracy to the merchant class would occur very slowly; in France, it occurred very quickly, culminating in the French Revolution of 1789. The emergence of capitalist class relations included the transformation of work into a form of commodity, that is, labor became something to be bought and sold in a market. Thus, capitalist class relations were synonymous with the emergence of labor markets.

A third feature that characterizes the emergence of capitalism was the *growth of new financial systems*. Some early indications of this phenomenon are evident in the ways in which the emerging merchant class lent extensive sums of funds to the feudal aristocracy, in large part to finance wars. However, the growth of financial markets occurred in a variety of other forms as well, including the transition from a barter based system to a cash based one and the corresponding use of money as a medium of exchange. The birth of financial markets also saw the birth of banking institutions, which in large part arose out of the tradition of gold traders in northern Italy, and the emergence of double entry accounting, letters of credit, stock markets, and bonds.

A fourth characteristic that defines capitalism was a series of territorial or geographic changes. The

Figure 7.1 Fifteenth-century Italian city-states

Table 7.1 Defining characteristics in the development of European capitalism

- Large-scale presence of markets, commodity production, private ownership of means of production, profit maximization
- Ascendance to power of the merchant class, decline of aristocracy
- Financial markets, banks, stocks, bonds, accounting
- Long-distance trade relations in many goods, not just luxuries
- New ideologies such as rationalism, individualism, secularism
- The Protestant ethic
- Growth of scientific method, revolutions in astronomy, germ theory of disease
- Rise of the nation-state, nationalism, liberal democracies
- Colonialism and overseas expansion

spatial structure of capitalism is synonymous with *uneven development*, that is, the emergence of rich and poor regions and countries, growing and declining cities, places that enjoy investment while others suffer disinvestment. The process of uneven development occurs at several spatial scales. At the first and broadest scale, the global one, capitalism created uneven development in the form of colonialism, in which Europe dominated the rest of the world for 500 years. A second spatial scale involved the reorganization of the European continent into a series of national territories, giving birth to the nation states of Europe. The third spatial scale saw the increasing bifurcation between the cities and the countryside, as reflected in the Enclosure movement in Britain, the growth of cash cropping, and the privatization of commonly owned land. A fourth spatial scale occurred within cities, where land itself became a commodity, something to be bought and sold for a profit. The growth of capitalist land markets involved the growth of rent as the defining characteristic of how land use would be organized, as well as a bifurcation between home and work: no longer would large numbers of people work and live at the same place.

A fifth characteristic that defines the emergence of capitalism was the explosive growth in *long-distance trade relations*. Long-distance trade had, of course, existed long before the emergence of capitalism, including traders such as Marco Polo. However, under feudalism most trade occurred only among luxuries and precious goods; gold, silks, and spices (particularly when meat spoiled easily) were all traded only because their high value could cover the transportation costs involved in bringing them over long distances. In feudal Europe, trade existed only in a very peripheral way, whereas in capitalism, trade became central to the emergence of market relations. Thus, there arose new trading routes over sea and land, giving rise to the "Age of Exploration." The growth of international trade had enormous impacts on the structure of local production in different parts of Europe. Trade allowed communities to specialize in the production of a handful of goods, developing and making use of comparative advantage. Thus, trade allowed for the interdependence among places as each community gave up the necessity to produce all of its own goods for itself. Increasing trade among communities increased the general wealth of the European population significantly: a family in London, for example, had access to oils from Greece, wines from France, silks from Italy, metals from Germany, timber from Scandinavia, and furs from Russia.

A sixth defining feature of capitalism was the emergence of *new ideologies* and intellectual cultures. Indeed, the growth of new ways of thinking was indicative of the fact that social and spatial relations do not change without changes in the consciousness of the people who inhabit those areas. Some of the earliest ideological changes associated with the emergence of capitalism included the Renaissance during the sixteenth and seventeenth centuries. During this period, there was an explosive growth of science, including astronomy (e.g., Copernicus, Kepler, and Galileo, who invented the telescope), chemistry, physics and mathematics (e.g., Bacon, Newton, Leibnitz), and in the nineteenth century, biology (e.g., Darwin and Van Leewenhoek, the inventor of the microscope) and the germ theory of disease. However, capitalism also created a much more widespread series of ideological changes that hinged largely on the emergence of a secular individualism, the emergence of rationality as a defining form of human experience, a faith in technological and social progress, and a general belief that human beings had control over nature and over their own destiny. Another sign of the emergence of capitalism can be seen in the Protestant Reformation which began in the early sixteenth century in Germany, with its emphasis on individual salvation. The sociologist Max Weber argued that Protestantism gave rise to industrial capitalism in northwestern Europe in his famous book, *The Protestant Ethic and the Spirit of Capitalism*. In the eighteenth and nineteenth centuries, capitalism also gave rise to the Enlightenment, including such significant figures as John Locke, Adam Smith, Proust, Pascal, Voltaire, Rousseau, Descartes, Kant, Hegel, Nietzsche, Goethe, Weber, and Marx.

A seventh characteristic that defines the emergence of capitalism includes the *rise of the nation-state* as the predominant political entity in the world. A nation consists of a group of people with a common language, history, territory, and perception of the world, or sense of themselves as a nation. Nations, of course, preceded capitalism. Under feudalism, political organization was dominated by feudal empires such as the Frankish, Byzantine, Holy Roman, Austro-Hungarian, and Hapsburg Empires. However, with the emergence of capitalism, nation and state increasingly became synonymous. The ideology of nationalism, including its emphasis on self-determination, autonomy, and sovereignty contributed to this trend. The growth of nationalist ideology in turn reflected the growth of an increasingly mobile labor force, rural to urban migration, the spread of mass literacy, the expansion of an intelligentsia, the formation of state bureaucracies, and political parties, all of which contributed to the experience of modernity and the homogenization of culture. The nations of Europe, as we know them today, emerged in particular following the Treaty of Westphalia in 1648, which saw the disintegration of the Holy Roman Empire, and the Congress of Vienna in 1815, following Napoleon's defeat. Thus, capitalism is not only a process of market making, but also a process of state making. The relations between the state, or government, and markets are one of the most significant features of contemporary capitalist society.

An eighth feature that defines the emergence of capitalism may be seen in the growth of a *new global system*. Capitalism was, from its beginning, an international system, that is, it created uneven development on a worldwide basis. The particular mechanism that allowed capitalism to do this was *colonialism*, which allowed Europe to conquer the rest of the world for five hundred years. Europeans had guns, ships, and horses, as well as diseases that gave them a significant advantage over most of the rest of the world. In the sixteenth century, Spain and Portugal conquered most of Latin America and parts of Africa. England, France, the Netherlands, and Belgium also established large empires throughout the world, while late colonial powers such as Germany and Italy acquired relatively few foreign possessions (Figure 7.2). Thus, there would seem to be an intimate interrelationship between the emergence of colonial empires on the one hand and European nation-states on the other. Ultimately, Europe transplanted the nation-state overseas, creating roughly 180 such states today.

The Industrial Revolution

The development of capitalist social relations in Europe and elsewhere accelerated rapidly with the growth of the Industrial Revolution, which in large part gave western society its modern form. It is significant to note that industrialization began in the late eighteenth and early nineteenth centuries, long after the advent of capitalism. Three hallmarks define an industrial society: inanimate energy, rapid technological change, and rising productivity. Each of these will be examined briefly.

Inanimate energy, in contrast to animate energy, refers to the application of energy derived from sources other than human and animal muscle power. The emergence of industrialization is defined, above all, by the application of inanimate energy to the production process. The first form of inanimate energy to be used in early industrial Europe and North America was running water. Early textile factories, for example, relied heavily on streams to propel looms. The second form of inanimate energy to be used extensively was wood: with the invention of the steam engine by the Scottish engineer James Watt in 1769, wood became a valuable resource to power steam engines throughout Europe. However, as Europeans quickly cut down many of their trees, deforesting much of their continent in the process, wood became increasingly scarce and expensive. As the price of wood rose, eroding profits, they switched to coal. Europe is blessed with extensive coal deposits which underlie many of the early areas of industrialization, ranging in a large belt from Wales through London, under the English Channel, across Belgium and northern France, through the Ruhr River area of western Germany, and into Silesia in southern Poland and northern Czechoslovakia. In the nineteenth

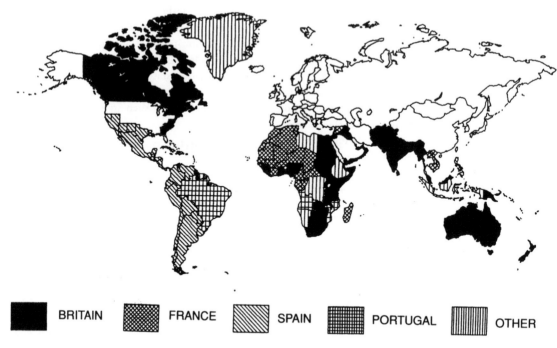

BRITAIN   FRANCE   SPAIN   PORTUGAL   OTHER

Figure 7.2 European colonial empires at their maximum extent

and early twentieth centuries, early industrializing Europe switched to other forms of fossil fuels, including petroleum, natural gas, and later, oil shale. In short, industrialization would not have been possible without the widespread and abundant prevalence of fossil fuels.

A second and very important characteristic of emerging industrial society was the rapid rate of *technological innovation*. Capitalism is a very dynamic system and it rewards constant change in the production process, i.e., new technologies that increase productivity. Productivity may be measured as the output divided by the inputs required. Productivity increase implies reducing the inputs required per unit of output. The Industrial Revolution saw numerous technological innovations appear and reshape the way in which work was done. For example, the spinning jenny, cotton gin, and steam engine arose in the 1760s; the power loom replaced the hand loom in 1785. Other revolutionary innovations included the printing press in 1811; the railroad in 1820; the

electric generator in 1831; vulcanized rubber in 1839; the telegraph in 1844; refrigeration in 1851; the gasoline engine in 1859; the open hearth furnace for steel production in 1866; the telephone in 1876; rayon in 1884; and the automobile in 1892. All of these innovations represented new ways of converting inputs into outputs more efficiently, that is, they raised productivity significantly.

As technological innovations raised the rate of efficiency of production, the supply of many goods increased much more rapidly than the demand. The market price of many goods began to decline, increasing the purchasing power of wages and raising the standard of living for many people. For example, the cost of producing food began to decline significantly, raising the quality of diets and improving the status of health and life expectancy. The increase in purchasing power that coincided with the Industrial Revolution dramatically reconfigured the patterns of everyday life and the social relations of industrializing Europe.

The Geography of Industrialization

The Industrial Revolution was more than a social, economic, and political phenomenon, it was also distinctly a spatial one. Industrialization occurred first in northwestern Europe in the late eighteenth century, particularly in Britain, the world's first industrialized nation (although a very early textile industry was also found in Liège, Belgium). The emergence of Britain as the first industrialized country may be attributed in large part to the prior commercialization of English agriculture during the Enclosure movement, the growth in long-distance trade relations (e.g., the Hudson Bay Company, the British East India Company), and Britain's abundant supplies of coal. In Britain, as in many places, industrialization began with the textile industry; cities such as Manchester, Leeds, and Birmingham became known as the "workshops of the world." Later, British and Irish cities that produced ships, such as Glasgow, Liverpool, and Belfast, would be added to these (Figure 7.3).

From Britain, the Industrial Revolution diffused a half century later to the rest of the world. However, for 50 years, in the late eighteenth and early nineteenth centuries, Britain enjoyed a significant period without international competition, a factor that largely explains the size and power of the British Empire. In the nineteenth century, northern France, particularly the lower Seine River valley, began to industrialize quickly, as did Belgium, the Netherlands, and the Ruhr area of western Germany. Simultaneously, the Industrial Revolution diffused to North America, where it saw the growth of the textile industry in New England. Southern Europe and Russia industrialized relatively late, and lagged considerably behind northwestern Europe in this process, a fact still evident today in their relatively lower standards of living and urbanization. In the 1870s and 1880s, following the Meiji Reformation and two centuries of isolation, Japan joined the ranks of the world's industrialized nations, the only member outside of the Euro–American circle to do so.

The Industrial Revolution was reflected in the growth and decline of different industries at different places at different moments in time. Among these, the most critical was the growth of the *textile industry*: everywhere, industrialization has been led by the emergence of textile production, including Britain, France (the cities of Lille and Lyon), the U.S. (New England), Japan, and Russia. In the twentieth century, newly industrializing countries such as Korea, Taiwan, Hong Kong, Singapore, and China have seen a rapid growth of their textile industries. The reason that the textile industry is synonymous with early industrialization has to do with its low start-up costs, the ease with which firms enter and exit the industry, its reliance upon unskilled labor, and the fact that it requires little by way of an infrastructure.

In contrast to textiles, later stages of industry, including the iron, shipbuilding, and railroad industries, arose much later. The period from the 1820s through the 1870s is frequently known as an epoch of heavy industry. Likewise, the period from the 1870s through the 1920s saw the rapid growth of the steel, rubber, and glass industries. In the twentieth century, the automobile, plastics, and petro-chemicals industries have also grown rapidly. In the 1980s, the electronics, or "high tech" industries have also surged in employment and output. Each of these industries was the "high tech" of its day, celebrated as dynamic, technologically sophisticated, and a source of new jobs. Thus, with the rise of each manufacturing sector, the Industrial Revolution set into motion a series of changing comparative advantages centered around the changing fortunes of different industries, as discussed in Chapter 5.

Other characteristics of the Industrial Revolution included the emergence of the factory system, urbanization, population effects, and a new international system. The factory system brought together large numbers of workers and machines for the first time in human history. The emergence of the factory system reflected the increasingly capital-intensive nature of work with industrialization. Unlike labor, machinery is an ideal tool for repetitive tasks involving little variation in routine. The factory system was augmented by the growth of the assembly line and interchangeable parts in the 1890s. Factories were

Figure 7.3 Centers of British industrialization

instrumental in the creation of an industrial work-ing class and the corresponding deskilling of many artisans. With the factory system the entire experi-ence of time itself changed dramatically; no longer did many people work to the rhythms of the seasons, but the rhythms of the machine, and the clock, bell, stopwatch, and calendar all became significant demarcators of the experience of time. Like space, time became a commodity. It is significant to note that the factory system's formation of an indus-trial working class was resisted by many workers at the time. For example, early French textile workers were known to take a heavy wooden clog, known as a "sabot," and jam it into the machinery, from which comes the term "sabotage." In Britain, resistance to the factory system was reflected most dramatically in the growth of the Luddites, who blamed machinery for their long working hours and low wages; the term "Luddite" has since become generalized to mean anyone opposed to technological progress.

The Industrial Revolution was also reflected in the rapid growth of cities, that is, widespread urbanization. As early as 1400 A.D., Europe had become covered with a fine mesh of new communities, known in England as "new towns" and in Spain as "villanovas." Unlike feudalism, in capitalism most work was concentrated in urban areas, a reflection of the inter-industry linkages and agglomeration economies that bound many capitalist firms together. Thus, cities became pools of capitalist investment as well as labor. Simultaneously, capitalism unleashed a dramatic mechanization of agriculture, reducing the need for rural labor and leading to the depopulation of many rural areas. As rural opportunities declined and urban ones soared, capitalism initiated a wave of rural to urban migration, which was widespread in Europe and the U.S. in the nineteenth century and is common in most Third World countries today. The growth in urban population in industrializing societies may be illustrated in an urbanization curve (Figure 7.4) which depicts the concentration of large numbers of people in urban areas during the course of industrialization.

Demographic Changes Associated with Industrialization

Industrialization also had demographic consequences, a phenomenon known as the **demographic transition**. Industrialization changed the natural growth rates of societies by altering their birth and death rates significantly. Death rates dropped as contagious diseases were brought under control, in large part because of improved diets, increasingly widespread antibiotics, and lower infant mortality rates. Industrialization changed the birth rate as well: as larger numbers of people came to live in cities and were no longer engaged in agricultural work, the economic value of children gradually declined. Children ceased to be significant sources of agricultural labor, or social security for their parents during their old age, and the lower infant mortality rates insured that higher numbers of children survived infancy, decreasing the necessity to have many offspring. In particular, as large numbers of women became employed in industrial labor markets, the value of women's time, and hence the opportunity cost of having children, rose dramatically. Thus, the decline in the birth rate had little to do with birth control and everything to do with the changing costs and benefits of having children. Industrialization also induced a shift from the extended to the nuclear family, giving rise to much smaller average families. Industrialization in this way allowed at least a short-term escape from the Malthusian scenario of overpopulation (see Figure 7.5).

Global Industrialization

Technological changes and the drive for market expansion spawned a new international system, including global markets, internationalized production systems, and a new hierarchy of nation-states. Specifically, Britain, having defeated France in a battle for global supremacy that ended with the Napoleonic Wars, became the world's pre-eminent power. In this respect, the timing of industrialization is important. Early industrializers such as the U.S. and Britain faced little international competition; they relied heavily on

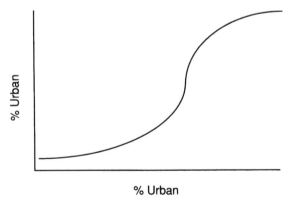

% Urban

Figure 7.4 The urbanization curve

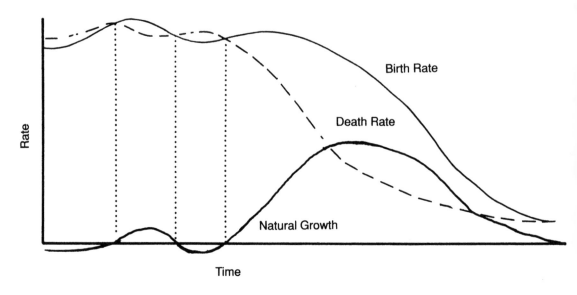

Figure 7.5 The demographic transition

low-technology industries such as textiles, and often adopted a *laissez-faire* attitude towards state intervention in the economy. However, with later industries, which relied heavily on government aid, in particular to build an infrastructure and skilled labor force, government intervention became more widespread. Late industrializing countries, in contrast, such as Germany, Japan, the Soviet Union, as well as many newly industrializing Third World countries today, faced stiff international competition. For example, when Germany began to industrialize in the late nineteenth century, British textile producers were already well entrenched in the global market. Consequently, many late industrializing countries condensed the early stages of industrialization which centered around light industry and moved rapidly into heavy industry. Many late industrializers, therefore, relied very heavily on the state, including the provision of an infrastructure, international protection, and a variety of other forms as well. Late industrializers, in contrast to early industrializers, had to carve their way into existing international markets. However, some late industrializers enjoyed advantages not available to early industrializers, including access to the latest technology.

In summary, the Industrial Revolution changed the entire economic landscape. Factories, mills, bridges, canals, and railroads of the industrial era became the new cathedrals of the Industrial Age. The natural environment was radically modified, draining swamps, dredging rivers, building earthen dams, opening up new continents for agricultural exploitation. The new society dramatically changed cities, blasting new boulevards through old neighborhoods, transforming urban areas from quiet centers of feudalism into hustling, bustling, crowded, teeming, dirty centers of commerce. Industrialization, in short, modified the agricultural landscapes of Europe into landscapes of modernity, landscapes of teeming cities, wealth, and poverty, created an international market system characterized by new trading patterns, and ushered in new experiences of time and space.

BUSINESS CYCLES AND UNEVEN DEVELOPMENT

One of the fundamental properties of capitalist development is that it is inherently uneven over time and space. As we have seen, uneven development in space takes various forms, including rich and poor countries, growing and declining regions, booming suburbs and decaying central cities, prospering urban areas and poverty-stricken rural ones. Uneven development in time occurs through cyclical oscillations in economic activity, in which employment, output, investment, profits, and wages rise and fall in short- and long-term business cycles. To a large extent, uneven development reflects the degree to which firms invest and disinvest in different places at different moments in time. This section examines the reasons that unevenness over time and space are integral to capitalism.

Business Cycles and Economic Change

From the very beginning, capitalism was marked by intensive periods of growth and decline. Such intervals of growth and stagnation are frequently referred to as boom and bust periods, or recessions and depressions. In the U.S., severe recessions or depressions occurred in 1819, 1837, 1857, 1873, 1893, and 1907. In the boom period of the 1920s, many politicians proclaimed that "New Era" politics had "conquered the business cycle"; this fantasy soon collided with the hard facts of the Great Depression of 1929–39.

The Great Depression marked the most spectacular of all of the crises of capitalism in the western world. It began with the stock market crash of October 29, 1929, and increased in severity in the following years. The U.S. Federal Reserve mistakenly tightened the money supply, forcing interest rates up, and the U.S. Congress initiated a series of tariffs against foreign imports, igniting an international trade war. With industrial economies already inextricably linked, the Depression soon became worldwide. Unemployment in Europe and the U.S. rose officially to roughly 25 percent, and unofficially may have reached as high as 50 percent. As a result, large numbers of people suffered serious declines in their incomes, with numerous repercussions, including eviction from their homes when they failed to pay their rent or mortgages; many became homeless and were reduced to living in parks and public shelters. Bread lines and soup kitchens became common in many cities, and housewives were frequently seen scavenging through garbage for that evening's meal. Indeed, the Great Depression was a period not only of economic stagnation, but also of severe social instability, one permanently etched into the consciousness of the generation that lived through it. The end of the Depression, partly a result of Franklin Roosevelt's New Deal programs, largely coincided with the armament of the U.S. in preparation for World War II.

Following World War II, a period in which the U.S. enjoyed unprecedented international hegemony, many western nations witnessed relatively low unemployment, and a rapid growth in incomes, and theorists and policy makers believed that it was possible once again "to conquer the business cycle." Even during the boom period of the 1950s, however, the specter of recession reared its head in 1959. With the international oil crises of the 1970s, recession, economic stagnation, and unemployment – often coupled with deindustrialization – appeared repeatedly, particularly in 1974 and 1979–82. In 1989, following the rapid growth of the 1980s, recession once again plagued the economic landscape. It is evident, however, that post-World War II recessions have generally not been as severe as those earlier in the century, a phenomenon some observers attribute, among other reasons, to improved public policy and control of national money supplies.

The stubborn persistence of boom and bust periods despite the best attempts of policy makers to mitigate them indicates that they bear further investigation. Boom periods are generally marked by significant increases in employment and output, rising profit levels and wage rates, and are frequently periods of relative social stability. Bust periods, in contrast, are often periods of social turbulence, in which large numbers of firms face declining profits and declining output, frequently going into bankruptcy, laying off large numbers of workers, forcing up the unemployment rate and forcing wages down.

Alternative Explanations of Business Cycles

Why do these boom and bust periods occur? The reasons underlying the existence of business cycles are complex, and while there is little uniformity of opinion, several leading theories have been advanced. One of the first explanations for the cyclical behavior of capitalism is offered through the Marxist perspective. Marx and other Marxist theoreticians argued that one cannot explain capitalist development over time and space simply by looking at the behavior of the individual firm. Emphasizing the labor theory of value, they argued that capitalists must inevitably overaccumulate surplus value, leading to a tendency to produce more than can be consumed, and forcing prices, and therefore profits, downward over time. Crises therefore are endemic to capitalism, part of its normal structure and operation. This means that there can never be an "equilibrium" in the economy, but that instability lies at the heart of capitalism: boom and bust periods are not abnormal but a normal part of commodity production. Therefore, Marx argued, production under market societies is inherently anarchic and self-destructive. However, during downturns in the business cycle, firms can take advantage of new opportunities open to them, including the introduction of new production techniques and new products, the exploration of new market areas, and the shift to new and more profitable locations (e.g., via deindustrialization and relocation overseas). Therefore, crises are functional to the normal operation of capitalism because they lay the condition for new rounds of production and accumulation. Marxists often argue, therefore, that capitalism is not simply crisis *ridden*, it is inherently crisis *dependent*. Marxists also argue that these crises will exacerbate existing class contradictions, culminating in a series of conflicts in which capitalism will destroy itself.

One of the most famous theories of business cycles, named after the Russian economist Kondratieff, was presented in Chapter 6. Kondratieff examined the history of capitalism and found that capitalism underwent periods of significant expansion and stagnation roughly every 50 years (thus giving rise to the term "Kondratieff cycles"). He linked these periods of rapid change to the emergence of new technological innovations and new industries: the emergence of the textile industry; the emergence of steam power and the railroad industry, and as a result of that industry's backward linkages, steel and iron production; the emergence of electricity, shipbuilding, and the automobile industry; and the emergence of the electronics industry, petro-chemical products, and service activity. The third Kondratieff cycle was particularly important because it marked the transition from craft/petty commodity capitalism to industrial monopoly capitalism. Each of these Kondratieff cycles was characterized by the growth of different industries (each the "high tech" sector of its day), significant improvements in productivity, major changes in the world system, including the rise and fall of different national powers, and new economic landscapes, including changes in regional comparative advantages and new forms of urbanization. For example, in the years since the oil crises of the 1970s, the world economy has seen the relative ascendancy of Japan, a slow deterioration in the western nations' competitive position, the microelectronics revolution, and globalized financial markets linked by worldwide telecommunications systems.

A third theory commonly invoked to explain business cycles was named after the economist Simon Kuznets. Kuznets argued that while Kondratieff cycles are helpful in understanding long-term changes in economic activity, they do not explain short-term fluctuations in output and unemployment. He argued further that short-term boom and bust periods are frequently superimposed on long-term oscillations, and linked these short-term fluctuations to changes in the investment behavior of firms. Investment fluctuates far more rapidly than any other component of gross national product. Kuznets speculated that investment is a function of expected future demand, that is, each firm must assume that the current *change* in the demand for its output will continue into the foreseeable future (what neoclassical economists frequently refer to as "the accelerator hypothesis"). During upturns in the business cycle, firms must expect that the demand for their output will continue to rise, and therefore they must invest as heavily as possible in new technologies in order to remain competitive. A firm that does not invest as heavily as possible during growth periods will suffer a decline in market share, and ultimately be out-competed by its competitors. During downturns in the business cycle, firms must anticipate that the demand for their output will continue to fall into the foreseeable future, and therefore must cut their losses as rapidly as possible by laying off workers and by disinvesting as rapidly as possible. In this way, the individual rationality of each firm's behavior produces a collective irrationality, which Kuznets believed explained why investment rates fluctuated so wildly. Kuznets argued that firms tended to engage in massive surges of investment and disinvestment roughly every twenty years, leading to the formation of so-called Kuznets waves. Further, he speculated that highly capital-intensive industries, which relied significantly on heavy investments in machinery and equipment, should exhibit the most cyclical pattern of investment and disinvestment over time: prime examples of these include the steel, automobile, construction, and lumber industries. In contrast, employment in the service sector tends to be less sensitive to cyclical oscillation than in manufacturing.

A fourth and final theory of cyclical behavior was offered by the British economist John Maynard Keynes in the 1930s. Keynes' theories of cyclical behavior and the government's role in dealing with these were heavily utilized by the U.S. government under Franklin Roosevelt in the 1930s, giving rise to Keynesian economic forms of political intervention. Keynes theorized that the government's role in business cycles should be to spend and tax counter-cyclically. During downturns in the economic cycle, government should spend heavily, even if it meant that tax revenues exceeded expenditures; hence, Keynesian spending became synonymous with deficit spending. Thus, in the U.S. in the 1930s, the government spent very heavily on public works projects, such as the Tennessee Valley Authority and the Civilian Conservation Corps. During upturns in the business cycle, Keynes argued that the government

should tax more heavily than it spent, accruing a surplus that could then be used to stabilize employment during future downturns. In short, the government's job was to stabilize the economy, using fiscal policy such as taxes and spending programs to walk a fine line between unemployment and inflation. Such a theory indicates that government employment is generally less cyclical than employment in the private sector; thus, locations with significant amounts of government employment should suffer fewer and less severe oscillations of the business cycle than those places with relatively less public sector employment.

The Geography of Business Cycles

At this juncture, the temporal rhythms of capitalist development can be related to its spatial or geographical pattern. Uneven development in time and space are outcomes of one process, or two sides of the same coin. It is important, therefore, to relate the periodicity of capitalist development in time to the uneven landscapes of capitalist development in space. We have seen how uneven development in space is produced in large part through the formation of a comparative advantage in production, that is, the specialization around a limited group of industries. Through this process, cities, regions, and entire nations are linked to a broader market through their export base. The export base, as we have seen, has important impacts on the local structure of production through its direct, indirect, and induced effects, and in particular through its networks of backward linkages. However, in a temporal perspective, it is critical to note that a comparative advantage is not a static phenomenon, but changes continually over time. This means that comparative advantage can be created and destroyed in individual areas, in particular through cyclical fluctuations in investment and output. For example, when capital seeks out lower cost locations and cheaper labor, it may end the comparative advantage of a particular region in a particular industry: a striking example is offered by the textile industry of New England, which had resided in that area for a century, only to have evacuated to the U.S. South in the 1930s and 1940s. Thus, capitalism continually creates new places through the process of investment while simultaneously dismantling old places through the process of disinvestment.

From the perspective of an individual region, different business cycles may create and destroy different comparative advantages at different moments in time, i.e., the same region may see the rise and fall of different waves of investment at various moments in time. This process is well illustrated in Figure 6.2, which depicts the primary industries in different regions of the U.S. throughout four Kondratieff cycles (see p.103). Note how flows of capital successively shaped different parts of the country around different types of investments. New England, for example, lost its comparative advantage in textile production in the 1930s, but eventually gained a new comparative advantage (in electronic goods) in the 1970s. Similarly, the South was transformed from an agricultural region centered around tobacco and cotton to a lightly industrialized region beginning in the 1920s, including textiles, apparel, and furniture (thus the South's gain was New England's loss). The industrialized Great Lakes states, which developed primarily around the steel industry, diversified to include the automobile, rubber, and chemicals sectors, only to lose them later on. The Prairies states have traditionally remained agricultural, although some cities (e.g., Minneapolis and Omaha) have acquired a significant service base. The Pacific Northwest, which was constructed around the lumber industry after the transcontinental railroads were completed in the 1880s, eventually acquired new bases in aerospace (the Boeing Corporation) and, recently, computer services (e.g., Microsoft). Finally, the Southwest, which developed originally around tin and copper mining, grew rapidly in the 1920s, largely as a result of the aerospace and electronics industries (much of which was propelled by federal defense expenditures) and motion pictures, particularly in Hollywood. In several regions recently, such as the South and the Southwest, tourism has become a principal source of nonlocal revenues. Note, however, there is no guarantee that a new set of industries will

emerge in an area; it depends upon relative factor input costs and their productivity.

Thus, from the point of view of an individual place, business cycles represent different waves of investment at different moments in time. Each "wave" affects local economy and local landscape in a variety of ways. First, it affects the inputs into the production process, including the amount of capital and labor and land, and the patterns of local linkages to local firms supplying goods and services. Different industries will have different patterns of interregional backward linkages and leakages. Second, different industries will affect the local infrastructure in varying ways, placing different demands on local roads, bridges, sewers, and communication lines and thus affecting the tangible, built environment. Third, different industries will create different types of local labor markets, generating different types of jobs and skills, and paying different wage and salary levels, creating different propensities to unionize, and perhaps locating in different areas within the same region (e.g., suburbs vs. central cities).

Finally, different industries will affect local family structures and gender roles, i.e., business cycles affect not only the sphere of production but also the sphere of reproduction. The British geographer Doreen Massey offers a striking example of this phenomenon in her study of Wales, in which coal mining, long a patriarchal activity in which men earned family incomes and women stayed at home, was replaced by electronics, in which women frequently worked, leaving men to tend the domestic sphere. Finally, by altering patterns of work, wages, and family structures, business cycles also affect local housing patterns, residential neighborhoods, and even local cultures and ideological climates.

Thus, each business cycle, or each wave of investment, leaves a distinct imprint on a region that survives for a considerable length of time. It is critical to note here that each imprint is affected by the residues of the earlier cycles of investment: each wave of investment by firms not only responds to and modifies the vestiges of the past, but is in turn modified by them as well. For example, in New England, semiconductor firms frequently utilized the old infrastructure and even buildings of textile, shoemaking, and chocolate making factories. The theoretical implications of this process are important: it implies that as the comparative advantage of an individual place changes over different moments in time (that is, over different business cycles), each place will tend to accumulate a series of imprints of different waves of production over time. An analogy to this process can be made to the Roman practice of a palimpsest, a piece of parchment that would be written on several times before being discarded: eventually a palimpsest came to reflect the imprint of many different people writing different messages at successive moments in time. Analogously, individual regions reflect successive rounds of investment at different moments in time. Thus, over their historic trajectories, individual regions can acquire a uniqueness that reflects the various rounds of investment in their history, each of which has a different impact in time. In this way, the same industry may have different impacts upon different regions because its effects (i.e., on the infrastructure, labor and housing markets, etc.) are influenced by the vestiges of earlier rounds of investment in each place.

THE RESTRUCTURING OF PRODUCTION

As we have seen, economic landscapes under capitalism are produced under the general logic of profit maximization, and are structured and continually restructured as the production process changes. Uneven development is an intrinsic part of capitalist industrialization (and deindustrialization), and reflects firms' changing locational needs in the search for the highest rate of profit. Put differently, the creation and dissolution of comparative advantage is a consequence of the powerful pressures of the market to innovate and seek out the highest rate of return. Frequently the technological and locational changes alter the distribution of economic activities through different business cycles, which reflect periodic oscillations in investments, outputs, profits, employment, and wages over time. A useful way of summarizing this complex body of issues is through the notion of industrial restructuring, which may broadly be defined as the tendency of firms to change their technological inputs, outputs, markets, labor costs, and locations over time.

Industrial Dualism

In an examination of the historical restructuring of industry, it is useful to differentiate firms into two broad categories that were commonly found in the nineteenth-century industrial metropolis and still characterize the urban division of labor in the twentieth-century "post-industrial" city. The first category includes *small-scale, labor intensive firms*, which frequently produce their output, in limited quantities and serve very specialized markets. In the nineteenth century, these firms included jewelry, button, garment, leather, shoe, gun, and watch and clock producers in most industrialized cities of the world. Such firms are typically very labor intensive in nature, that is, they employ relatively large numbers of people per unit of output, and wages and salaries comprise a substantial share of their output. Workers in such companies are often quite skilled, as the knowledge necessary to carry out the production process resides in the living worker, not dead machinery. Because these sectors are relatively easy to enter and exit, these industries are very competitive, that is, many small producers face one another in very competitive relationships. These firms tend to be bound together in intricate networks of input–output relations, that is, a great deal of subcontracting occurs among them.

This form of production is called **vertically disintegrated**, in which each firm tends to be specialized at a different stage of the production process. Such firms tend to face frequent changes in the design of their output, frequent changes in the nature of demand, and hence tend to produce in very short production runs and do not rely extensively on economies of scale. Because their markets are often unstable, these firms face considerable uncertainty, which further discourages heavy investments in expensive capital. In order to minimize their transportation costs and maximize their access to one another, such companies frequently locate near one another in city centers. Such agglomeration gives them maximum access to the labor market and to each other, as well as to specialized pools of information that are frequently found only within very narrowly confined areas in which many firms of the same industry are located (e.g., Madison Avenue in New York City). Therefore, such firms tend to be willing to pay very high rents such as are commonly found in city centers. Thus, the pattern of land values found in metropolitan areas arises from, and in turn affects, the urban division of labor.

At the other end of the production spectrum are *large, capital-intensive firms* that produce a standardized output in large quantities for predictable markets; hence, such companies rely extensively on economies of scale to reduce unit production costs. Because their markets change relatively little and slowly, these firms have little need to access the specialized pools of information generally obtainable only within agglomerated complexes of corporations. Employing large numbers of workers, these industries are often unionized. For example, in the nineteenth century, steel, flour, sugar, and meat packing companies exhibited these characteristics. Such types of firms tend to have very heavy fixed capital investments and often locate along transport lines, such as railroads or coastal areas in order to minimize transport costs when possible. However, since they are not as dependent on centralized locations as their small, vertically disintegrated counterparts, they typically will seek out locations on the urban periphery where rents are low.

These observations indicate that centralized locations on the landscape, i.e., central cities, have a comparative advantage in labor-intensive types of production, while peripheral areas and suburbs often have a comparative advantage in capital-intensive forms of production. The differences between labor- and capital-intensive forms of production may be seen by observing the labor cost as a proportion of total cost in the following industries. In services, roughly 80 percent of total expenditures are for wages; in textiles, 50 percent; in steel, 20 percent;

in electronics, 15 percent; and in petroleum, the most capital intensive of all, only 4 percent of aggregate costs are for labor. In each case, different production techniques – different combinations of labor and capital – produce different profit-maximizing locational strategies. However, no single production technique – and the geography associated with it – is stable for long, but changes as firms seek out new ways of maximizing profits.

One persistent feature of capitalist production is that it tends to become increasingly capital intensive over time, that is, firms tend to substitute capital for labor in the production process. Dramatic examples of this process are found in agriculture, in which mechanization has reduced the proportion of the labor force of most western nations to roughly 2 percent of the total; in manufacturing, the assembly, machines, and robots all serve as examples; in services, the copying machine and the computer are other forms. Capital intensification is profitable because new machinery raises the output per worker, improving efficiency and productivity levels; further, machinery, unlike workers, never complains, goes on strike, takes a vacation, or demands benefits. For these reasons, since the dawn of the Industrial Revolution, firms have employed machines to perform standardized, repetitive types of tasks.

Corporate Growth and Change

The heart of the restructuring process involves the relative mix of capital and labor that firms use at different moments, the degree to which they internalize or subcontract out functions, and the different geographies produced in the process. The changes that occur as firms switch back and forth between small-scale, labor-intensive techniques and large-scale, capital-intensive ones are important. It should be emphasized that there is no reason to think that firms change in one direction only (e.g., from small- to large-scale production), as commonly implied by traditional product cycle theories. Firms can also become smaller and more vertically disintegrated over time, as has happened in many industries in the late twentieth century.

Let us begin with the process of vertical integration and its geographic repercussions. As firms grow in size, a series of events occurs. First, total output rises, allowing firms to produce in larger production runs and derive economies of scale to reduce unit production costs. As the average firm size in the industry increases, inefficient firms are weeded out, and the industry becomes more oligopolistic (often resulting in a handful of giants that dominate the sector's total output). Second, with capital intensification, productivity increases, meaning the number of workers per unit output declines (note that the firm's total employment may actually rise, but not as quickly as total output). This means that the employment multiplier effects from spending in the industry will gradually decline over time. Third, the mix of workers' skills necessary to engage in production changes: there is considerable debate as to how new machinery changes aggregate skill levels in the labor market. Some observers point out that the introduction of new machinery obliterates many skilled professions, leading to a net deskilling; however, new technologies also generate new skills and occupations (e.g., computer programmer); the net effects of this process have yet to be determined empirically. Fourth, as the firm begins to transport as well as produce goods in large quantities, it begins to achieve economies of scale in the transportation process as well as production. That means that transport costs per unit output decline over time; for example, it is cheaper per ton to move large quantities of coal than it is to transport small quantities. Contrary to popular impressions, this means that large firms are frequently more locationally footloose than are small firms. Fifth, the firm tends to become more vertically integrated, that is, many of its linkages with ancillary firms providing parts and services, which were previously subcontracted to external suppliers, become internalized in-house. For example, a firm may make its own parts rather than subcontract them out to a supplier, or hire its own lawyer rather than employ a law firm, or establish its own print shop. In large production runs, the costs of such activities can be distributed over a larger quantity of output, making it more efficient and cost-effective to do so.

Vertical integration inevitably generates important repercussions in the economic landscape. In restructuring the labor process, firms change their optimal location just as they change their optimal mix of inputs. Frequently, such firms will flee the expensive rents and expensive labor costs of very centralized areas, often abandoning the center city for cheaper and more efficient locations in suburban areas. Thus, as the production process becomes more capital intensive over time, inner cities begin to lose their comparative advantage in production, and manufacturing begins to relocate out to suburban areas.

The mirror image of vertical integration is vertical disintegration, in which firms shed off many functions formerly performed "in-house" by subcontracting them out to suppliers (and in the process increasing their inter-industry linkages). Vertical disintegration typically occurs during periods of unstable and rapidly changing markets, products, and production processes, and represents an adaptive strategy by firms to minimize uncertainty. By disintegrating, a firm may eliminate part of its functions (e.g., an automobile maker may get rid of a division that produces, say, ball bearings or windshields) and purchase those goods and services from smaller, specialized suppliers. During downturns of the business cycle, the firm is not obligated to retain these functions or pay its workers; it can simply terminate its contract with the supplier. As a result, the firm becomes smaller, more specialized, and its pattern of inter-industry linkages (and multiplier effects) changes. Correspondingly, so does its optimal location, i.e., it will often seek out more centralized locations where it can realize agglomeration economies. In the process, the firm increases the costs of transportation per unit output, and to minimize this cost, will seek out locations near other firms even if it must pay higher rents in order to do so. In short, vertical disintegration induces a concentration of firms together in space. In the late twentieth century, vertical disintegration has played a significant role in the rise of "post-Fordist" production techniques and the formation of specialized districts of production such as Silicon Valley in California or Italy's Emilia-Romagna region.

Does this discussion of restructuring have any applicability to services? Many types of services, particularly headquarters and administrative functions as well as business, finance, and professional services, exhibit similar characteristics to those of vertically disintegrated manufacturing firms, i.e., they are often relatively labor intensive and rely upon extensive linkages with other firms, frequently in the form of face-to-face communications. For this reason, while center cities have lost most of their manufacturing employment, they have gained significant numbers of jobs in financial and professional services. The most dramatic growth of financial and business services has been in large cities, frequently in downtown areas, where they have encouraged gentrification that has had dramatic effects on local neighborhoods and housing markets. Bound together in dense networks of linkages, many of which are very specialized, the complexes of service firms that dominate most inner cities areas today exhibit many of the same features that complexes of small manufacturing firms did in the late nineteenth century: they are labor intensive, vertically disintegrated, have small production runs, and demonstrate intricate inter-firm linkages. In short, the location of services in center city areas largely conforms to the expectations of the theory articulated above. Conversely, those services that have become the most routinized and capital intensive, such as back office and data entry operations, are often the ones that have most readily abandoned center city areas for suburban and peripheral locations. The notion of restructuring, therefore, applies to *all* forms of commodity production, including manufacturing and services. The landscape of any urban area, therefore, will be formed through successive episodes of vertical integration and disintegration, typically manifested as business cycles. The exact nature and outcome of this process will vary with the nature of the particular industries involved, the pace of technological change, and the conditions of the global market, and each metropolitan area will reflect a unique combination of different "layers" of investment, some of which will be vertically integrated and others will not.

The restructuring of production and its geographical outcomes occurs at several distinct spatial scales. First, at the intra-metropolitan scale, restructuring leads to the closure of center city manufacturing plants and the flight of industry to suburban areas. As manufacturing firms abandoned center cities throughout the industrialized world, multiplier effects (the result of backward linkages) came into play and led to higher rates of unemployment and depressed land values. At the same time, however, the growth of services dramatically transformed inner city labor markets from primarily blue collar to overwhelmingly white collar forms of work. Second, at the regional scale, the process of industrial restructuring accelerated the deindustrialization of the U.S. Frostbelt or Rustbelt, i.e., the industrial Midwest and New England. New manufacturing plants – more capital intensive and vertically integrated than production in the traditional industrial belt – generally fled to nonmetropolitan areas, most commonly in the Sunbelt of the West and South. Finally, at the third scale, the international one, the restructuring process has seen a flight of manufacturing from westernized nations to Third World countries where labor costs are considerably cheaper, particularly East Asia and Mexico (the maquiladoras). At the same time, labor markets and land uses throughout the industrialized nations have increasingly been given over to the production of services of various forms.

DEINDUSTRIALIZATION

A widespread phenomenon throughout the industrialized world is the steady decline of manufacturing employment, a process commonly labelled deindustrialization. Translated into the health of local areas and the everyday lives of the people who inhabit them, this process is manifested in plant closures, layoffs, unemployment, and, often, rising rates of poverty. Because the loss of industry is such a widespread and important aspect of recent economic and geographic change, it deserves further attention.

Broadly speaking, **deindustrialization** may be defined as the loss of comparative advantage in manufacturing, particularly in heavy industry. Deindustrialization is a process that occurs unevenly over time and space, frequently reflecting the restructuring of different industries over time. As the world's first industrial nation, the U.K. was also the first to deindustrialize (Martin and Rowthorn 1986) and has witnessed the decline of its textile and apparel, steel, automobile, shipbuilding, and other industries. These losses have been largely, but not entirely, concentrated in northern Britain, i.e., the Midlands and Scotland, where many of the earliest and largest cities of the Industrial Revolution were concentrated. The evacuation of manufacturing there has led to widespread unemployment and a significant decline in local standards of living, lower rates of social mobility (especially for working-class youth), and rising social problems such as crime common to many economically stagnant regions. In contrast, the Greater London metropolitan area has witnessed a sharp increase in employment centered around financial and business services, leading to a severe North–South dichotomy in the U.K. Similarly, nations such as France (particularly the lower Seine River area and Lille), Germany (i.e., the Ruhr), Canada (southern Ontario), and even Japan have suffered the erosion of much of their manufacturing capability. Deindustrialization has occurred through the decline of different industries in different moments in time, including steel, automobile, rubber, textile, meat packing, shipbuilding, chemicals, and others. In the 1980s, even such celebrated industries as electronics began to exhibit signs of deindustrialization in the industrialized nations.

To appreciate the nature and extent of deindustrialization in the U.S., it is helpful to look back at the 1950s, which in many ways were something of a "golden age" in terms of the standard of living for many people. After World War II, the U.S. had virtually no international competitors, for most potential rivals had had their industrial base and infrastructure devastated by the war. The U.S. comparative advantage in heavy industry was reflected in the health of the Manufacturing Belt. During this period, the U.S. produced roughly 40 percent of the world's industrial output, and in particular enjoyed unprecedented hegemony in such industries as steel and automobiles. Beginning in the 1960s, however, the U.S. began to suffer mounting international competition, and by the 1970s suffered plant closures, rising unemployment, and a deteriorating trade balance in manufactured goods. As Figure 7.6 indicates, however, the erosion of manufacturing jobs during the 1960s, 1970s, and 1980s was gradual; indeed, there is often a tendency in the popular media to exaggerate the rate of industrial job loss. Nonetheless, during this period, the Manufacturing Belt, in which the bulk of job losses were concentrated, was steadily transformed into a "Rustbelt" (Figures 7.7–7.10). The states and cities within the region suffered serious economic hardships similar to those found in the U.K.

Causes of Deindustrialization

What are the causes of deindustrialization? Three principal reasons are offered here. First, it is apparent retrospectively that firms in nations suffering the worst deindustrialization did not sufficiently invest in new production techniques, leading to relatively *low rates of productivity growth*. Instead, many firms squandered their profits in high dividends and wages, paying generous returns to stockholders and workers. Nations with relatively high rates of savings as a percent of GNP generally enjoy high rates of productivity growth; after World War II, this trend characterized Japan, Germany, and, to a lesser extent, Sweden, Italy, Belgium, and the Netherlands. The U.S. and U.K., in contrast, saved only about 15 percent of their GNP and had a relatively low rate of productivity growth of roughly 1.5 percent annually. Capitalism places enormous pressure on firms to invest or to be driven into bankruptcy, and the insufficient savings and investment of American and British firms were a prime reason for their lack of competitiveness in the face of mounting foreign competition, including many firms and nations using "state of the art" production techniques.

Figure 7.6 Employment in selected U.S. manufacturing sectors, 1963–91

Second, when many multinational firms did make investments, they engaged in *foreign direct investment* (FDI) out of their home nations. In the 1950s and 1960s, for example, large numbers of U.S. multinational firms invested in Europe; by the 1970s, Europe had been replaced in large part by growing investments in the Third World. By the 1980s, U.S. multinational corporations were earning roughly one-third of their gross profits from overseas operations, frequently "milking" their domestic plants in order to subsidize their investments abroad. To a lesser extent, the same process was true for British, French, and even German firms as they sought to gain access to the global market. Thus, the internationalization of many industries coincided with the deindustrialization of the first world.

Third, it is apparent that the deindustrialization was exacerbated in large part by a steadily *deteriorating trade balance*. Competitors in East Asia and southern Europe, frequently relying upon cheaper labor, began to out-compete U.S. and British firms in a variety of industries. Shortly after World War II, with the Marshall Plan, in which the U.S. subsidized the construction of the economies of many West European nations, the nations devastated by World War II began to re-establish themselves as

significant producers on the international stage. By the 1960s, Japan and West Germany became formidable competitors with the United States in the production of steel and automobiles. By the 1970s and 1980s, these European producers and Japan were joined by several Third World countries, particularly the "Four Tigers" of East Asia as well as Mexico and Brazil. However, even France and Germany, which had prospered as manufacturing in the U.K. and the U.S. faltered, began to suffer an erosion in industrial jobs by the late 1980s.

A popular, but erroneous, interpretation of deindustrialization is that it is a result of the growth of services. It is true that in the post-World War II era, when employment in manufacturing in many western nations began to decline, employment in services grew significantly. However, there is no convincing logic to the position that the growth of services is responsible for deindustrialization. The rising demand for services, however, does not explain the declining demand for manufactured products. Why, for example, should the increase in employment in a law firm or hospital lead to a decline in employment in a steel mill or rubber factory? If anything, because services consume some manufactured goods, the growth of services should increase employment

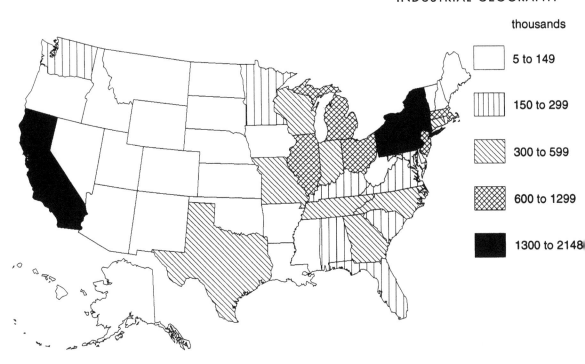

Figure 7.7 Location of U.S. manufacturing employment, 1960

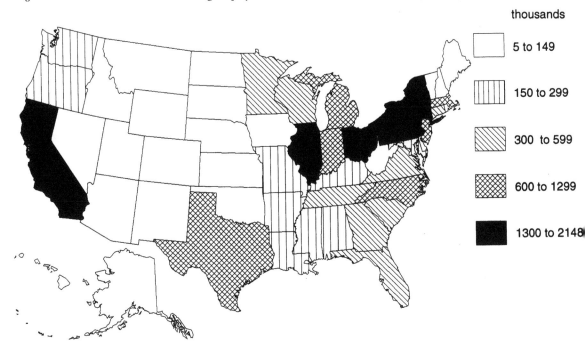

Figure 7.8 Location of U.S. manufacturing employment, 1970

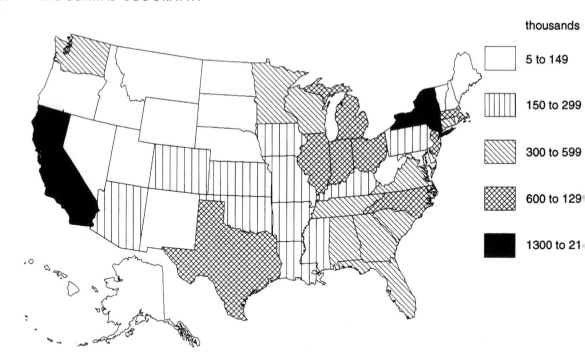

Figure 7.9 Location of U.S. manufacturing employment, 1980

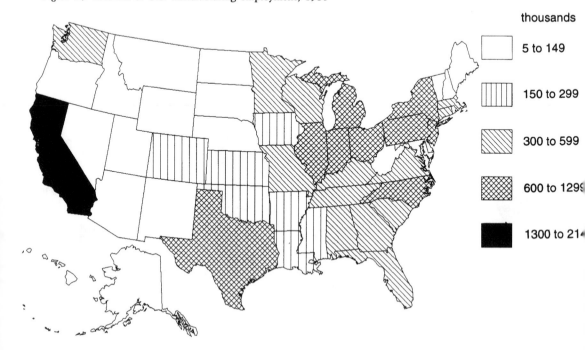

Figure 7.10 Location of U.S. manufacturing employment, 1990

in industry. Indeed, the loss of manufacturing has decreased employment in many services, particularly those "tied" to industry (e.g., shipping). Geographically, the highly uneven pattern of employment growth of services and manufacturing reflects relatively few connections between these two sectors. In some places, such as Pittsburgh or London, the decline in the steel industry was replaced by a growth of business services; in others, such as Detroit or Manchester, U.K., the decline in automobile production was not compensated by the growth of services, leading to steady decline. In cities such as Los Angeles or Tokyo, both services and manufacturing increased in employment simultaneously in the 1970s, creating a booming local economy.

It is important to note that the deindustrialization occurred differentially in terms of output and employment. In terms of output, manufacturing has remained roughly 22 percent of the GNP of western nations since the 1950s. In terms of employment, however, manufacturing has accounted for a steadily decreasing share of total employment; in the U.S., it dropped from roughly 45 percent of total jobs in 1950 to about 18 percent by 1990. The fact that western manufacturing firms can generate a constant or even rising output with fewer workers is a reflection of rising rates of industrial productivity.

The Geographies of Deindustrialization

Like all economic processes, deindustrialization occurs unevenly over space. It is useful to interpret the geography of deindustrialization at three distinct spatial scales. First, at the *intra-metropolitan scale*, the deindustrialization of the inner city coincided with rapid employment growth in suburban areas. As manufacturing plants in the center of urban areas began to close down after World War II, inner city tax revenues dropped as corporate and personal income taxes shifted to outlying suburban areas. The deindustrialization of the inner city carried with it numerous multiplier effects, which increased local levels of unemployment and caused nonbasic sectors

such as real estate to suffer as well. Deindustrialization, in short, caused much of the poverty found in inner cities of the U.S. and the U.K. Simultaneously, as inner city labor markets failed to provide adequate incomes, those who could left for the suburbs. Those who could not exit often became reliant upon government assistance. Many minorities in center city areas, who had resided for generations in working-class neighborhoods, suffered disproportionately from this process. From these comments, one may argue that the deindustrialization of the center city was a primary factor leading to the creation of the black ghetto.

A second spatial scale at which deindustrialization occurs is *intra-nationally*. In the U.K., this is evident in the split between the Midlands and the southeastern region centered on London. In the U.S., the parallel process results in the bifurcation between the Rustbelt (or Snowbelt), extending from New England along the southern Great Lakes, which has traditionally comprised the core area of U.S. manufacturing, and the Sunbelt, the nation's southern and western regions. As Figures 7.11–7.13 show, the Rustbelt states, in addition to California, have retained the bulk of U.S. manufacturing employment in the three decades after 1960. Consequently, in the years since World War II, most of the deindustrialization of the U.S. economy has occurred in the Rustbelt states. In the 1960s (Figure 7.11) manufacturing job loss in the Rustbelt was relatively modest, while growth in the Sunbelt was explosive. By the 1970s (Figure 7.12), however, industrial job loss in a wide belt of states ranging from Connecticut to Illinois was matched by very rapid job gains in the South and West. Finally, by the 1980s (Figure 7.13), when the entire U.S. economy continued to lose such jobs, deindustrialization continued throughout the Rustbelt but job gains in the Sunbelt were quite small. In both the U.S. and the U.K. this process occurred as different industries shifted from the national core to the periphery or overseas (e.g., the automobile industry and shipbuilding from Liverpool, wool from Manchester, textiles from Leeds and York, or in the U.S., the evacuation of the textile industry from New England

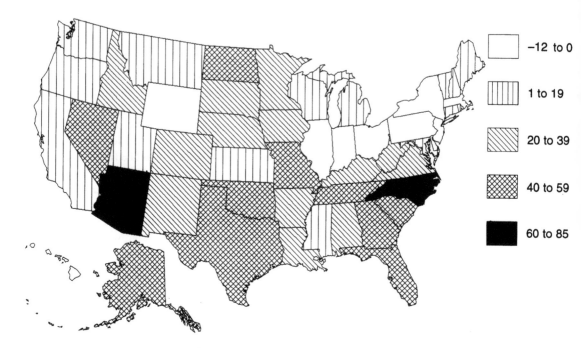

Figure 7.11 Percent change in total manufacturing employment in the U.S., 1960–70

to the Carolinas in the 1950s and the steel, automobile, rubber, and other industries to Georgia, California, or other states in the 1960s, 1970s, and 1980s). As with the intra-metropolitan scale, the inter-regional shift also induced significant migration: southeastern Britain has gained population while that of the Midlands has remained constant, and in the U.S., most of the population now lives in the Sunbelt.

A third spatial scale at which deindustrialization occurs is *internationally*, i.e., the relocation of manufacturing from the First World to the Third World, where labor costs are often one-fifth to one-tenth as expensive as in industrialized nations. In the 1970s and 1980s, the internationalization of the automobile, steel, plastic, shipbuilding, textile, and other industries was reflected in the growth of production overseas, particularly in the East Asian nations of South Korea, Taiwan, Hong Kong, and Singapore as well as in Mexico and Brazil. Frequently this process took the form of the flight of branch plants, or so called runaway shops, from the high labor costs of

the U.S. or the U.K. to the low labor costs of the Third World. In Mexico, it appears in the growth of the *maquiladoras*, or assembly plants clustered along the U.S. border, in which the assembly of electronic goods, toys, and many other products occurs.

All three of these scales are examples of how capitalism perpetuates uneven development, i.e., by creating and destroying places simultaneously through the spatially uneven process of investment and disinvestment. The suburbanization of manufacturing, the relocation of industry from the Rustbelt to the Sunbelt or the British Midlands to the Southeast, and the flight to the Third World are all indicative of the constant search for profit-maximizing opportunities. In each case, some places gained manufacturing employment (i.e., suburbs, the Sunbelt, the London region, selected Third World nations) while others lost (i.e., inner cities, the Rustbelt, the Midlands, most industrialized nations). Thus, the deindustrialization of one set of places is synonymous with the industrialization of

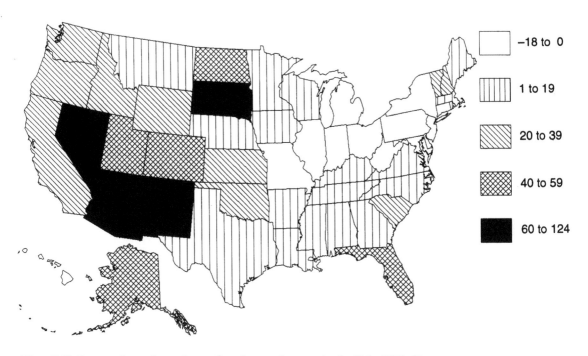

☐	−18 to 0
▥	1 to 19
◪	20 to 39
▦	40 to 59
■	60 to 124

Figure 7.12 Percent change in total manufacturing employment in the U.S., 1970–80

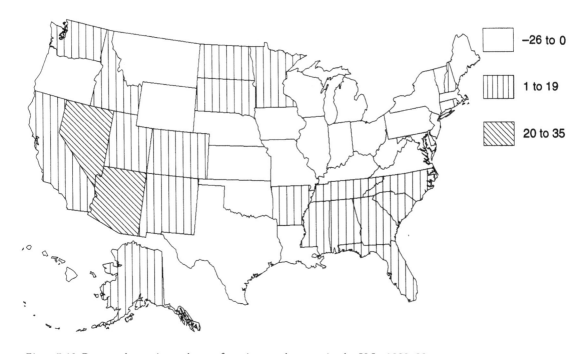

☐	−26 to 0
▥	1 to 19
▨	20 to 35

Figure 7.13 Percent change in total manufacturing employment in the U.S., 1980–90

others, part of the ongoing reproduction of uneven development that is an intrinsic part of capitalist geographies.

For people living in places that have been deindustrialized, the social and economic consequences can be severe. As a local steel, automobile, shipbuilding, textile, or rubber factory closes, the backward linkages and multiplier effects come into play. Layoffs and unemployment rates begin to rise, and labor unions find themselves declining in membership. The backward linkages extend to the retail trade sector, reflected in bankruptcies and declining numbers of shops, and in declining real estate values as well. For unemployed workers who formerly had jobs in manufacturing, deindustrialization can be catastrophic, particularly elderly workers with few other skills and few other opportunities in the labor market. It should come as no surprise, therefore, that deindustrialization is frequently accompanied by foreclosures, evictions, homelessness, divorce, depression, suicide, alcoholism, domestic violence, and other social problems. Frequently, deindustrialized areas are the source of out-migration as people seek greener pastures in labor markets elsewhere. However, for those that remain, deindustrialization can also be reflected in low rates of confidence/self-esteem, causing many to give up the search for jobs altogether so that those that are unemployed become unemployable.

Deindustrialization has also had significant consequences for the social mobility of different immigrant groups. In the nineteenth and early twentieth centuries, when most of the immigrants to the U.S. originated from Europe, manufacturing generated large numbers of jobs at relatively high wages and a significant amount of upward social mobility. However, for immigrants entering the U.S. or western Europe in the post-World War II decades, originating largely from the Third World, deindustrialization has been accompanied by relatively low rates of upward social mobility. Thus, the socioeconomic success of different ethnic and racial groups is closely linked to the timing of their entry into changing national economies. Immigrants to the U.K., Germany, Canada, or the U.S. today thus face very different circumstances (i.e., lower economic growth and social mobility) than did those a century earlier.

Despite the large number of job losses in manufacturing that the western world has endured, there is no guarantee that deindustrialization will continue indefinitely. There is no law of economics that holds that the U.K. or the U.S. cannot once again produce, say, large numbers of shoes. Manufacturing is still a significant source of employment in many First World nations and is unlikely to disappear altogether tomorrow. Indeed, the crises of the 1980s made wage rates in parts of the U.K. and the U.S. competitive internationally (below those in Germany or Japan) and spurred many firms to increase their rates of efficiency. In the late 1980s, U.S. manufacturing firms invested heavily in new technologies and increased their productivity markedly. However, within the context of a changing global economy, in which much manufacturing has relocated to the Third World, there is little reason to expect a prolonged resurgence of industrial employment in the industrialized West. This means, then, that the western nations must be actively involved in the search for new employment opportunities in new industries, constantly seeking to create new comparative advantages even as they lose existing ones.

THE GLOBALIZATION OF PRODUCTION

One of the most important features of late twentieth-century production is the steady formation of a global economy. The history of capitalism is replete with examples of this phenomenon: after all, capitalism was, from its inception, an international system, and globalization was a fundamental attribute of 500 years of European colonial domination over the rest of the planet. However, in the last 50 years the pace and extent of globalization has been unparalleled historically; many observers hold that the current trend is indicative of the "Fifth Kondratieff" cycle of the late twentieth century. The signs of globalization are not difficult to detect, involving massive growth in international trade, investment, production, and consumption. Any trip to the grocery store should provide ample evidence. The process of globalization has made local and national economies increasingly interdependent upon one another and has unevenly generated new patterns of costs and benefits around the world.

Historical Overview

The emergence of a global economy must be seen in the context of that other major institution, the nation-state. The political structure of the contemporary world-system – including roughly 180 nations, in large part the consequence of colonialism – is central to understanding how politics and economics overlap, intersect, and occasionally collide in the formation of industrial landscapes. Different "regimes" of international production are generally accompanied by the rise or fall of different national powers. During the Industrial Revolution, for example, the world's pre-eminent economic and political power was Great Britain, which enforced the famous Pax Britannica for more than a century.

Perhaps the most effective starting point for summarizing the current growth of a global economy is World War II, which ended the long-standing system of European hegemony. Unquestionably the winner of this devastating conflict was the U.S., which emerged with its cities and infrastructure intact and those of its allies and opponents in ruins. In 1950, the U.S., with 6 percent of the world's people, produced 40 percent of its goods and services. U.S. domination in heavy industry, concentrated largely in the Manufacturing Belt, was unparalleled. With very high levels of income and rapid rates of economic growth, the U.S. standard of living was the envy of the world. Internationally, the U.S. was by far the most powerful nation in the world, with an enormous military (largely arrayed against and justified by the threat of the Soviet Union).

As the world's hegemonic power, U.S. management of the world trade system occurred in various ways.

In the aftermath of World War II, the United Nations – under U.S. leadership – established the Bretton Woods system for the management of global finance, in which exchange rates were pegged to the U.S. dollar (whose value was in turn pegged to gold). The World Bank and International Monetary Fund likewise played key roles in selectively funneling resources to various national economies for short-term and long-term assistance. In 1947, the chief industrialized powers initiated the General Agreement on Trade and Tariffs (GATT), which has played a leading role in regulating international trade, including reducing political pressures for protectionism. Started as a means to facilitate trade between the U.S. and its European allies, GATT steadily expanded to include most world economies.

In the 1970s, the international oil crises profoundly shook the world economy. The Organization of Petroleum Exporting Countries (OPEC) raised the price of oil tenfold (from $3.00 to more than $30.00 per barrel). Because petroleum is such a critical commodity to western economies (recall that the Industrial Revolution would not have been possible except for easy access to fossil fuels), these price increases had devastating effects in the western world, which plunged into recession while inflation rates climbed (a phenomenon known as stagflation). The petro-crises forced a series of long-term changes in the western world, accelerating the deindustrialization of many national economies (including the U.K.), halting economic development in much of the Third World, and, through the recycling of petrodollars, laid the foundation for the debt crisis of the 1980s. Further, the oil crises allowed many firms (particularly automobile producers) to make heavy inroads into the large U.S. market, a boon to the Japanese and Germans. Plagued by slow growth of productivity, deindustrialization, federal budget deficits, and numerous other problems, the U.S. competitive role in the world economy suffered a steady and painful decline (as did that of several other nations); as a share of total world output, the U.S. economy declined to roughly 25 percent by the late 1980s. Thus, the oil crises initiated a period in which the unsurpassed U.S. domination of the world system began to come to a close; in retrospect, future historians may look back upon this period as the beginning of the end of the Pax Americana.

The period since the 1970s has been marked by two important features. First, there has been a steady erosion of the U.S. and U.K. competitive standing in the world system, a feature caused both by the insufficient domestic investments of firms and by the rise of competitors elsewhere, particularly Japan and the newly industrializing countries (NICs) of East Asia, which have enjoyed the most rapid rates of growth in the world. This phenomenon has resulted in large trade deficits for countries such as the U.S. and the U.K. Under the Reagan Administration in the 1980s, the U.S. became the world's largest debtor nation, relying heavily upon foreign capital. Second, there has been a rapid growth of regional trade blocks, the most well known of which is the Europe 1992 reforms, which will eventually fully integrate the 15 economies of the European Union by eliminating all trade and investment barriers among them. In like fashion, the U.S. signed a Free Trade Agreement with Canada in 1988 and Mexico in 1992, resulting in a North American Free Trade Zone (which, unlike the European Union, will include a Third World nation). The growth of regional trade blocks has reduced protectionism for member states, but continues to threaten the revival of such measures on an international basis (raising, for example, fears of "Fortress Europe").

To comprehend globalization, it is necessary to examine the behavior of the agents that have created it, particularly multinational corporations. Next, attention turns to the newly industrializing nations.

Multinational Corporations

Without doubt, the primary agent in the post-World War II round of globalization has been the multinational corporation (MNC). MNCs have grown rapidly in size, employment, and power in the last half-century, and exert an enormous influence over the industrial landscapes of many nations. The names of large MNCs are household terms around the world: ITT, General Motors, Exxon, General Electric, Nestlé, Dow Chemical, Coca Cola, and Matsushita. Although large, transnational corporations have existed as long as capitalism (e.g., the Hudson Bay and British East India Companies), modern MNCs are the first in history to operate as an integrated unit in a global market, and are capable of switching vast quantities of resources across national boundaries. The total sales of many MNCs exceed the Gross National Product of most small nations (for example, General Motors earned $95 billion worldwide in 1988, larger than the GNP of Sweden, Indonesia, or Nigeria).

MNCs are involved in all sectors of the global economy. Agricultural MNCs (e.g., Cargill) dominate global trade in food stuffs, especially critical grains such as wheat. In services, large MNCs dominate financial markets (e.g., banks such as Citicorp or Dai Ichi Kangyo). When they establish manufacturing facilities in Third World nations (e.g., the Mexican *maquiladoras*), MNCs have become the agents through which the Industrial Revolution has diffused internationally. Contrary to common perceptions, however, most MNCs do not invest in poor, Third World nations, but in other highly industrialized ones: the bulk of U.S. MNC investments, for example, is concentrated in Europe, and most European MNCs invest in other European nations. Highly industrialized nations offer numerous advantages that translate into high rates of profit, including well-developed infrastructures and skilled labor. However, MNCs are also the principal agents of **deindustrialization** in western nations; for many communities, globalization has entailed economic devastation, plant closures, layoffs, rising unemployment, poverty, and out-migration. Hence, globalization is neither an inherently "good" nor "bad" process: its effects vary widely, depending upon the history and structure of production in local areas.

MNCs enter a host country's economy in a variety of ways, through the establishment of new plants, the acquisition of existing firms, or joint ventures with local companies. Frequently, host country governments place limits on the extent of foreign ownership in particular sectors. Many governments are highly distrustful of foreign MNCs, an acknowledgement that they operate not in the interest of any one nation but in their own best interest (i.e., profit maximization). The loyalty of MNCs, by definition, is not to any one nation, but to themselves in the context of many nations. This is not to say that the interests of MNCs are inherently opposed to those of nation-states, but that they are simply different; when cooperation with national governments is profitable, many MNCs willingly comply. However, there is little evidence to support the common, oversimplified assertion that MNCs will end the sovereignty of nation-states. National governments still play critical roles in terms of regulating trade, immigration, exchange rates, and investment.

When MNCs invest in a nation, both costs and benefits are accrued. The benefits of MNC investments are relatively straightforward: they generate jobs, skills, and incomes, and lower unemployment, and may help to alleviate poverty in an area. MNCs are also important agents of international technology transfer, and may improve rates of productivity. The costs of MNC involvement, however, are often not as clear, and may exceed the benefits. By competing with less efficient locally based craft producers, MNCs may force them into bankruptcy. Some MNCs are attracted to Third World settings to escape environmental restrictions, and thus release copious quantities of pollutants, often toxic in nature, into the local environment. Finally, MNCs have been known to meddle in the political affairs of nation-states, supporting politicians and parties that support their interests and mobilizing against those who do not (including, in the extreme case, supporting military takeovers).

The Ascendance of East and Southeast Asia

A fundamental feature of the world today is that most nations are poor; underdevelopment is the norm, development the exception. However, the exception to the largely bleak conditions faced by most people in the world is the newly industrializing countries (NICs), which have enjoyed the most rapid economic growth in the last 20 years. Most, but not all, of the NICs are located in East Asia, the fastest growing region in the world economy. The spectacular growth of NICs – often, initially, with foreign economic assistance and investment – illustrates export-led growth at its best, as these nations acquired a comparative advantage in many forms of manufacturing based on cheap, well-educated labor, cooperative governments, and high rates of savings and investment. The "sunrise" industries of NICs – textiles, electronics, automobiles, steel – generally coincide with the "sunset" industries of the West. To a large extent, this phenomenon represents the restructuring of production discussed earlier, an outcome of firms' perpetual search for optimal rates of profit and the consequent restructuring of the production process.

Unquestionably the giant of East Asia is Japan. Alone among many nations in the world, Japan reveals a unique history, one that made it the world's only non-western industrial power. Isolated from the world for 250 years during the Tokugawa shogunate, Japan was never colonized; when forced to "open up" to the world in the 1860s, it embarked on a rapid program of industrialization. Even the devastation of World War II only momentarily halted this ascendancy, for following the war, a demilitarized Japan invested in state-of-the-art production technologies, turning the disadvantage of an embarrassing surrender into a powerful advantage. Today, Japan has the world's second largest GNP, following only that of the U.S. It is, for example, the world's largest producer of automobiles and motorcycles (Toyota, Nissan, Honda, Mazda, Subaru, Isuzu, Suzuki, and Mitsubishi are common names everywhere). Japan also has powerful and efficient firms that produce steel, textiles, and plastic products in prodigious quantities. Its fishing industry roams the world. Japan's electronics firms, based on the southern island of Kyushu, are the world's largest producers of televisions, computers, scientific equipment, stereos, microwaves, and telecommunications equipment. Japan is the world's largest banker, with the largest ten banks in the world by assets, its largest stock market, and vast quantities of overseas

Figure 7.14 East Asia, including Japan and the "Four Tigers"

investments. With abundant capital, Japanese firms have invested heavily abroad, erecting automobile and electronics plants in Europe and the U.S., buying vast quantities of commercial real estate, and purchasing foreign stocks and, in the case of the U.S., federal government debt. Internationally, Japan became the largest creditor nation on earth. Thus, the decline of the Pax Americana was accompanied by the rise of a "Pax Nipponica."

Why has Japan done so well? Several reasons are evident. First, Japan has a high rate of savings (20 percent of GNP, compared to 5 percent in the U.K. and 2 percent in the U.S.), which permits high rates of investment in new capital stock and maximum levels of productivity. The nation's infrastructure is well developed and well maintained, with state-of-

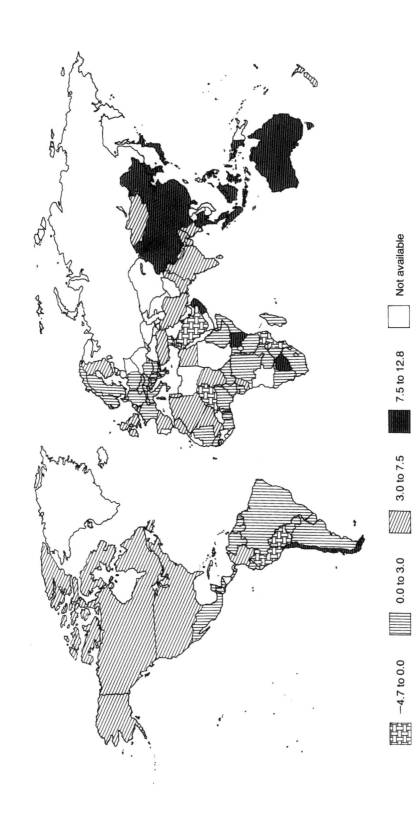

Figure 7.15 Average annual percent change in GNP, 1988–90

-4.7 to 0.0   0.0 to 3.0   3.0 to 7.5   7.5 to 12.8   Not available the-art train systems and an excellent telecommunications network. Japan is not saddled with large military expenditures, spending only 1 percent of its GNP on defense, compared to 6 percent in the U.S. The Japanese government and industry work closely with each other, as evidenced by the famous Ministry of International Trade and Industry (MITI), which subsidizes research and encourages collaboration among large firms. Japan has a rigorous educational system which stresses scientific skills, and, consequently, has large numbers of engineers, physicists, and mathematicians. Finally, Japanese corporate culture stresses a Confucian loyalty on the part of its workers; strikes are rare in Japan, and workers toil for long hours and take few vacations, lending confirmation to the stereotype of a nation of "workaholics."

Despite Japan's prominence, the most rapidly growing NICs are the "Four Tiger" nations: South Korea, Taiwan, Hong Kong, and Singapore (Figure 7.14). With variations, the story is remarkably similar in these nations. Each received considerable U.S. assistance and investment after World War II. Each aggressively encouraged foreign investment, and saw tremendous growth in exports and imports, leading to large trade surpluses, particularly with the U.S. In the 1980s, the average annual rate of GNP growth in South Korea was 10.6 percent, in Taiwan, 7.5 percent, and in Singapore, 3.3 percent. In comparison, GNP growth in the U.S. or the U.K.

frequently hovered around 1 percent. In South Korea, economic growth centered first around textiles, then around electronics, steel, and automobile production. In Taiwan, manufacturing was concentrated in household goods and garments. In Hong Kong, garment production was prominent next to household goods, toys, shipping, and finance. Singapore has largely shifted its economy from garments and electronic goods to finance and business services. The consequences of this rapid growth include the formation of a large, urbanized middle class: today, the standard of living in nations such as Taiwan or South Korea does not differ appreciably from many parts of southern Europe such as Greece.

In addition to the Four Tigers, other NICs wait in the wings, exhibiting among the highest rates of economic growth (Figures 7.14 and 7.15). Thailand, for example, enjoyed GNP growth of more than 6 percent annually in the late 1980s. Malaysia, with a growth rate of 6.3 percent annually in the 1980s, has emerged as the world's largest producer of semiconductors and refrigerators, and will soon begin exporting automobiles. Even impoverished Indonesia began to develop a textile industry. Finally, China, under the market reforms of Deng Xiaoping, has initiated several Special Economic Zones targeted for government assistance and foreign investment, mostly located in the south (e.g., in Guangdong and Fujian provinces) in the vicinity of Hong Kong.

SUMMARY

Any casual survey of the changing historical geography of capitalism will reveal a pattern of constant upheaval and change. However one puts it – restructuring, business cycles, the gain and loss of comparative advantage, globalization – it is clear that the one fundamental facet of the capitalist landscape is its tendency to acquire new forms. As noted, this tendency in large part is reflective of firms' constant search for the highest

rate of profit, whether it be through technological change, new markets, takeovers, foreign investment, or relocation.

In this context, the dramatic changes of the late twentieth century comprise only the latest of a long series of chapters. Both the global economy and various regions within it have been profoundly reshaped by deindustrialization, deregulation, internationalization, and astonishingly rapid changes among nation-states (e.g., the declining competitiveness of the U.S., the rise of

Japan, the collapse of the Soviet Union). Clearly, these changes are interrelated and often mutually reinforcing, comprising various facets of a "Fifth Kondratieff" cycle whose effects have yet to be fully understood. What emerges clearly from these observations, however, is the growing inter-dependence of various regions and nations. The rapid growth of international trade has firmly linked together the fortunes of individual places. To some extent, this phenomenon reflects the progressive restructuring of manufacturing, which has taken the simultaneous forms of deindustrial-ization in the western world and the rise of the newly industrialized nations. Yet the growth of financial services is also part of the same general trend, but one accomplished largely through the introduction of telecommunications systems. All of this makes for a serious argument that anyone interested in industrial location in either an academic or an applied context must be prepared for yet further rounds of unanticipated change in the future.

SUGGESTED READING

Harvey, D. (1989) *The Condition of Postmodernity*. Oxford: Basil Blackwell.

Martin, R. and Rowthorn, B. (1986) *The Geography of Deindustrialisation*. London: Macmillan.

Massey, D. (1984) *Spatial Divisions of Labor*. New York: Methuen.

Scott, A. (1988) *Metropolis: From the Division of Labor to Urban Form*. Berkeley: University of California Press.

Storper, M. and Walker, R. (1989) *The Capitalist Imperative: Territory, Technology, and Industrial Growth*. New York: Basil Blackwell.

Vogel, E. (1986) "Pax Nipponica?," *Foreign Affairs* 64: 752–67.

Wallerstein, I. (1979) *The Capitalist World-Economy*. Cambridge: Cambridge University Press.

8

HOW COMPANIES ACTUALLY
MAKE LOCATION DECISIONS

This chapter emphasizes the close relationship between location decisions and the general class of investment decisions, presents a set of techniques useful in analyzing physical investment decisions, and sensitizes you to the dependence of location decisions on the assumptions made about current and future circumstances.

The enormity of the location decision is quite apparent now that we have introduced the theoretical principles of minimizing costs and maximizing profit, recognized the importance of competition via location, and noted the effects of evolving technology. Production methods, scale, suppliers, and markets must be known over a wide area and forecast into the future. The location and outputs of competitors and potential competitors affect the profitability of a particular investment; these must be estimated in the future. The availability and cost of employees must be predicted, as well as the potential (and likely unknown) gains from agglomeration. The time spent in compiling and analyzing the required information is expensive, whether the time is taken by staff, owner, or consultants of the organization considering the investment. Given the investment magnitude involved and the importance of fixed investment, the final decision nearly always requires the consideration of the chief executive and perhaps the organization's governing board. It is a wonder that organizations can ever make a good location decision in a timely fashion.

We can complicate matters even more, by reminding you that every capital investment decision is a location decision, at least implicitly.

Investment in fixed plant, or in equipment specific to a particular production process, helps to reinforce or to modify the status quo, including the locations of a company's operations. Should a company undertake a location analysis for each capital investment? Though most organizations allocate capital to their investment needs a year or more in advance, the decisions about how and where to make the investments are often made very quickly. This quick time frame reinforces the tendency toward "locational inertia." It is easier to make an incremental decision in support of the status quo than to consider reconfiguring the entire organization.

A more important source of locational inertia, of course, is the prior investment of an organization in its operations in a particular place. A work force has been assembled, and the local labor force outside the organization is understood. Supply and distribution networks are quite location specific, and are expensive to modify. For single plant companies, capital availability is location specific, dependent upon relationships with local banks and credit agencies. Finally, to the extent that existing physical facilities have not been depreciated in real or in financial terms, they provide a short-run cost incentive to make incremental investments to support the status quo.

When a company considers substantial new investment, the location decision often becomes more explicitly recognized. Substantial new investment is not uncommon: examples include new construction to increase capacity by a large percentage, the replacement of central equipment with new equipment that forms the core of substantial process change, or investment in people and facilities to serve an important new market. In these cases, a locational decision may be forced by the unavailability of space for expansion, a poor fit between the proposed production process and the current supplier or labor-market relationships, or a recognition that the greatest market for increased production lies sufficiently far from the current locations to warrant an investment in the potential market. In such cases, the investment location must be married to a location decision – what we have argued throughout this book becomes explicitly obvious to the actors involved.

Successful investment planning requires a foundation in strategic planning. Strategic planning takes place at two levels in a complex organization: at the organization-wide, or corporate level, and at the level of each particular business or product line in which the organization is involved. The former, **corporate strategy**, entails decisions about the lines of business in which the organization is involved. When U.S. Steel purchased Marathon Oil Company, executives made a dramatic shift in corporate strategy, motivated by a search for steady profits. When Sears, Roebuck developed a credit card, it entered a very different line of business, motivated by profitability and encouraged by its huge information base on consumers' income, purchasing, and credit patterns. More relevant for facility planning is **business strategy**, a set of decisions concerning the way in which the organization will conduct a given line of business. Developing a business strategy requires assessment of all the strengths and weaknesses of the organization: costs relative to competitors, reputation, technology, ability to innovate, suppliers, strength of ties to clients and markets, capital availability. Elements of a business strategy include the organization's types of products, pricing, and market focus. Implementing strategic decisions requires coordination of purchasing, work force, production, and marketing decisions – including the location of suppliers, production, and marketing. Without explicit coordination, facility location can easily work at cross purposes to other aspects of business

strategy: the local labor force may have cost or quality characteristics that do not match the production characteristics required; marketing strategy may require proximity of production to key markets or clients; particular supply linkages may take on crucial importance for technical or delivery coordination. Therefore, determination of corporate and business strategy is an important background and guide for any investment decision, especially one with as many variables as a location decision.

An investment-cum-location decision has many components, each of which is inspected in the following sections. At each juncture, decision making is assisted by the recognition that the proposal is to freeze investment capital in a particular process at a particular place, to make what is infinitely pliable into something quite fixed, with implications for the future asset base, cost structure, competitive position, and geographical distribution of the organization. Thus, the capital-budgeting procedure described toward the end of the chapter is critical, because it compares the stream of returns from this investment with the stream of returns from alternative uses of the investment capital.

PRACTICAL CONSIDERATIONS

Alternatives to Substantial Investment

Investment capital comes at great expense. At minimum, that expense is the interest rate at which the organization can raise funds, from an intermediary like a bank or from individual investors. Most organizations have many potential uses for financial capital: to buy more supplies, to extend credit to customers, to replace equipment, to subsidize development costs, or to pay off a high-interest loan. Because capital is limited, the relevant cost of capital is greater than the interest rate paid to the lender (or the return expected by the investor). Rather, the company must recognize that the true value of capital is equal to *the return on the most lucrative investment that the organization could make.* This is called the **opportunity cost** of capital. For any organization facing more investment choices than the capital it can borrow or buy, the opportunity cost of capital is higher than the nominal interest rate.

Recognizing this very high cost of capital (for some organizations, as high as 30 percent even when nominal interest rates are 10–15 percent per annum) leads managers to look for low-investment ways to resolve organizational needs. A very common organizational problem is *insufficient capacity*. Before considering investment in additional production lines or building space, there are several ways to increase capacity.

1 *Increase efficiency*. Slight modifications in the routing of materials or information, in the placement of equipment, or the assignment of tasks can increase production with little fixed investment. Identify bottlenecks, the points in the order-processing, materials handling or processing, or packaging and distribution stream at which delays occur, evidenced by a backlog waiting to be processed or insufficient work at the next stage. People and equipment may be usefully redeployed to reduce the bottleneck.

2 *Increase labor*. The most common way to obtain a substantial increase in output is to add a production shift, using the existing plant and equipment for a longer portion of the day or week. Constraints on additional shifts include the hours of the day, the availability of workers, and the need for "down time" for maintenance, repair, or adjustments. Common alternatives to adding shifts are intensification of work (requiring more production in the same number of hours for wage workers, or the same salary for salaried workers) and extending the work day (via overtime or compensatory time). Employees must agree to such changes, however, or the increased output will be minimal.

3 *Contract production outside the facility*. A particular process or component can be purchased from another facility, within or outside the corporate organization. This entails a loss of control over production conditions, and a potential increase in total costs. Contract costs may be greater than internal production because of the profit that another organization may require. On the other hand, an external supplier (or another facility within the corporate organization) may have lower production costs because of its specialization, scale, experience, or location. If these lower costs are reflected in the potential supplier's delivered price, contracting is a cost-saving measure. Considering contracting options makes organizational management ask important questions about the organization's strengths and efficiencies relative to potential suppliers: a contracting arrangement can prove beneficial in the long run. These questions are often called the decision to make versus buy, and they are an important part of the investment-cum-location decision-making process. Beyond the cost considerations, contracting is beneficial when increased output is temporary or uncertain, or when capital constraints prohibit investment.

Each alternative to fixed investment faces its own constraint. There are limits to the increases in output that can come from minor adjustments to work rules, equipment configuration, or production scheduling that limits significant output increases. Beyond a certain point, adding shifts or increasing overtime brings diminishing returns, because equipment may need down time or repair time, and because there are only three eight-hour shifts in a day. With respect to capacity or total subcontracting, each company has an advantage over contractors in certain elements of production, or else the company should not be involved in production at all.

In addition, there are problems that cry out for fixed investment: technological change that renders existing facilities obsolete; labor-wage increases that render current production methods uneconomic; or increased, price-sensitive demand in markets remote from current production facilities. For whatever reason, companies do occasionally "bite the bullet" and engage in expensive and risky fixed investment. At this juncture, locational considerations generally become explicit.

Location Considerations Depend upon the Investment Motivation

Let's think of three possible reasons for a decision to invest large sums of financial capital into fixed, physical capital: capacity constraints; changed production methods; or additional product lines. Each of these reasons suggests fixed capital investment in the company's facility or facilities. Several locational questions arise. Should the investment be made within (or alongside) the existing facility? This is referred to this as *in situ* investment. Should the investment be made in a new location, keeping the older facility in operation (establishing a **branch plant**)? Should the new facility supersede the old (**relocation and closure**)? Or should the new investment be used as an opportunity to separate different activities that had occurred in the existing facility (**production segmentation**)? Common sense, along with some of the principles we've covered so far, suggest how these questions should be answered. Each answer is contingent upon the motivation for the new investment.

In situ investment

Investing in an existing operation is usually cheaper than establishing a new or a branch facility, for several reasons. First, the locational search is either eliminated or shortened, and time is expensive. Assessing possible new locations takes time because of the number of options to be inspected and the number of variables whose values are unknown. Before this quick and cheaper option is taken, several questions arise.

1 How much time does the organization have before the new investment must be in place: is there time for a full location analysis?
2 How likely is it that an alternative location will offer substantial benefits to the organization? Is the current location peripheral to needed inputs or markets? Is it unusually expensive? Are similar organizations growing in the local area?

3 Are there reasons to move or divide operations, distinct from the benefits of any particular location: will a move motivate a needed reorganization of logistics or responsibilities? Should the organization develop flexibility from having operations in more than one political jurisdiction or labor market?

Second, establishing a new facility is more expensive than adding space and equipment to an existing operation, unless essentially all the improvements on the current site (structures, equipment) must be replaced. There are other options between investing in the current location and developing an entirely new location. If the organization can identify and purchase an existing facility in an advantageous location, the cost may not be greater than *in situ* investment. Third and finally, the organization must consider internal economies and diseconomies of plant scale, which reduce or increase the unit cost of maintaining operations in one facility. These economies include physical overhead and facility management. Potential diseconomies include the difficulty of managing a larger facility, and the generally poorer labor relations within large facilities.

Branch plants

If existing capacity is a problem because demand has grown nearly evenly in all the company's geographic markets, then the required location analysis is similar to the analysis that led to the establishment of the existing facility. Several factors are likely to have changed, however. Most importantly, the company's increased output may make a second facility feasible or desirable because scale economies may have been exhausted with the current facility. Management of employees and materials logistics becomes difficult after a facility exceeds a certain size. Even if the current location were still optimal for one plant, it probably isn't optimal to place two plants in the same location. The current plant's location is a given, however, so the second plant's location needs to focus on a market area that is not well serviced by the existing facility. As a new, segmented market area, the location of the second plant should follow the total cost and spatial competition considerations that were presented in Chapters 2 and 3. Other factors have probably changed since the earlier facility was established, or may be expected to change over the life of the planned facility. How are labor relations in the existing facility? Are they so cooperative that they need to be carefully transplanted to a new facility, or are they so tense that a new facility should "start from scratch" in a new local labor market? Have the company's key suppliers or transport modes changed? What about the production process, and the corresponding need for particular kinds of employees? Are some technical employees so critical to the implementation of the production process that the company's success would be jeopardized if they were moved to a new facility – or if they refused to move? Any one of these considerations affects the similarity between the past and the current investment location decision.

If the increased demand reflects expansion into a remote market, the increased capacity might be considered in the form of a branch operation. This is especially likely if the remote market is international, and if there are trade barriers (such as tariffs or quotas) that prevent serving the international market by exporting to it. However, when the company estimates the size of its demand in the remote market, it must account for competitors' reactions. A company can focus on building market share in a specific product line or geographic area in a quiet, evolutionary way, so as not to arouse or alarm competitors. However, building or buying a facility dedicated to that product line or geographic market will be noticed by all competitors. Market share influences a company's ability to price, advertise, and hold on to its distribution networks (e.g., shelf space in retail stores, loyalty of the best sales agents and manufacturers' representatives, etc.). Therefore, each competitor will move to counteract the new facility – with a nearby facility of its own, with price-cutting in the market around the new facility, or with a defensive marketing campaign (including pricing, promotion, and advertising) across all markets in which they compete. Such reactions are especially likely when there are several competitors with substantial market shares. Therefore, this phenomenon of market-defending in reaction to a new facility has been termed **oligopolistic reaction**.

The extent and type of expected oligopolistic reaction determines the size and profitability of opening a facility in a new market area, thus influencing the decision to invest in such a facility. How can the reaction be predicted? Competitors' motivation to react is increased if the market area is currently very profitable for one or more competitors (allowing them the flexibility and motivation for predatory price-cutting), and if the market is stable in size (not declining and therefore not worth fighting over, nor growing rapidly and thus worth sharing with a new, local competitor). Competitors' ability to react is increased if they are part of large, multi-locational companies that can afford to defend a particular market area. Competitors' likelihood of reaction is greater if similar actions (by the company considering the investment or by others) have yielded reactions in the past.

Production segmentation

New, additional product lines are likely to have different locational needs from the existing lines, whether the differences are in inputs, labor orientation, market location and elasticity, or location of competitors' operations. The best location for each production process should be identified separately. However, certain costs of management, and perhaps of inventory or equipment, can be shared if the operations share a facility. The benefit of optimal locations for each of the activities must be compared to the benefit of sharing the same facility. For small companies, the determining factor is likely to be insufficient managerial staff and expertise to operate multiple facilities.

Relocation and closure

If the existing production process and accompanying equipment are obsolete, then the existing facility cannot continue to operate in its current form. Substantial investment must occur *in situ* or the facility must be closed. Given the dynamism of technology, markets, and regional characteristics, it is likely that the existing location is not the one that would be chosen today. The local labor force or local institutions may not support the technology, the skill requirements, the technical linkages, or the capital/credit markets that the facility requires. Competition may have increased the cost pressure on the operations, so that a lower cost location seems attractive. Over the operational life of the current facility, the center of gravity of the relevant markets has certainly changed, as has the competitive pressure on particular geographic markets (because of subsequent facilities added and planned by competitors). For any of these reasons, the need for substantial new

fixed investment can motivate a search for a totally new location. In addition, establishing a new facility in a new location makes substantial organizational change easier: the mix of employees and managers changes; the new location can occasion new or no labor-union rules; the characteristics, skill mix, industrial experience, and corporate experience of employees, contractors, temporary workers, and even government institutions can be selected by virtue of selecting the new location.

However, locational inertia comes into play, in the form of a complex question. Would (a) the returns on new investment in a new, more optimal location exceed (b) the proportional returns on what may be substantially less investment in the current location? As we will see in the section on financial analysis, below, this depends on the relative size of investment required in a new versus the current location, the disruption of moving, the costs of suboptimal location, the opportunity cost of capital, and the time horizon inspected.

CASE STUDY

Industrial de-location: selecting plants for closure

Disinvestment requires as careful an analysis, including locational analysis, as does investment. Plant closure may be motivated by *decline* in the company or the industry in which the plant operates, resulting in a decision to drop a line of business, to exit a geographic market, or to trim total production capacity by closing less profitable operations. Alternatively, plant closure may be motivated by *growth*. A need for new investment in equipment, space, or layout should prompt an assessment of all options, including the return on investment in a new location. An investment decision that prompts a location analysis may yield a decision to close an existing facility. Further investment in an existing facility may not be warranted because of insufficient room for expansion, poor configuration of the facility, or poor location with respect to projected markets or a preferred labor force. The firm may decide to close the old operation in favor of a new operation in a new location. In practice, an organization (a firm, a corporate division, or a governmental unit) facing rapid growth seldom closes existing facilities, until the period of rapid growth subsides and consolidation can occur.

In some of these examples, the decision to close a *particular* facility reflects factors other than the facility's location: its condition, its production configuration, or its product line. In other cases locational characteristics of the facility slated for closure are key aspects of the decision. Space for expansion, distance from growing markets, local tax payments, and local labour market characteristics are all possible factors in a closure decision. The labour pool available locally must be assessed in terms of skill and company-specific familiarity, weighed against wage levels, work rules, or even age. (It is often argued that the relative youth of workers hired during the 1980s in new, especially Japanese-owned automobile assembly plants in the U.S. and U.K. gives these companies a cost advantage, because their retirement pension liabilities are much lower than auto makers with older facilities. Health insurance also costs more for older workers, in a system like the U.S. where large employers buy group health insurance for their employees.) Depending on the geographic scale of the facility's market area (worldwide for an integrated steel plant, regional for a steel "mini-mill," local for a small bakery or bank branch), some of these locational characteristics are either very localized, leading to short-distance relocations (such as the dynamics of neighborhood markets that would affect the decision to close a bakery or a bank branch) or quite region-wide, leading to long-distance relocations (such as labor relations or wage rates that might affect the decision to close a mini-mill). Labor relations are an especially important locational characteristic leading to plant closure, in circumstances where the firm operates other, similar facilities.

The Sheller-Globe Corporation was a large, diversified manufacturing company, the result of a 1967 merger of Sheller Manufacturing and Globe-Wernicke. In 1982, its automotive parts division operated plants for automobile steering wheels in Grabill, Indiana and Brampton, Ontario (Figure 8.1). The

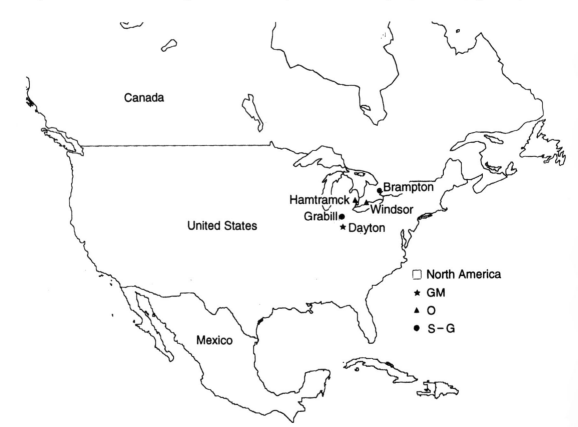

Figure 8.1 North American steering-wheel plants, 1982
Source: MacLachlan, 1992

Brampton plant was smaller, but was the only facility in the U.S. and Canada that produced polyurethane steering wheels, a newer product that found demand in larger and luxury cars. The older technology, thermoplastic steering wheels were produced at Sheller-Globe's two plants, as well as by Olsonite, another auto parts manufacturer, and by General Motors (GM). This oligopolistic market structure is important to the case, as is the oligopsony among the buyers. North American steering-wheel manufacturers had only three customers: GM, Ford, and Chrysler. Another important context is the 1965 Automotive Trade Act (the "Auto Pact") between Canada and the U.S., which removed most trade barriers for new cars and parts between the two countries.

In 1982, Olsonite gained the technical ability to produce polyurethane steering wheels. To maintain its North American monopoly on that product, Sheller-Globe decided to purchase Olsonite's automotive parts operations. At year's end, Sheller-Globe thereby gained two additional steering-wheel plants, in Hamtramck, Michigan and Windsor, Ontario; the Hamtramck plant was closed immediately because of its condition and to reduce excess capacity. The Brampton plant was dedicated to polyurethane steering-wheel production, and was operated at capacity in expectation of increased demand for the newer product.

In 1984, expecting even greater demand for polyurethane steering wheels, Sheller-Globe decided to reactivate a plant in Niles, Michigan (purchased as part of a larger corporate acquisition) that it had mothballed in 1980. Though this facility had to be re-equipped to produce steering wheels, it was newer than the Brampton plant, had expansion room, and could be operated under a union contract that stipulated wages averaging $7.50 per hour. The average wage at the Brampton plant was $14 per hour.

By 1984, however, Sheller-Globe's 65 percent share of the Big Three's steering-wheel purchases was being threatened by two new entrants and an additional plant. Izumi, a Japanese company that exported limited amounts of polyurethane steering wheels to Ford and Chrysler, opened a plant on Long Island, New York – at the encouragement of Ford, which wanted to reduce its dependence on Sheller-Globe. Centoco, a Canadian company, decided in 1984 to use its plastic molding expertise to start making thermoplastic steering wheels in Tillbury, Ontario, not far from Windsor, where Sheller-Globe operated a former Olsonite plant. Centoco's strategic and locational decisions were well justified: it faced encouragement (in the form of contracts) from GM and Chrysler; it hired former Olsonite managers and technicians who were dissatisfied following Sheller-Globe's acquisition of the Windsor plant; and it even purchased some equipment from the Windsor and Hamtramck plants. Finally, GM began producing polyurethane steering wheels in a new *maquiladora* facility in Matamoros, Mexico – exporting equipment and supplies into Mexico without tariffs, performing labor-intensive work in that low-wage facility, and paying U.S. tariffs only on the value added in Mexico. Within a year, Sheller-Globe's market share had shrunk to 45 percent.

During 1985, demand for polyurethane steering wheels fell (as the Big Three restricted their use to more expensive cars), even as production capacity increased. The next year, Sheller-Globe closed its Brampton plant. Falling market share combined with a shrinking market motivated the closure of a polyurethane-wheel plant. The smaller size, higher wages, and smaller site of the Brampton plant motivated its selection as the plant to close.

In 1987, the Big Three lost eight percentage points of their North American market share to imports and to Japanese companies that assembled cars in North America, using parts (including steering wheels) imported from Asia. In addition to this indirect form of import competition (the Big Three weren't importing from Asia, but North Americans were buying cars that used imported steering wheels), Sheller-Globe faced greater domestic competition on the basis of price, quality, and product technology. Izumi and GM's Inland Division won more Big Three contracts formerly held by Sheller-Globe. Centoco developed a hybrid wheel that wrapped polyurethane around a thermoplastic core, and which could be made at a cost between that of the two earlier products. By May 1987, Sheller-Globe announced the closure of the Windsor thermoplastic plant. Again, closing *a* plant was motivated by falling shares in an increasingly competitive market; closing the *particular* plant was motivated by higher wages, higher non-labor costs, and concern over the more militant union (despite the integration of the Canadian and U.S. auto industries, workers were represented by the Canadian Auto Workers in Canada and the United Auto Workers in the U.S.).

This case shows the complex set of influences on plant-closure decisions: acquisition and excess capacity, technological change, import competition, and the actions by oligopsonistic buyers to encourage new suppliers to compete with a monopolistic supplier.

Sources

MacLachlan, I. (1992) "Plant closure and market dynamics: competitive strategy and rationalization," *Economic Geography* 68(2): 128–45.
Stafford, H.A. and Watts, H.D. (1991) "Local environments and plant closures by multi-locational firms: a cross-cultural analysis," *Regional Studies* 25(5): 427–38.

COMPARING POTENTIAL LOCATIONS

Beyond the considerations of internalizing versus purchasing an activity, of investing at current locations versus at a different location, and of acquiring a previously operating facility versus building a new facility, there remains the fundamental question of where. Most of the models surveyed in Chapters 2 and 3 assumed a continuous field of possible locations, with distance from suppliers, competitors, and markets as the key measures. Alternatively, a company can compare a fixed number of discrete locations for costs or potential profits. Let's look at each method in turn.

Comparing an Infinite Number of Locations

The only way to consider every potential production location across a defined area is to assume that all relevant considerations vary as a function of distance from some landmark point(s). Think of the cost-minimizing model we built in Chapter 2. If we considered only transport costs, and assumed that these costs were a function of distance from the production site to input sources and market locations, then we could "zero in" on the cost-minimizing production site. One approach for accomplishing this was discussed in Chapter 2. This entailed allocating x and y coordinates for each input source and market center, calculating the averages of the squares of these coordinates (the center of gravity), and then searching near the center of gravity for the calculus-based conditions for a transport cost minimum. The use of x and y coordinates ignores the importance of specific transportation routes. Once we recognize that transportation does not occur equally easily in all directions, but is channeled into existing rail, road, and waterway networks, the featureless two-dimensional plane of Cartesian geometry loses a great deal of utility. How much utility is lost depends on the geographic scale being considered and the extensiveness of transport networks. To select a potential site across the entire United States, with its extensive network of road and rail, Cartesian coordinates do relatively little damage to the transport analysis. Selecting a transport-cost-minimizing location within a small, rural area with a limited road network obviously requires consideration of the specific routing of roads rather than the selection of a simple weighted center of a two-dimensional plane. Location within a country with sparse transportation infrastructure (road, railroads, and ports) also requires attention to the specifics of the network. This is a big drawback if available routes are few and limited, as in a poor or unpopulated setting.

The biggest drawback to such a continuous view of space and distance, though, is its emphasis on Weberian Step 1 considerations. Other important locational variables do not vary as a strict function of distance: the costs of labor, taxes, or energy; the likely reactions of competitors; or the local availability of required business-support services. These discrete locational variables cannot be considered in this way. If transport costs play a large role in determining profit levels for a facility, it makes sense to review locational options in an iterative fashion. Continuous space can be studied to determine the general area of a planned investment (for example, the large region within the U.S. to be inspected for a new plant to serve the entire U.S. market). Then, we must turn to specifics of transport network and non-transport considerations to determine the best location within that general area. At some point, then, we have to look at the costs and benefits of a discrete number of potential locations.

Comparing a Discrete Number of Locations

Given the intractability of actually analyzing two-dimensional space as a continuum of potential production locations, locational analysis generally restricts itself to a discrete set of points. Which points, and how they should be analyzed, is the subject of this section.

How can we select the potential sites to be compared? As we noted earlier in this chapter, an initial approximation considers each of the main input locations and market centers. In many modern industries, input locations are considered unimportant: production might be very market-oriented because of ubiquitous inputs, pure inputs, high

transport cost, perishability of the product, or the proximity that some service operations require to their clients. In these cases, substantial market locations are good potential locations. An additional complication ensues when the market covers a wide area, rather than consisting of discrete points. In such cases, the entire area is carved into sub-areas, each with a particular market potential and other local characteristics. The size (and therefore the number) of sub-areas depends on the fineness of the analysis desired. Any of these criteria may yield a very large number of potential locations. The analyst should use some set of additional criteria to screen the set of locations. Relevant criteria will reflect the nature of the process and the organization: a minimum or maximum population or local labor market size; presence or absence of large organizations in the same industry (depending on whether agglomeration or oligopolistic reaction are more important considerations); availability of specific transport infrastructure (an international airport, a containerized cargo seaport, or a confluence of at least two limited-access motorways). These screening criteria must be carefully selected, because they delimit the locations that will receive a more complete analysis.

Analyzing input sources and market locations using linear programming

The decision maker has a goal of *optimization*: to minimize the total costs of transport. The decision maker knows the relevant *constraints*: the level of output required of the plant, the locations of the input sources and the market destinations, the relative weights of each input per unit of product, and the transport rates per unit of weight and distance. There is a well-established set of methodology for optimizing under specific constraints. If we assume that transportation costs are a linear function (i.e., a fixed multiple) of distance, the assumptions of Step 1 of the cost-minimizing model relate well to the requirements of linear programming. **Linear programming** is any of a number of ways to determine the combination of inputs that minimizes or maximizes a function of those inputs. In our case, total transport cost is a function of the transport inputs to and from each market destination and input source. Because we know the proportional weights of these shipments, we know the rate at which one distance-based cost is substitutable for another – for example, the rate at which location at one input source increases the transport costs to markets.

A linear programming approach first allocates all production to one point: perhaps the source of the most expensively transported input. The total transport costs are calculated. The next set of calculations determines how much transport could be saved by allocating some production to other locations. The computer calculates the pattern of these cost substitutions, one transport movement being substituted for another. This pattern of opportunity costs (the transport cost of allocating production to one location rather than another) is used to determine the optimal production location or locations. The potential production sites considered are the sources of each input and each market center: intermediate solutions are not usually considered.

Recall from Chapter 2 that transport costs are seldom linear multiples of the distance traveled. Rather, rates per unit of distance generally decline as greater distances are to be travelled, reflecting the fixed costs and the terminal costs associated with a transport movement. This curvilinearity of transport costs concerns us in two ways. First, it encourages the selection of endpoints as production locations: the location of a key input, the location of the market, or the location of required trans-shipment (e.g., at a port where inputs are moved from water-borne to land-borne transport). Locating production at one of these points eliminates a set of extra transport movements, with their associated fixed and terminal costs. The fact that our programming approaches consider only endpoints is less of a problem. Second, however, if we want to analyze the effect of non-linear transport costs on optimal location, we need to use non-linear forms of programming.

This method is not difficult to implement in practice. The decision maker must know the production function and must be able to identify input sources and market centers. Linear programming can actually accept multiple possible input sources, allocating input needs from each in the most efficient manner. Similarly, linear programming can be used to locate more than one facility, as in the case where a branch facility needs to fit into a company's network of existing plants (which are sources of supply that are held constant in the analysis). The production function does not have to have the fixed coefficients we used in Chapter 2. Substitution among inputs is allowable, as long as the substitution is linear (e.g., walnut wood substituting for pine wood in a fixed proportion to make furniture).

It is more difficult to include in our programming analysis those locational factors that do not vary as a function of distance: labor availability and cost, energy costs, taxes, oligopolistic reaction, to name a few. Thus, any programming or similar logistical analysis must be augmented by less exhaustive comparison of potential sites' costs and markets.

Comparing distribution costs from a number of sites

There are cases where production costs do not vary greatly across potential locations, because the major inputs are ubiquitous. Material inputs are considered ubiquitous if they are literally omnipresent (like sand, air, and water), or if they are purchased at a uniform delivered price (as are many consumer items and small pieces of equipment). Labor inputs are considered ubiquitous if the kinds of workers needed are not rare (e.g., general production workers or clerical staff) and are hired at a nationwide minimum wage or a nationwide trade union wage. For other activities that do face some cost differences according to location, these differences are overshadowed by the costs of transporting the product to market combined with very price-elastic demand. Examples include bottled soft drinks, ready-mix concrete, and cardboard boxes. Finally, still other activities face requirements of rapid but irregular delivery to clients, so that proximity to one or more key clients is more important than any cost considerations. Certain specialty metal fabricators or printing operations fall into this category.

We began this section by assuming that a company has decided that its capacity shortfall requires a new facility. If the company intends to maintain its existing facilities, the new one must fit rationally into the existing network of capacity and demand. Which potential facility location will result in the lowest total market transport costs? To analyze this, we must consider the company's system of market centers C_1, C_2, and C_3 (which could be its own distribution warehouses, or its chief purchasers) and facilities P_1 and P_2, including *one* of the potential sites P_A. It makes sense to assume that the company knows where its current facilities are, and how much each can produce. To proceed, the company must know where its products are shipped, and how this is likely to change if a new facility adds to total capacity. Knowing the location of each market center and production facilities, and knowing the amount to be shipped to each center and the capacity of each facility, it is possible to determine the shipments from each facility to each market center that minimize the company's market-distribution transport bill. It is obvious that each facility should supply its nearest market center. After that allocation is made, however, how should each facility's excess capacity be allocated, and how should each market center's unmet demand be satisfied? The systematic method of determining this optimal shipment pattern is a linear programming application known as the **transportation problem**.

After deciding on the optimal pattern of shipments from P_1, P_2, and P_A, the analyst must discover the optimal pattern of shipments from P_1, P_2, and P_B, and then the optimal pattern of shipments from P_1, P_2, and P_C. Then, the total distribution transport cost bills are compared from the three different configurations, each with a different location for the facility to be added. The new-facility location should be selected which yields the lowest-cost configuration. Figure 8.2 illustrates this simply and graphically. In this case, existing facility P_1 is too

small to satisfy all the demand at C_1, and P_2 is too small to satisfy the demand at C_2, though the excess demand at C_2 is not as great as at C_1. In addition, a third, smaller demand is present at C_3. Figure 8.2 suggests three possible locations for a new, larger facility, P_3, which has sufficient capacity to satisfy the system's excess demand. Each possible location yields the optimized product flows illustrated in Figure 8.2. The least-cost alternative suggests the appropriate location for P_3.

In Chapters 2 and 3, we distinguished industries according to their input requirements and market locations. Given our definition of an industry as the set of establishments that use similar technology to produce similar products, we might expect that all establishments serving the same geographic market would have the same optimal location. How can we explain the variation in production locations chosen by companies in the same industry? Now we can see the context in which fixed investment decisions are made. Each decision has to fit into a pre-existing set of company facilities and competitors' facilities, and each decision faces a different set of pre-existing facilities and competitors' facilities.

Comparing revenues at different sites

Recall the measure of *market potential*, presented in Chapter 3 as the sum of markets served from a given location, where each market is diminished by the extent to which f.o.b. delivery costs dampen demand. This measure can be used to compare the amount of sales that could be expected from each of a set of potential production locations. If we ignore oligopolistic reaction as well as the possibility of production cost differentials across potential locations, the optimizing firm selects the location with the greatest market potential. As mentioned in Chapter 3, this location is likely to be near the densest market area to be served by the facility.

The industrial belt of Canada and the United States, surrounding the lower Great Lakes, forms an important market area for many industrial products (fabricated metals, automotive parts, even

(a)

(b)

(c)

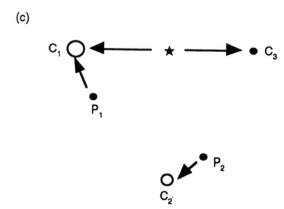

Figure 8.2 Optimized product shipments from three alternative locations for a new production facility

manufacturing process consultancy). This multi-state and two-province area (see Figure 8.3) could be divided into a set of states and provinces, or a set of metropolitan areas, or in any other way for which adequate data exists for an analysis of market sizes and distribution costs. Each market sub-area has a market center: the largest city and transportation node in the sub-area, or perhaps the geographic center of a large sub-area with widely dispersed internal activity. Each market center is analyzed as a potential production point and as the destination for products bound for each sub-area. Distances are measured between these market centers, and each center's market size is the size of the market in the sub-area.

So far, this market potential measure ignores competing suppliers, except insofar as the price elasticity of demand (assumed to be the same everywhere) reflects purchasers' alternatives. What about spatial competition, the fact that transport costs provide a competitive edge to a seller close to a particular market? Spatial competition can be modeled by multiplying the market potential at each market center by some measure of relative distance of competitors from the market center. This has the practical effect of systematically reducing the market potential in sub-areas with competitors, encouraging a revenue-maximizing location away from competitors' location.

By merely listing all relevant location considerations, a decision maker can add all the costs and benefits accruing to a facility at each potential location. The location with the lowest net costs (after subtracting benefits) is, in some sense, the best location. What should be included in such a listing? Table 8.1 provides an example. Clearly, alternative locations should be compared for the cost of material and energy inputs, including the charges for transporting them, the cost of employees (wages, salaries, benefits, payroll taxes, unemployment insurance, training costs), construction or purchase and remodeling costs, taxes on corporate property, income, and inventory, and public incentives for new investment. Personal income taxes and housing costs are relevant, to the extent that they represent a cost differential

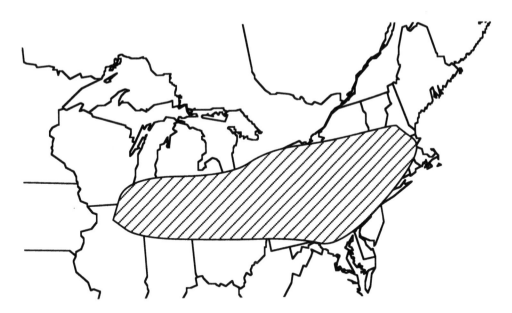

Figure 8.3 The Canadian–United States industrial belt

for current employees whom the company wants to relocate to a new location. These considerations are measurable, and can be added (or subtracted) for each potential location.

Other considerations are less quantifiable, such as the availability of workers with particular skills, the appropriateness of an existing building, the quality of life, or the likelihood of competitive reaction to the company's investment. These considerations can be checked off in a yes–no or acceptable–unacceptable fashion. Preferably, the analyst devises measures for each consideration. If skilled workers are unavailable, what would be the cost of encouraging skilled workers to migrate? What would be the cost of training local workers? Can local or national government authorities subsidize the cost of training? If the buildings being considered are not well suited for the planned facility, how much production would be lost to inefficiency, or how much would remodeling cost? Does the local quality of life represent a cost for which relocated managers will have to be compensated, or a benefit that will make relocation easier and potentially cheaper for the company? Competitive reaction can be estimated as the revenue lost to competitors' price

reaction to the investment, multiplied by the probability of such a reaction (where a certain reaction has a probability of 1.0).

How can this information be obtained? There are several inexpensive sources of locational data: national censuses, local economic development offices, and local utility companies. Because of their broad coverage, national governmental data sources offer the benefits of impartiality and uniformity, but often provide insufficient detail. Local public and quasi-public sources are eager to provide information that might encourage the location of additional investment within their jurisdictions, but may not provide full disclosure. (However, everyone involved recognizes the negative reaction that would come from explicit falsehoods.) At greater cost, organizations can subscribe to private information sources (some of which, however, rely on the same public data sources that could be accessed directly). Finally, some information requires site visits, especially for less quantifiable considerations.

The required information may be difficult to obtain, but this difficulty is no greater than for alternative location analysis methods. Given a creative

Table 8.1 A sample industrial location check list

	Location A[1]	Location B
Costs per unit output		
Delivered materials		
Employee occupation 1		
Employee occupation 2		
Employee occupation 3		
Electric power		
Product distribution		
Marketing costs[2]		
Overhead expenses		
Annual property tax		
Annual corporate income tax[3]		
Annual corporate payroll tax		
Annual inventory tax		
Annual public-utility assessments		
Amortized expenses		
Property acquisition costs		
Construction costs		
Major equipment costs		
(Investment credit or subsidy)		
Personnel relocation expenses[4]		

Notes: 1 Only those potential locations that satisfy all basic, non-quantifiable requirements (availability of critical workers, appropriate land-use and physical environmental designations, basic transport infrastructure) should be considered within such a check list.

2 If the degree of price promotions or quantity discounts varies across locations because of differing competitive environments, those costs (or reductions in average revenue) could be listed here. These costs can be estimated in any of several ways, including estimating the likelihood of different degrees of competitive reaction and multiplying each cost impact by its probability.

3 Assumptions about the predicted level of corporate income must be standardized across possible locations. Some jurisdictions tax corporate income according to the profit reported by facility, others according to the proportion of corporate revenues within the jurisdiction, others according to the proportion of corporate employment within the jurisdiction.

4 These can be estimated according to the varying necessity to relocate managers and technicians and the varying expense of encouraging and moving these personnel.

approach to the estimation of costs and benefits, the check list method encompasses most of our locational concerns in one process. However, this method does not consider two critical concerns. *Generating the list of potential locations* is a crucial first task, for the check list cannot compare unidentified locations. Some of the approaches presented above can be used to generate a set of potential sites. For example, a logistical analysis focused on transportation costs could determine which large region is optimal for the new investment. If transport costs loom very large in the competitiveness of firms in the given industry, the particular location within the region matters somewhat less. Several potential sites can be selected within that region, perhaps based on the availability of suitable buildings or building lots. These sites can be compared in our check list fashion. Alternatively, the availability of key employees may be the most important consideration, so that the only potential locations are those with substantial operations in the same or related industry. These potential locations could be compared via a cost-benefit check list.

The second critical consideration ignored by simple check lists is the *differences in timing of costs and benefits*, across different potential locations. The organization will pay for differences in investment costs (a new building required at one site, slight remodeling required at another, and a substantial public subsidy of investment at a third site) early in the life of the investment. Differences in operating costs will become more important after the facility is up and running. Differences in the useful lives of the various proposed locational solutions (a new building versus a renovation versus a low-investment option) will not become apparent in the short run. These temporal concerns are the focus of the following analysis.

Capital-Budgeting Approaches

Given the avowed corporate goal of profit maximization, and measuring profit as a return on invested assets, we must compare investment alternatives according to their total return on investment. The previous paragraph notes one complication to any such comparison: the variation in the time over which profits will be made from an investment. A comparison of alternative projects' return on investment is also complicated by the large number of potential investments a company can consider in a given year. Capital-budget analysis, using the present discounted value of future profits, is one way of overcoming both difficulties. In a capitalist economy, any economic organization faces similar opportunity costs of its capital resources, whether we are analyzing government agency expenditures of tax receipts, or voluntary organizations' expenditures of contributed funds. Capital budgeting is critical in any such investment decision.

First, the analyst considers only a subset of the possible investment alternatives facing the company. In the case of a location decision, a large company has many other needs in addition to resolving the capacity constraint or new-market opportunity that motivates the location decision. Instead of new fixed investment, the company could pay off debt earlier to save on interest payments, distribute more dividends to stockholders to increase the market value of the stock, engage in more product development to yield a high-profit new product, or acquire a key supplier to cut costs. To make economic sense, the capacity- or market-oriented investment must provide a profit that equals or exceeds the profit that could be made from these alternative investments. The company should estimate the maximum likely profit rate from any of these investments. This target profit rate is the *opportunity cost of capital* used in any of the investments.

The opportunity cost of capital is used to discount the profits expected in future years. The value of money depends on when it is to be received or paid. The first and obvious reason for this is the inflation of prices (or deflation of the value of money) so common in modern economies. A second, intuitive explanation for the time-dependent value of money is the interest income that can be earned on money that you or I hold until we must spend it. However, this interest is earned because our idle money can be used to finance investment that returns more value than it cost: manufacturing equipment, product development costs, students' educations. Given the range of investment options and returns on $500,000 invested today, that money may be worth more than a promise of one million dollars five years from now. Indeed, if we assume that the opportunity cost of money is 15 percent per year, a promise of one million dollars in five years is only worth $497,000 today.

In analyzing the current value of future costs or revenues, the key variable is this assumed opportunity cost. In our analysis, this is called the discount rate. In capital-budget analysis, the discount rate can be set to equal the perceived opportunity cost of capital. A high discount rate vastly reduces the value of profits in the distant (e.g., ten-year) future, as we can see from the formula:

$$V = \sum_t \frac{1}{(1 + r)^t} X_t \qquad (7.1)$$

where V = present discounted value of a stream of net cash flow from a particular project,

$r =$ discount rate being used, and

$X =$ net cash flow from the project in year t.

The discount rate serves as a proxy for all the investment alternatives facing the company. Using the discount rate in this way, a project analysis can concern itself with only the locational decision at hand, rather than analyze all alternative capital expenditures. The stream of year-by-year cash flow can be forecast for each locational option, as a function of the costs and revenues incurred. Look at Table 8.2 which compares three hypothetical investment options for a manufacturer. The project analysis considers only the projected costs and revenues from investing in increased capacity for an existing product. Investment option A is the expansion of an existing company facility for higher capacity (*in situ* investment). Option B is the purchase of an existing plant from a competitor that has gone out of business. Option C is to build a new facility near a growing market area. Each locational option has its own mix of costs and revenues in each year. Option C, along with additional new-location options, may be derived from a check list analysis, discussed in the previous section. We can compare the total cash returns to each option, discounting each future year's expenses or revenues by the annual discount rate. The highest valued stream of net revenue is financially the best option.

Study the notes following Table 8.2 which illustrate some of the assumptions that must be made by the analyst. Construction, purchase, and equipment costs are straightforward. Public incentives may be listed publicly by local development authorities, or some negotiation may be required. Moving inventory and equipment and relocating one or more key employees costs money – often much more than estimated in Table 8.2. Our estimate of sales revenue is fraught with uncertainties. Option B illustrates one uncertainty: the possibility that competitors will react to one company's new investment by discounting their prices within the contested markets, reducing the effective price any producer receives. For this analysis, not only the likelihood of oligopolistic reaction needs to be estimated, but its impact on average sales price and its duration (e.g.,

until the market grows to absorb all the capacity, or until some high-cost producers exit the market) must be estimated. The cost of sales, here encompassed in one line, contains all the concerns for locational differences in costs that we covered in Chapter 2.

In addition, this variable requires an estimate of cost variation because of production technology, scale of the facility, and the proportion of capacity at which each facility is likely to operate in each year over the planning period. The average costs used in the table vary across the options, but also vary within each option over time.

Table 8.2 lists only the costs and revenues that are directly attributable to each form of capacity increase. In an actual case, this is complicated by the need to account for the total operations, especially in option A which entails adding capacity to an existing plant. We have ignored the re-investment that may be necessary toward the end of the five-year period, as well as the depreciation expenses that prepare the company for that eventual re-investment. There are innumerable financial considerations that do not appear in our hypothetical case, such as depreciation and its effect on property and income tax rates. Also, note that national taxes do not appear on this table. In the U.S., the deductibility of state and local taxes from profits taxed by the federal government reduces the net cost of state and local taxation: that is not considered in our simple case.

The net cash flow for each option in each year appears in boldface in Table 8.2. For option A, the total for the five years listed is $1,478,750. For option B, the stream of net earnings totals $3,034,000, including some large capital expenses and operating losses in years 1 and 2. Option C yields a net cash flow of $5,647,500, clearly the most lucrative flow despite the large capital investment ($2 million) in year 1. This reflects the lower unit costs of option C once it reaches capacity, as well as the oligopolistic reaction to option B that is expected to lower its average revenue in years 1 and 2.

However, the $2.7 million net revenue that the analysis estimates for option C in each of years 4 and 5 is worth somewhat less than that today.

Table 8.2 Capital-budgeting approach to comparing three capacity-building investments

	Option A Expand current plant	Option B Buy existing plant at a different location	Option C Build new plant near a growing market
Additional capacity	1,000 units/year	2,000 units/year	3,000 units/year
Additional net revenues (or costs)			
Year 1	**($371,250)**	**($1,030,000)**	**($1,850,000)**
PP&E costs[1]	($500,000)	($1,000,000)	($2,000,000)
Incentives[2]	25,000	100,000	200,000
One-time costs[3]	(10,000)	(30,000)	(50,000)
Sales revenue[4]	250 units @ $5,000 = $1,250,000	500 units @ $4,000[5] = $2,000,000	—
Cost of sales[6]	(1,125,000)	(2,100,000)	—
Property tax[7]	(2,500)	—	—
Income tax[8]	(8,750)	—	—
Year 2	**$462,500**	**($400,000)**	**$440,000**
Sales revenue	1,000 units @ $5,000 = $5,000,000	2,000 units @ $4,000 = $8,000,000	1,000 units @ $5,000 = $5,000,000
Cost of sales	(4,500,000)	(8,400,000)	(4,500,000)
Property tax	(2,500)	—	(10,000)
Income tax	(35,000)	—	(50,000)
Year 3	**$462,500**	**$1,488,000**	**$1,677,500**
Sales revenue	1,000 units @ $5,000 = $5,000,000	2,000 units @ $5,000 = $10,000,000	2,500 units @ $5,000 = $12,500,000
Cost of sales	(4,500,000)	(8,400,000)	(10,625,000)
Property tax	(2,500)	—	(10,000)
Income tax	(35,000)	(112,000)	(187,500)
Year 4	**$462,500**	**$1,488,000**	**$2,690,000**
Sales revenue	1,000 units @ $5,000 = $5,000,000	2,000 units @ $5,000 = $10,000,000	3,000 units @ $5,000 = $15,000,000
Cost of sales	(4,500,000)	(8,400,000)	(12,000,000)
Property tax	(2,500)	—	(10,000)
Income tax	(35,000)	(112,000)	(300,000)
Year 5	**$462,500**	**$1,488,000**	**$2,690,000**
Sales revenue	1,000 units @ $5,000 = $5,000,000	2,000 units @ $5,000 = $10,000,000	3,000 units @ $5,000 = $15,000,000
Cost of sales	(4,500,000)	(8,400,000)	(12,000,000)
Property tax	(2,500)	—	(10,000)
Income tax	(35,000)	(112,000)	(300,000)

Notes: 1 Property, plant, and equipment costs include the costs of constructing an addition (option A), purchasing and remodeling the existing plant (option B), and land and construction costs of the new plant (option C).
2 The available public incentives for private manufacturing investment amount to 5 percent of the cost of option A, and 10 percent of the cost of options B or C.
3 These are costs of moving equipment, costs of relocating managers and key technicians for options B and C, and revenue forgone due to down time in the reconfiguration of the plant (option A).
4 Options A and B achieve full production in the fourth quarter of year 1, while option C requires 18 months for construction and start up. In addition, the market area to be served by option C is growing, but will not

use the new plant's full capacity until year 4. (In a more complex analysis, option C's excess capacity could be shipped to another market area in years 2 and 3, and the additional transport costs could be reflected in higher cost of sales from option C in those years.)

5 Option B is in a competitive market area, and the company expects competitors' reactions to reduce prices to $4,000 per unit in years 1 and 2.

6 The cost estimates recognize the greater scale economies of the larger operations, and the greater production efficiency of the new facility built specifically for the company (option C). However, option C does not reach capacity until year 4. Until it reaches full capacity, its average costs are higher, and decline as full capacity is approached. It is in these cost-of-sales estimates that locational differences in production costs (because of input transport, energy costs, labor costs, agglomeration economies, transport to distant markets) are manifested.

7 For options A and C, the property tax is 0.5 percent of the property's value. No tax is owed on option C until production begins. Option B includes a ten-year exemption from property tax liability.

8 The corporate income tax (exclusive of nation-level taxes) is 7 percent of profit for options A and B, and 10 percent for option C. For simplicity, we will calculate profit as sales revenue minus cost of sales, rather than including depreciation and other complications.

Similarly, the $2 million that the company would have to spend to build the facility has to be committed in year 1, and thus looms large in the time-dependent value of money. Table 8.3 compares the current value of the profit streams from each option, without discounting, using a 15 percent discount rate, and using a 30 percent discount rate. In addition, Table 8.3 shows the result of carrying on the analysis for a ten-year period, during which time the large profit streams from options B and C grow larger. We can learn a great deal from Table 8.3. First and foremost, option C appears to yield the largest stream of net revenues, regardless of how they are discounted. Selecting a higher discount rate increases the company's emphasis on investments that reach positive cash flow quickly. This makes sense if the company's opportunity cost of capital is very high: if the company is borrowing money at very high interest rates, or if the company has potential investments in other activities that promise very high yields. Selecting a short time horizon penalizes investment options whose "payback" is later. This makes sense if the investment outlook is very volatile. For example, technology and markets change so rapidly in some industries that a fixed investment may be worth little or nothing in five or seven years.

In any logical or quantitative analysis, however, the results depend totally on the conditions and relationships that are assumed. In this case, let's see what would happen if we assumed no oligopolistic reaction to option B (see Table 8.3). If the company could sell the output of the acquired plant at the same $5,000 average price that was assumed elsewhere in this example, the five-year, undiscounted stream of net earnings from option B would rise from $3 million to $5.4 million, awfully close to the $5.6 million undiscounted stream of net earnings from option C. Indeed, when we ignore the possibility of price discounting in the market served by

Table 8.3 Comparing net revenue streams

	Option A	Option B (Without competitive reaction)	Option B-2	Option C
Five-year analysis				
Undiscounted value	$1,478,750	$3,034,000	$5,394,000	$5,647,500
Value, with 15% discount rate	825,373	1,370,849	3,208,883	2,702,422
Value, with 30% discount rate	485,105	570,045	2,050,281	1,267,158
Ten-year analysis				
Undiscounted value	3,791,250	10,501,000	12,834,000	19,097,500
Value, with 15% discount rate	1,596,186	3,851,412	5,689,446	7,186,774
Value, with 30% discount rate	788,492	1,546,129	3,026,365	3,031,719

option B, its discounted stream of net earnings exceeds that of option C in the five-year analyses.

In using this method of capital budgeting or any other analytical method, many estimates must be made to reflect the many different possible conditions and relationships that affect the outcome of the investment decision. Location decision making is clearly not an easy task.

FINAL CONCERNS

Organizing the Decision-Making Process

The complexity of the fixed investment decision is a substantial problem for a small company. Entrepreneurial companies generally have small or no administrative staffs, so that no one has the time and resources to oversee a complex procedure that does not arise very often. The task of assembling sufficient information about many different potential locations is especially difficult for a small company whose contacts and experiences are geographically limited. Thus, sheer size (or the lack thereof) hinders the locational search. In addition, the feasibility of a move is circumscribed by the local nature of small-firm financing (typically dependent upon a relationship with a local banker or venture capitalist) and labor (sometimes dependent on the entrepreneur's family and friends for below-market-rate labor and expertise). These organizational, information, and financial reasons combine to inhibit the locational searches of small companies. Most small companies in need of fixed investment for new products or additional capacity select pre-existing or nearby locations.

Larger companies have a very different set of problems with investment decision making. The complexity of the decision is often matched by the complexity of the organization. Personnel in charge of manufacturing, logistics, marketing, investment programming, and real estate development all take an interest in the decision, because it affects the performance of each of their areas of responsibility. It must be clear who within the organization is in charge of the information gathering and negotiating process, and it must be clear who is empowered to make a final decision. A great deal of information may be available to different people within the company, based on their experience with far-flung markets, suppliers, labor forces, and the like. Compiling this information across the different branches of the company can be as difficult as compiling it from outside the organization. For all these reasons, a fruitful locational search needs the support and imprimatur of the top manager(s) who will ultimately approve the investment (or not). Despite the complexity of the decision and the organization, the information and personnel resources of a larger company generally allow it more flexibility in locating new fixed investment.

Companies often turn to external consultants to assist in the decision-making process. Consultants that specialize in fixed investment analysis and in location studies have ready access to the range of information required about many different locales. This is especially helpful to a company planning an international investment, where the range of choices and the diversity of information sources are much greater than for domestic investments. Consultants also serve as useful third parties or intermediaries in negotiations between companies and local/national governments, over issues such as public incentives, tax rates, availability of hard currency to a foreign-owned company, and ability of a foreign-owned company to send profits and royalties to the parent company.

Seldom, if ever, do companies simply accept the recommendation of an external consultant. The consultant's most valuable services are information gathering, preliminary negotiation, and the organization of a logical pattern of decision criteria. For example, in the capital-budgeting analysis of the previous section, a consultant might defend or critique the set of potential location options (perhaps three options were presented by the client company, with an additional three added by the consultant); explain why the capital-budgeting methodology was chosen, rather than a logistics-driven cost-minimization analysis or a market-and-competition-driven analysis of revenues and capacities; provide much of the information required for the analysis; and analyze the results of different assumptions, such as the speed with which a new plant reaches capacity, or the degree of competitive price reaction to a company's investment. While the consultant may end the analysis with a recommendation, it is the presentation of choices, tradeoffs, and information that is valuable to the client.

The same value can be attached to the work of an internal group working toward an investment recommendation. The final decision maker needs the range of information and choices, along with a defense of the procedures used and the relationships assumed, before committing funds.

Ending the Process

Given more information and better analysis, most decisions could be improved. However, investment analysis carries substantial opportunity costs. First, the cost of staff time or consultant time could be used on other projects, or on the investment itself. Second, if the investment is a worthwhile use of scarce resources, it is better made sooner than later.

How can an organization decide when to stop the decision-making process, make the decision, and begin the investment? One common approach is to set a deadline for the decision. The deadline might be prompted by a meeting of the board of directors, or by a time beyond which the added capacity will not be available for the peak selling season.

Another approach is to charge the process with criteria for coming to an end. For example, the methods in this chapter focus on the comparison of pre-selected locations and investment options. If option C appears to be the best option, that does not suggest that a yet unidentified option D, or G, or X is not even better. It is only in the continuous world of x,y coordinates that our analytical methods lead us to "the optimum" solution. What we can do, however, is carefully select a range of options for analysis, determine the opportunity cost of the search, and to end the search when the difference in the forecast profitability of the best and the second-best options makes further search cost uneconomic. When one can forecast that additional time and money invested in the search would yield a low return in terms of better location, stop the search process.

SUMMARY

This chapter makes the intimate relationship between investment and location even clearer. The various methods for analyzing location decisions are essentially investment analysis methods; the investment alternatives include no action, action on the current (set of) sites, or action at new locations. As many as possible of the myriad considerations that we've presented earlier in the book enter such an analysis in the form of variables across alternative actions.

Organizational action is generally in response to decision makers' perception of some problem or unusual opportunity, and generally is weighed against the costs of action, the costs of inaction, and the general goals of the organization. The nature of the problem that prompts an investment decision affects the locational strategy considered. Capacity shortfalls, production difficulties, uncompetitive costs, and changing markets each suggest different locational responses. The costs of action or inaction depend in part on changes in the environment of the organization. Forecasting the future is a tricky business, however. Given the importance of assumptions about future states to the analysis of alternative actions, a great deal of industrial location decision making reduces to decisions about the future.

SUGGESTED READING

Beckmann, M.J. (1980) "Continuous models of transportation and location revisited," *Papers of the Regional Science Association* 45: 45–53.

Fotheringham, A.S. (1983) "A new set of spatial interaction models: the theory of competing destinations," *Environment and Planning A* 15: 15–36.

Haynes, K.E. and Fotheringham, A.S. (1984) *Gravity and Spatial Interaction Models*. Beverly Hills, Calif.: Sage Publications.

Hoover, E.M. (1967) "Some programmed models of industry location," *Land Economics* 43: 303–11.

Jeske, J.W. (1965) "Linear programming as an aid in the solution of plant location problems," *Industrial Development* (January).

Schmenner, R.W. (1982) *Making Business Location Decisions*. Englewood Cliffs, New Jersey: Prentice-Hall.

Wagner, H.M. (1969) *Principles of Operation Research with Applications to Managerial Decisions*. Englewood Cliffs, New Jersey: Prentice-Hall.

Werner, C. (1985) *Spatial Transportation Modeling*. Beverly Hills, Calif.: Sage Publications.

9

THE ROLE OF GOVERNMENTS

s chapter interprets government involvement in ustrial location in three ways. First, we note the 's in which national governments establish the damental conditions for the operation of market non-market relationships, including property ts, labor regulation, and interest rates. Second, make brief mention of industrial activity that is ed or at least directly influenced by national and l governments even in mixed economies: public ices and defense-related operations. The bulk of chapter is devoted to national and sub-national ernmental attempts to influence the location of ate investment, under the rubric of regional cy and urban policy.

E STATE, THE MACROECONOMY, D PRIVATE INVESTMENT

vernments set the ground rules for economic ivity. The functioning of any society requires wers to basic questions of rights, ownership, and ropriability. Are land, labor, or material resources ned in common or by individuals? What about ownership of ideas? Can ownership be trans- ed, according to what rules of exchange? Is the ue of something what the market will bear? vernments also provide a forum for public debate r these basic rules. As the collective expression – the collective domination – of the people of a ional territory, government provides the economic tem with legitimacy, giving the economic actors ietal permission to trade skills, time, resources,

and money. In an immediate sense, the police powers of the government can be brought to bear on those who disobey the legal restrictions on economic (and other) behavior.

Each of these constructs has actual limitations. Public debate and societal permission are limited by the degree of responsiveness of governments to the people they serve. Governments themselves must work to achieve and maintain legitimacy in the eyes of the national population, or fail to do so at the peril of their long-term existence. The larger and more heterogeneous the nation, the more difficult it is to achieve societal permission for any measures or rules. Nominally sovereign governments, even those fully supported by their populations, must make decisions with recognition of a host of economic and political pressures. National sovereignty is limited by the international connections of the modern economy: currency exchange rates, commodity trade, population migration, capital movement, and infor- mation flows. Nations cannot set their economic systems and policies in a vacuum. The international context impinges on the commodities it can export and import to support internal production and liveli- hood, on the legal or illegal movement of people and capital, and on the movement of technology and other information. One of the most straightforward mechanisms of international connection is the multi- national corporation. Such a corporation can choose where to incorporate itself, where to own facilities, and among what countries to trade within itself and outside itself. As such, it is a resident of several countries and a citizen of none. Even in capitalist economies, where private profit is a primary motivating force, public actions through governments have very substantial influence on the geographic pattern of industrialization and industrial change — on who locates what, where. First, let's examine some leading theories of why governments do what they do. Then let's turn to the many different roles that governments can play in industrial location by looking at the varied influences on industrial location.

Theories of the State

Explaining how and why governments derive their enormous power to shape a society and a geography is no easy task. Much of the history of political science as a discipline, and indeed, many other disciplines today, has been given over to this topic. In order to appreciate the role of government in theoretical terms, it is first necessary to review the contending theories of the role of the state. While there are virtually as many theories of the state as there are theorists, it is possible to group them into three broad categories: liberal, conservative, and Marxist. There are, of course, certain similarities that run through these perspectives (e.g., both the classically liberal and conservative notions emphasize the primacy of the market to one extent or another), and there are also significant differences *within* each of these camps, yet the differences among them serve to highlight the contrasting ways in which this vital topic has been broached by different schools of thought over time. This brief review is intended only as a brief introduction to help reveal the role of the state in the creation of industrial landscapes.

Liberal theories

Perhaps the most widespread perspective on governments among academics and the lay public alike is the liberal view, which emerged gradually during the Enlightenment and the ascendancy of capitalist social relations in Europe during the seventeenth, eighteenth, and nineteenth centuries. One of the principal thinkers involved in the construction of this theory was John Locke, the British political scientist. The liberal notion encompasses a wide variety of views of the state, some of which contradict each other, but can be nonetheless characterized by several unifying themes.

Essentially, the liberal theory of the state holds that it is the collective product of individual choice. Drawing upon consensus views of society (e.g., Rousseau's "social contract"), this perspective holds that the state is created because individuals are willing to sacrifice some of their autonomy to the government in order to procure the benefits that derive from state actions. The state is thus produced and collectively reproduced in the minds of its citizenry, and legitimized only when individuals grant it power through elections and voting. This perspective thus assumes equal access to power ("everyone can vote"), and minimizes the structural significance of the state in terms of class, race, or gender relations.

The consequences of this view are important. Because the state is held to arise from the collective desires of its citizens, and its authority is periodically legitimized in the form of elections, government is held to represent the "collective interest." Thus, state actions are explained as necessary for society as a whole, even if they contradict the wishes of any specific individual. This perspective is important, for example, in explaining the government's right of eminent domain in the U.S., in which the state is entitled to appropriate the property of any land owner (with due compensation) if it is judged necessary to do so. Government actions are particularly important in the event of market failures (e.g., negative externalities). Thus, the state is held to intervene not in favor of any specific social group or class, but in favor of society as a whole. Other theorists, notably Marxists, however, question whether there exists an identifiable "public interest" as assumed by this perspective. A vexing problem with the liberal view is that widespread state intervention in many nations has led to the creation of large, unresponsive, and inefficient bureaucracies, which often as not may be self-serving rather than serving the public that pays the taxes necessary to support the government.

Conservative theories

Although there are profound political differences between contemporary liberals and conservatives, in fact they share a great deal in common. Both perspectives arose during the Enlightenment, both hold the individual is the ultimate unit of social analysis, both tend to minimize the role of class and other social relations, and both subscribe to a consensus view of society, i.e., as a homogeneous collection of individuals. The differences between the liberal and conservative views, however, lie in the degree to which they emphasize the rights and obligations of the state in a market-based society. Indeed, as noted in Chapter 7, the relations between states and markets constitute one of the most important issues of political debate today; further, they vary considerably among different nations. As noted previously, early industrializing nations (e.g., the U.S. and Britain) are typically dominated by modes of political thought that de-emphasize state intervention, while countries that have undergone more recent industrialization (e.g., Germany, Japan, and newly industrializing Third World nations) have political climates generally more conducive to state intervention in the economy.

In general, conservatives tend to uphold the virtues of the market, particularly private investment and consumption, to a markedly higher degree than do liberals. While conservatives do not constitute a homogeneous whole, they tend to believe that only markets generate wealth. States, on the other hand, only consume it. Thus, most conservatives would subscribe to the dictum that "the state that governs least governs best." More radical versions of conservative thought argue that *all* state actions are inherently inefficient and ultimately counterproductive. Because the public sector is not bound by the iron dictates of competition, it has little incentive to perform efficiently. Thus, conservatives frequently oppose state intervention (with the critical exception

of national defense, about which more will be said later), including, for example, various forms of public planning and the provision of public goods and services, arguing that the private sector can offer these more efficiently. For these reasons, conservatives also tend to oppose taxes on the grounds that they transfer resources from efficient, private sector actors to inefficient, public sector ones.

However, like all perspectives on the state, the conservative view suffers from inadequacies: faith in the market, for example, does not explain why private investors have been historically unwilling or unable to provide public goods like roads or low-income housing, in which there is generally little profit to be made or the benefits of investment cannot be captured through market prices. Further, markets frequently generate negative externalities such as traffic congestion or air and water pollution, problems that only state intervention is ultimately capable of mitigating. The history of capitalism has in fact been typified by increasing state intervention over time, as the public sector has enveloped larger spheres of activity formerly performed at home (e.g., education, medical care, child care, etc.).

Marxist theories

A third body of thought concerning the role of the state arises from the Marxist school of political economy, which is significantly different from the two approaches outlined above. The Marxist perspective on the state owes its origins primarily to the works of Lenin, not Marx, who casually dismissed the state as the "executive committee of the bourgeoisie," and predicted that it would eventually "wither away" with the rise of a communist, classless society. Lenin in contrast, provided an explicit theorization of th state in terms of class relations, i.e., as the instrument by which one class dominates another. Thus far from being a politically neutral actor such a depicted in both the liberal and conservative views the Marxist camp holds that state policies ar products of the struggle for power. Unlike the libera claim that power is uniformly distributed amon individuals, Marxists emphasize that power is inevitably unequally distributed among social classes. Even elections are facades of equality that mask the real, structural role of the state as a social institution acting on behalf of the ruling class. There can be no mythical "public interest," only a series of incompatible interests of different social groups. State intervention for "society as a whole" in fact amounts to disguised intervention on behalf of the powerful. However, there is a significant variation within the Marxist school of the degree to which the state is subservient to the interests of capitalists, ranging from those who argue that the state is a simple tool of the ruling class to those who accord it a measurable degree of autonomy, noting that under particular circumstances, the state may even be used to advance the interests of the working class in its struggle for power. More recent perspectives, therefore, hold the state to be an arena in which different social groups contend for power. With these comments in mind, and noting that there are problems as well as aspects of truth to each of them, let us turn to some specific aspects of state intervention in capitalist economies.

Macroeconomic Policies

As we have said repeatedly, industrial location decisions are essentially decisions to invest or to disinvest, allocating funds today for returns expected in the future. National governments can greatly influence the investment climate facing firms, by making funds easier to obtain, by determining the after-tax returns from investment, and by making the future look more or less promising.

Governments have two principal vehicles for affecting the general economy in the short run: fiscal policy and monetary policy. **Fiscal policy** is the combination of government expenditures and taxes. Most often, government spending and taxing decisions are made separately, with attention being paid to the political consequences of each spending or taxing program. However, attention needs to be paid to the "bottom line" – deficit or surplus? The perceived success of Keynesian fiscal policy in

the U.S. in the 1960s has led to a general acceptance of the premise that downward spirals in the business cycles can be mitigated by governments' deficit spending. By definition, this increases current demand for goods and services in greater measure than current, disposable income is reduced by taxes. In the opposite situation, rapid expansion of the economy can supposedly be dampened by government surpluses, to avoid the inflation that comes as industrial capacity and labor supplies are fully utilized. However, government surpluses have been rare in recent decades, regardless of the growth rate of national economies. Persistent labor unemployment even during economic upswings has made it politically difficult to restrict economic growth via government surpluses. In addition, when governmental budgets are established by item-by-item decisions on programs, with revenues developed in a separate, parallel, process, there is a substantial tendency toward deficit spending at the national level in many western industrialized countries.

Monetary policy entails decisions by sovereign governments to print additional money – or, more fundamentally today, to issue credits to the nation's banks. By buying government bonds from banks, a government provides banks with more credits that can be used to make loans. Because banks can lend multiples of their net assets, this is a powerful way of increasing the amount of money in the private economy. By this mechanism, and by the interest rate at which the central bank lends money to large lending banks, the government can influence the amount and the interest rate of funds available for investment. If funds are increased more rapidly than investments can be made, general price inflation can result from increasing amounts of currency "chasing" a more slowly increasing supply of goods and services.

How do these two types of macroeconomic policy affect corporate investment decisions? The answers are complex, and potentially contradictory. At the beginning and in the midst of an economic recession, falling levels of aggregate demand reduce the forecasts of return on productive investments. Therefore, fewer proposed projects promise sufficient returns to justify dedicating funds to them. However, stimulatory governmental policy (fiscal or monetary) may indirectly lead companies' analysts to forecast a more rapid return to economic growth, to the extent that a project's revenue projections are related to the national economy. Because of increased government purchases, stimulative fiscal policy may directly lead to increased governmental demand for a given company's product. This predicted or actually increased demand may make a proposed investment seem worthwhile. With respect to monetary stimulus, reduced interest rates payable on borrowed funds lower the threshold at which the return on a proposed project justifies its investment. This reduction in the threshold occurs whether the alternative to the project is not to borrow funds, to hold on to available funds, or to invest in an alternative project. Why? Because borrowed funds cost interest, which responds to monetary policy, while available funds can be invested financially for a return that also responds to monetary policy.

However (and this is where the effects of policies can be contradictory), if corporate managers suspect that the government's stimulatory policies will lead to greater inflation rates, the results are uncertain. The effects and the timing of inflation are difficult to predict. Inflationary expectations may lead managers to discount the current nominal interest rate for borrowed funds, and to expect that future output can be priced at a premium. These responses should increase investment. A fear that inflation will increase input costs more rapidly than product prices (for example, when much of a company's product is exported to countries facing lower inflation rates), or that demand will be moderated by consumers' reduced real purchasing power, will restrict investment decisions. The uncertainty and unevenness of inflation are the sources of much of its economic damage.

So these are some of the ways in which government macroeconomic policy affects corporate investment decisions. How does this affect the *location* of investment? Recall that *in situ* expansion or refitting usually entails a smaller investment outlay than purchase of another facility or construction/leasing of a new

facility. When production changes or expected demand requires that capacity be expanded or modified, macroeconomic variables such as interest rates and expected total growth or decline enter into the calculation of the maximum feasible investment outlay. Expansionary government policies increase the ability of firms to engage in more costly investments, such as new facilities. This reduces locational inertia and increases the mobility of industrial activity. Major shifts in the distribution of industrial activity – from the industrialized to industrializing countries, from the U.S. Northeast to the South and West, from the U.K. Midlands to overseas and the Southeast – are facilitated by expansionary macroeconomic policy.

The *direction* of the geographic shifts is affected by macroeconomic policy. Stimulative fiscal policy generally reduces labor unemployment, increasing wages and the difficulty of hiring and retaining employees. Employer responses include capital intensification and location of new facilities in regions or countries less affected by stimulative economic policy. The lower interest rates of monetary stimulus ease the investment in more capital-intensive production. Capital-intensive production changes the labor complement of production and affects its locational needs. While capital intensification can take many forms, one prominent form is a need for less skilled workers, along with a few highly trained employees or contractors to design or program adjustments in the new equipment. An operation slated to be housed in a totally new facility using more capital-intensive techniques may thus need a different employee mix. Workers at the original operation may be unwilling to modify job descriptions. If capital costs facing a corporation do not vary according to the location of the investment, then the capital and labor needs of the new facility argue for a new location, where new workers can be hired under new job descriptions, and where less-skilled workers can be hired for lower wages and benefits. Such a relocation solution is only one option facing the corporation, and its choice is affected by considerations presented throughout this book.

Government Roles in Technological Change

Chapter 6 dealt with the ways in which technological change affects industrial location decisions. A central theme of this book is that industrial location and locational change entails a matching and a dynamic interaction among *what* is being produced, *how* it is being produced, and the characteristics of places *where* it might be produced. A technology is a systematic way of determining what will be produced how. Technological change is an improvement in the understanding of what can be produced, and by what means. Thus, technological change affects the full range of locational needs; industrial reaction to place characteristics; and the ways in which a given industrial operation affects the characteristics of its geographic environment.

The ways in which governments can affect technological change are equally all-encompassing. Kline and Kash organized these effects under four rubrics: government influence on the factor, organizational, and international "climate" for innovation; government surveying and communication of emergent technologies in other industries and other countries; government coordination of standards and directions for technological development; and governmental operations to fill gaps in the commercial development and dissemination of technology.

With respect to the "climate" for innovation, Kline and Kash noted that the resources spent on the development of new technologies have uncertain returns, and governmental regulation of taxes, depreciation schedules, securities markets, and procurement influence the availability of resources that can be spent in such uncertain ways ("patient capital"). Governments also play crucial roles in the development of relevant skills in the labor force, from the provision of public education to the support of graduate study and the support (or lack of support) of workers who wish to get mid-life training. Organizational flexibility, communication, and security are requisites of the kind of resource commitment and continuous re-structuring that support innovation. National governments set the rules for corporate structure, cooperation, and ownership, influencing these requisite characteristics. National governments also set and negotiate the rules for international trade, with the goal of securing access to the foreign markets that are so important to earning a return on technology development. The market for technology itself is influenced by government protection of intellectual property, via the granting and regulation of patents, copyrights, and trademarks. If regulation of intellectual property rights is too stringent, competition may be reduced. If regulation is too lax, there is little incentive for innovation, and little technological progress is made within or imported into the country.

There are several famous examples of national governments that have taken pains to identify emergent technologies in universities, government laboratories, and overseas, and then to disseminate these technologies for adaptation and adoption in military, commercial, and mixed activity. The Manhattan Project that developed the United States's atomic bomb during World War II relied heavily on the coordination of efforts by academic researchers. Academic, military, and commercial expertise were combined in the Apollo space program, culminating in the first successful manned missions to the moon in the late 1960s and early 1970s. The expertise gained through this complex project found its way into a myriad of new commercial and military products and processes. The United States Agricultural Extension Service established small stations throughout the rural U.S. during the twentieth century, testing farming techniques and sharing these minor innovations with nearby farmers. The Japanese Ministry of International Trade and Industry has continuously surveyed emergent technologies, and suggested their application within key Japanese companies.

These projects entailed both identification and development of new technology, but also the coordination of development. Coordination is needed to assure that component efforts are compatible, that standards are set, and that duplication is used as a thoughtful strategy rather than a wasteful by-product. Also note the importance, at least in the cases mentioned above, of governmental agenda setting: a nuclear weapon before the end of World War II; a man on the moon before 1970; improved agricultural productivity and rural living standards; sector-specific dominance of world markets. Finally, note that the government may form specific organizations to increase coordination of resources, "filling the gaps" in the pre-existing technology infrastructure. The location of these organizations can have a critical influence on the sub-national geography of innovation and of leading-edge technologies. The next section presents the ways in which these direct government location decisions are made.

These national governmental activities have several effects on the geography of industrial activity. To the extent that they facilitate the development of new product and process technologies within their national boundaries, they encourage the national location of new and of revitalized industries. To the extent that they facilitate the development of improved communication and organizational techniques, they may encourage a greater spatial division of activities, enabling production of particular goods and services to concentrate at particularly advantageous points within the country and outside the country. To the extent that government initiatives are unevenly dispersed or that locations vary in their ability to take advantage of these initiatives, the government actions have direct regional effects.

Sub-national or local governments have somewhat less influence over the extent and nature of technological change, because of the mobility of technological information or equipment. Indeed, local governments should spend at least as much of their limited resources on improving the inward mobility of new or improved technologies as on new-technology development. The development of new technologies is a risky and uncertain prospect. However, technologies can diffuse (or, in a more active sense, can be transferred by organizations) widely and quickly. The benefits-to-costs comparison for inward diffusion may be greater than for local new-technologies initiatives. Local and national characteristics affect the geography of the diffusion, via the factor and organizational climates for technological improvement. Thus these same rubrics,

especially climate setting, communication, and development/dissemination can categorize the efforts of local governments. Because of the varied nature and success of these efforts, they have a direct impact on the distribution of industrial activity.

Education and training are crucial components of the local technological climate. In many countries, the delivery of education and training services is locally controlled, with some local discretion over curriculum and student access to new technologies. The basic skills provided to all students affect the general ability to absorb and manage new technologies. Beyond that, advanced programs can be targeted toward the technologies of greatest interest or relevance to the local or sub-national area. With regard to other aspects of technological climate, sub-national governments have become involved in infrastructural development for technology creation and transfer, through several vehicles. Public real estate development has created buildings (often called "incubators") or large tracts of well-serviced land ("technology parks" or "research parks") that encourage the location and agglomeration of establishments engaged in the development or exploitation of new technologies. Dissemination is encouraged via agglomeration, and via incentives for university researchers to interact with (or form) local establishments and enterprises. Local, state, provincial, and regional governments have established public entities charged with identifying problems in private businesses and infusing them with available information, methods, or consultant contacts. All of these methods attempt to increase the growth and competitiveness of local business, in the hope that resultant employment and tax revenue growth will remain in the region. While some of these efforts may be inefficient or even wasteful, the development and diffusion of improved products and processes is not a zero-sum game in which one region's gain is another region's loss.

LOCATION OF GOVERNMENT ACTIVITY

The government itself invests in productive activity. In private-led mixed economies like Canada, the

U.K., and the U.S., government fixed investment falls into three broad categories: public-service provision, public infrastructure, and defense-related services and manufacturing.

Public services and infrastructure

Public services tend to be provided according to central-place hierarchies. Specialized or scale-dependent activities (universities, bureaucratic control operations) are positioned in major cities, and lower level services (schools, post offices) are dispersed according to the market. Government provision of services brings with it *different locational considerations* than private, profit-oriented investments. First, the government is charged with providing services to all citizens, including the poor and the geographically peripheral. Ideally, remote markets and poor populations have as much weight as wealthy, central regions. Therefore, analytic tools that find the logistical center of a market area are even more useful in siting government facilities than privately owned operations. Second, the government seldom has substantial competition for its activities – few private companies want to compete with the financial capability of a government (which can subsidize operations from tax revenues), and many government activities are regulated monopolies or are money-losing by definition. Therefore, competitive considerations seldom enter into the analysis of location. Third, governments must seek popular support (even dictatorships require a minimal level of popular support to stay in power). Location of government-owned fixed investment is one way to affect popular support. A region or an interest group that threatens to undermine the current government may receive investment that yields higher wage, salary, real estate, and other income to the region or group. In a representative democracy, regions represented by powerful members of government may receive disproportionate shares of government investment. Finally, and related to the attempts by governments to gain widespread popular support, governments use the location of their own activities as a form of regional development policy,

discussed in the next section. Note how different these considerations are from the bulk of the issues raised so far in this book. In part this results from the non-manufacturing nature of most government fixed investment. The logistics of material inputs and distribution are not as salient for government services as for manufacturing facilities.

Public infrastructure (highways, air and water ports, water and sewage systems, and in some settings, electricity, telephones, and railways) is an important influence on the location of private investment, as we have seen. Its location depends upon the same considerations as the location of public service centers: universal service, limited competition, popular support, and regional policy.

Military spending

One important way in which governments are often involved in shaping economic landscapes involves spending for military purposes. Defense is commonly acknowledged to be a legitimate function of the nation-state, a quintessential public service, insofar as it cannot be provided for some residents and not for others. However, military spending also affects the distribution of economic activity in time and space. For example, many economic historians speculate that it was growing military spending on the eve of World War II that helped to pull the United States out of the Great Depression. In the post-World War II era, the Cold War between the U.S. and the Soviet Union deeply affected the economy of the United States. In the 1980s, the Reagan administration spent more than $3 trillion on a variety of weapons systems, ships, and troops.

Unofficially, military spending accounts for roughly 30 percent of U.S. federal spending, or roughly $300 billion annually. However, if a variety of related expenditures is taken into account, national defense may unofficially rise to roughly 50 percent of U.S. federal spending. For example, expenditures related to military purposes comprise roughly 15 percent of the National Aeronautics and Space Administration's (NASA's) spending and 75 percent of U.S. foreign aid. Nuclear research and development spending is not included within the Department of Defense budget, but within that of the Energy Department; similarly, spending for veterans' benefits is under the Department of Veterans' Affairs.

Military expenditures have enormous impacts on different industries and places in the United States. For example, in 1985, military contracts generated 93 percent of the demand for the output in shipbuilding, 66 percent of that in aerospace, and 34 percent of machine tools. Military spending also has intimate connections to "high technology" and electronics industries. Some of the largest contractors include corporations such as General Dynamics, McDonnell-Douglas, Rockwell, General Electric, Boeing, Lockheed, and United Technologies. With numerous well-funded lobbyists, these firms constitute a powerful political force shaping the nature and location of government expenditures: some political scientists talk about a "revolving door" of employment as military contractors, lobbyists, and politicians. Finally, it should be noted that expenditures for military purposes have also generated a variety of new products and innovations, including networked computer technologies, fiber optics, and lasers.

Military spending is qualitatively different from many other industrial sectors in several respects. First, the market for military goods is a monopsony (i.e., there is only one buyer, in contrast to a monopoly, in which there is only one producer). Further, the Department of Defense, as a non-profit maximizing public agency, is largely unconcerned about the pricing of military goods. Rather, the criteria that underlay the purchase and production of military hardware center around short production runs, the minimization of acquisition times, and an emphasis on quality. These characteristics make the demand for military products very price inelastic, that is, demand does not fall significantly when prices for these products rise. As a result, many firms producing goods and services for the military engage in enormous cost overruns in which the amount that they charge to the government greatly exceeds the

original contracted price. In addition, many firms become accustomed to the guaranteed profits they garner from the Pentagon and consequently face difficult conditions when they must compete in civilian markets.

Military spending exhibits a distinct geography in the United States that drastically affects the location of many private firms and the location of jobs. Department of Defense expenditures are the most regionally imbalanced category of federal expenditures, and are concentrated in a handful of states (Figure 9.1). California is by far the largest single recipient of military spending, especially the southern part of the state in which there is a large number of aerospace producers. Other areas that receive disproportionate numbers of military dollars include New England, particularly Massachusetts and Connecticut, parts of the South such as Virginia, Maryland, and Georgia, and sections of the Pacific Northwest, including Washington and Alaska. There is a tendency for military contracts to be heavily concentrated in the south and west of the United States. The reasons for this include the large number of air and naval bases in those parts of the country, the location of wide open spaces for use in Air Force training grounds, and aggressive lobbying by politicians from those states. However, because military spending is publicly financed, most states pay more in federal taxes than they receive in military contracts, i.e., military spending in a few states is heavily subsidized by all the rest.

While military expenditures generate benefits in the form of national security, many economists also point to significant economic costs of excessive military spending. First, there are opportunity costs in terms of the relatively few jobs generated by military expenditures. Of course, military expenditures create many jobs; any time the government spends money it generates employment. Military spending is a basic activity (in terms of economic base theory) for several areas of the United States. However, military spending is generally very capital intensive, especially expenditures in the form of prime contracts to aerospace, electronics, and related firms; therefore it generates relatively fewer jobs per

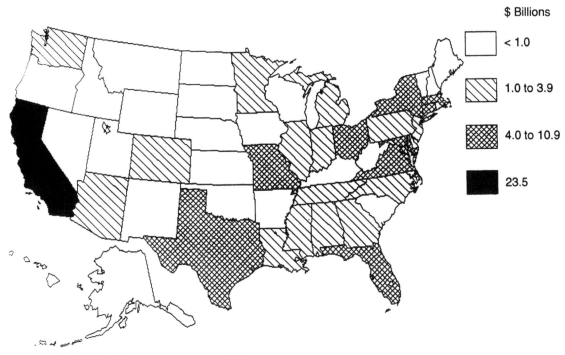

Figure 9.1 Total U.S. military prime contracts, 1990

billion dollars than would be created if the same funds were spent on other sectors. Second, there are opportunity costs in terms of the uses to which military expenditures are put. Many observers question whether the expenditure on military purposes would not be better spent repairing the nation's decaying infrastructure, insufficient housing, or inadequate educational system. Third, there are inflationary pressures generated by large numbers of military dollars: at best, military output is never used, but rusts in a warehouse. Military spending therefore generates incomes but does not increase the aggregate supply of usable output. By pumping more dollars to chase a constant supply of goods, military spending therefore contributes to the general rate of inflation. Some observers assert that the high inflation rates of the 1970s reflected the military expenditures of the Vietnam War in the 1960s. Fourth, there are significant effects on the nation's supply of skilled labor. In the United States, roughly one half of all scientists and engineers are employed in military-related industries and not

in the production of civilian goods. In contrast, Japan, with half the population of the United States, has more engineers in absolute terms and a much smaller military mission.

A fifth effect of military spending concerns its impact on research and development. The Department of Defense absorbs roughly 70 percent of all federal expenditures on research; an additional 17 percent of federal funding of space research goes to military purposes, and research for nuclear weapons is financed by the Department of Energy. Further, many universities depend upon military contracts, and have restructured their research programs to reflect the resources available for military research.

These issues lead some observers to argue that defense spending is a significant obstacle to U.S. economic growth in the late twentieth century. Evidence for this position is derived through an international comparison of military spending as a percent of GNP. In the United States, roughly 7 percent of GNP is devoted to military spending, while other wealthy industrialized nations, such as Japan and Germany, allocate roughly 1 percent.

The size of military budgets in the 1990s is a matter of deep political concern. The collapse of the Soviet Union largely eliminated the primary justification for the large defense budgets of the Cold War. However, military spending has become so deeply entrenched in U.S. politics that substantial reductions in the Pentagon budget are politically difficult; even liberal members of Congress often invoke "national defense" when cuts in military spending are aimed at *their* district. Indeed, the Pentagon is frequently unable to stop military spending programs even if it wanted to, as demonstrated by its frequently fruitless attempts to close outdated military bases in many parts of the United States. In short, military spending, like all government expenditures, is a politically charged process. Political struggles over the size, nature, and spatial distribution of public expenditures are important to how economic landscapes are created and changed, and any serious student of industrial location must have an understanding of the dynamics of government and power to appreciate the geography of economic activities.

REGIONAL POLICY

Much of our concern about the location of industrial facilities stems from their effect on their environs. Different effects are manifested at different scales: physical environmental effects at a relatively local scale, tax revenue effects at the scale of the taxing authority, trade balance and currency exchange effects at a national scale, and labor effects primarily at a local scale. Since many of these effects are local, focused over a metropolitan or other sub-national area, industrial location is an important target of governments' regional policies. We can define **regional policy** very simply as a set of public programs or decisions that influence the distribution of people, activity, income, and growth across a country.

Let's divide this broad concept of regional policy into two dimensions. The first relates to the primary purpose of the public programs and decisions: does the purpose have to do *explicitly* with geographic distribution across the regions of a country, or is geographic distribution merely an *implicit* result of some policy? Cutting across this division of explicit versus implicit regional policy is the level or scale of government that organizes the programs and makes the decisions. In a federal system such as the U.S., Canada, or Mexico, individual states or provinces may establish programs explicitly designed to affect their standing with respect to other states or provinces. Because these programs often set regions in competition with one another for industrial investment, population migration, or technological advancement, we will call these *competitive* regional policies. In any nation (federal or not), the central government may concern itself with the sub-national distribution of wealth, welfare, or growth. This concern is especially likely to develop where national governments are elected from local districts, and where regional politics have an ethnic component to bring them greater salience. We can call the set of national programs and decisions that affect regional distribution *centralized* regional policy. Table 9.1 presents some examples of explicit regional policy.

Explicit and Implicit Regional Policies

Explicit regional policy, whether competitive or centralized, can entail *direct* governmental improvement of regional welfare or indirect governmental *incentives* for citizens and organizations to improve regional welfare. British government policies to disperse national government employment across the U.K., which led to 20,000 jobs being moved outside London in the 1960s and 1970s, were direct actions explicitly geared to increase economic activity in the rest of the country and to reduce congestion in London. Government-owned manufacturing facilities, generally in heavy industries facing great locational inertia, have generally entered into regional policy discussions via decisions to keep them open in the face of reduced demand or inefficient organization. Direct, explicit regional policy does no

Table 9.1 Types of government regional policy

Attraction of outside investment (esp. branch operations)
Marketing: advertising; corporate visits; mailed information
Finance: industrial revenue bonds; tax incentives; land acquisition; loans
Operations: customized worker training; public infrastructure; expedited regulatory approval
Development of indigenous enterprise
Finance: government-capitalized loan pools; venture capital pools
Business assistance: small-business development centers; incubator facilities
Technology assistance: research-park development; research grants; R&D tax credits; technology-transfer assistance (university–industry and industry–industry)
Export assistance: export-operations training; foreign trade missions; export financing

Source: Bartik, T.J. (1991) *Who Benefits from State and Local Economic Development Policies?* Kalamazoo, Michigan: W.E. Upjohn Institute for Employment Research, Table 1.1

have to entail the location of a government-owned source of employment. Regional welfare can be improved by transfer payments to individuals and households, by education and health programs, and by national government grants to local governments that have small tax bases.

Because so much of the manufacturing and traded-services sectors are in private ownership in the U.S., U.K., Canada, and Japan, explicit regional policies often take the approach of providing incentives for private companies to maintain, expand, and relocate facilities in particular regions. Earlier in this chapter we presented ways in which national and sub-national governments encourage industrial expansion via technological improvements. Such efforts may be directed at particular regions, as part of explicit regional policy. Other incentives for desired private actions are presented below. Under a centralized policy, the favored regions are decided based on some formula that is politically acceptable to the country as a whole. Competitive regional policies entail each regional government establishing its own system of incentives, based on its own perception of its development needs.

Given that incentives attempt to influence investment-location decisions, incentives can be designed to relate to any part of the industrial location process. We can study the range of possible incentives as an analogy of this book's study of industrial location influences. Table 9.2 lists the principal incentives

offered by a sampling of U.S. states, with an interpretation of how the incentive affects the industrial location decision-making process. The differences across the states reflect political differences, differences in the leading industrial activity sought by each state, and differences in state policy makers' interpretation of the location decision-making process.

First, think of the investment process, and the ways that government involvement can affect the location of investment. Investment cannot take place without capital – resources made available to earn returns later. Financial capital is very mobile across regions, but lenders and borrowers must be brought into contact via intermediaries. Large corporations can raise investment funds globally via equity (stock) and debt (bonds and commercial paper) markets, for physical investments to be made anywhere. Merchant and commercial banks serve as intermediaries between companies that need investment funds and parties (individual and corporate) with funds to invest. However, managers of small businesses generally find themselves limited to local intermediaries, because the cost of long-distance intermediation – the judging of the loan opportunity, the bank's familiarity with the company – is prohibitive for small transactions.

New businesses, with very high aggregate failure rates, find it difficult to obtain investment funds. Some government incentives entail the provision of financial capital for investments to be made within

Table 9.2 Principal incentives for private investment in selected U.S. states

	California	Florida	Illinois	Mississippi	New York
Direct incentives					
Industrial development bonds (IRBs)	$10 million	Through local authorities	—	Arranged through banks	Through local authorities
Pollution control financing	Via IRBs	—	—	—	Via IRBs
Customized training	Reimburse cost for previously unemployed	State establishes training on site	Funds some costs	—	Reimburse cost for distressed workers
Enterprise zones	Various sites	—	Various sites	—	Various sites
Small-business financing	Below market-rate loans; loan guarantees	Through federal programs	Below market-rate loans	Below market-rate loans; loan guarantees	Loans and loan guarantees
Minority-owned business financing	—	Small loans and equity involvement	Small loans at low interest	Grants and loans	Loans and loan guarantees
Technical assistance	—	—	For small & minority-owned businesses	—	Small, minority & female-owned businesses
Business taxes					
Corporate income tax	9.3%	5.5%	4.8%	5%; 0.25% tax on capital stock	9%
Sales and use tax (statewide portion)	7.25%	6%, with sector-specific exemptions	6.25%, with sector-specific exemptions	6% retail sales tax;1.5–3.5% manufacturers' tax	4%
Property tax	1.1% market value	Levied by local municipalities	On real estate only	Levied by localities	Levied by localities
Inventory tax	None	None	None	None	None
Income tax credits	—	15% wage costs of new empl. from targeted groups = tax credit; enterprise zone tax credits	1% investment credit; 1.6% training credit; 6.5% R&D credit	Fixed-amount credits for new jobs, R&D, new HQ, child care, basic-skills training	5% investment credit; enterprise zone credit

a specific region. When these incentives are targeted at smaller or newer enterprises, they may enable investments that might not be possible otherwise.

Explicit attempts to lower factor costs are popular components of regional policy. Think of the costs involved in manufacturing: material inputs and their transport, energy inputs, labor inputs, financial capital, and (in some cases) shipment to markets. Favored targets for government policy are costs that are directly controlled by the government: state-owned transport services (railway companies or highway construction); state-run energy utilities; state-regulated utility, transport, or wage rates. For costs or charges that are not under the direct control of governments, subsidies can be granted to establishments that use particular factors. Because a subsidy reduces the effective cost of a factor, it has the simultaneous result of reducing aggregate costs (making location within the subsidized region more attractive) and increasing the relative demand for the factor. Thus, subsidies should only be provided for factors that are in surplus supply – factors whose market prices result in under-use within the region. Labor subsidy is a good case in point, when the prevailing wage levels are above the level at which unemployed workers are attractive to businesses.

Despite this distorting effect of factor subsidies, a very important incentive in the U.S. is the **industrial revenue bond**. By selling such bonds, a local or state government raises capital for use by a private industrial operation that will expand, locate, or renovate within the government's jurisdiction. The interest income earned by the bondholder (the lender) is not taxed by the federal government; thus the bondholder is willing to accept a lower interest yield than on privately-issued bonds. The industrial company pays the principal and below-market-rate interest. All parties are happy: the company obtains capital at a reduced cost, the local government has provided incentive for productive investment, and the bondholder receives a fair after-tax rate of return. The federal government loses tax revenue on the bondholder's interest income, so that the general level of federal tax rates increases (or future taxpayers will pay for the reduction in current federal revenue).

It is instructive to note two potential problems with such a happy arrangement. First, because individual states and localities are free to decide when to sell industrial revenue bonds on behalf of expanding businesses, this incentive for local investment is competitive in fact as well as in our terminology. All sub-national governments feel obligated to offer this form of financing to industrial operations, so that the incentive to invest within any particular region is reduced. (A locality with a more restrictive bonding policy provides a disincentive for industrial investment.) The industrial revenue bond provision in the U.S. tax code becomes a way for the U.S. as a whole to subsidize capital investment. Given the importance of capital investment for the implementation of new technologies and for the increase in labor productivity that allows real wages to increase, that may not be a bad national policy. However, the state and local governments that sell the bonds are usually motivated by a desire to increase labor employment within their jurisdictions. Herein lies the second problem with this policy: a subsidy to capital investment increases the ratio of capital to labor, leading to local investments that increase labor employment less rapidly than if the subsidy were factor neutral.

Incentives in the form of subsidies for the employment or training of labor in industrial companies generally match the policy aims of local governments more closely. Whether such incentive programs are developed as centralized national tools of regional policy or as ad hoc, regionally competitive tools, they can be tailored to the needs of individual regions. The most effective subsidy in a region with many unemployed machinists or machine operators or clerical workers would target those particular types of workers and their training.

An important government incentive for private industrial investment is reduction of the taxes to be paid as a result of the investment. Property taxes (or rates) can be negotiated in advance with the local authorities, rather than assessed at the usual level. Taxes on corporate profit or taxes on total sales can be negotiated downward, postponed, or eliminated for the first few years of the new operation. Note that the company proposing a fixed investment has great negotiating leverage, because it has a range of locational options. (This is also why companies having announced relocations or new facilities have more leverage than companies contemplating an investment *in situ.*) After the investment has been made, it is more difficult to negotiate tax reductions for the facility.

Many of these incentives entail subsidy or risk-bearing by the government. The government is supported financially by taxes on its current and future citizens (corporate and individual). Note that competitive regional incentive programs generally shift funds from one group to another within the region: from taxpayers in general to owners of and workers in the enterprises that take advantage of the incentives (most often, mobile enterprises or establishments), or from future taxpayers to current owners and workers. In addition, a sizable proportion of the employees in new or expanded facilities come from outside the sub-national region, transferred by their employer or (more often) attracted to the region by its economic growth. The regional beneficiaries of this expanded population are landowners, who see their property values increase with activity- and population-driven demand. These shifts need to be measured against the benefits to present and future taxpayers: benefits from having more tax-paying companies or employees, or from lower costs of public support of unemployed workers, or from increased property tax revenues, or benefits to future taxpayers from the establishment of an economic base that (after a subsidized start) will grow in the future. In the case of centralized incentive programs, the shift is from taxpayers in general (now or in the future) to companies, employees, or potentially taxpayers in the regions designated for subsidy. Such a shift of resources is worthwhile if the general population favors more even development patterns, if the wealthier regions are congested, or if less industrialized regions will develop into better locations for new rounds of industrialization than the regions favored by companies without public incentives.

Besides the economic incentive of cost and tax reduction, what we know of industrial location decisions suggests other mechanisms for industrial attraction. Agglomeration of similar producers and proximity of related producers, of suppliers and consultants, are powerful locational magnets for small operations and for technologically changing operations. Governments cannot easily create an agglomeration or a nexus of related activities, but careful targeting of tax, labor training, and infrastructural programs may catalyze such developments.

H. Brinton Milward and Heidi Newman compiled the direct state and national government investments used to attract six very large automobile-related facilities in the U.S. during the 1980s. These facilities were proposed and built by Nissan, Mazda, General Motors, Chrysler and Mitsubishi, Toyota, and Fuji and Isuzu, at costs ranging from $500 million to $5 billion. After the companies announced plans, dozens of state and local governments developed proposals for comprehensive incentive programs, including highway and railroad construction, worker training, general purpose loans, land purchase, improvement, and conveyance, waterworks, and facility-dedicated school system improvements. The incentive programs that were negotiated and accepted by the companies ranged from $33 million to $111 million in total value. Each of these facilities planned to employ thousands of workers directly, many from the local areas. The cost of the incentive programs per job created in the proposed facility ranged from $11,000 to $50,000. Interestingly, the per-job cost of incentives increased over the decade of the 1980s, as the process became increasingly publicized and as local governments worried that few other automobile companies remained to propose such large new facilities.

It is important to put these six very large investments and incentive packages in perspective, for they are rare. Most industrial attraction efforts target much smaller facilities, and increasingly target *in situ* investment by companies already in the region. Most companies considering a new facility have neither the time nor the staff (or consultant budget) to spend years compiling and comparing proposals. An important stage in the location decision-making process is information gathering. The time-pressed information-deprived way in which a handful of potential sites are investigated (by those relatively few companies that undertake multi-site investigations) suggests that getting the appropriate information to the appropriate person at the appropriate time is critical to industrial-attraction efforts. Much of the time and energy of local industrial development staffs is spent trying to disseminate information to corporate managers and consultants. We have seen how difficult it is to discover and compare wage rates, tax rates, perceived quality of labor force, and costs of supply linkages across places. Businesses benefit from assistance in these questions. However, local development officials are not disinterested, neutral sources of information. Thus are the information problems compounded.

Finally, a word on variations in regional policy around the world is in order. The objectives of regional policy vary from nation to nation, and may include economic development, reductions in regional disparities, minimizing unemployment, political stability, changing patterns of population or economic growth, or expansion of resource frontiers. Further, the politically acceptable limits of government intervention to achieve these ends will reflect differing national contexts, histories, and aspirations. In Canada, for example, the issue centers largely around Quebec, the poorer Atlantic provinces, the grain and oil producing plains provinces, and their relations to Ontario. In France, regional development planning reached its zenith in the 1950s under the agency DATAR, which successfully encouraged growth in centers outside of Paris. In Britain, regional policy has largely reflected the schism between the northern and southern parts of the country. In Australia, which has never been plagued by severe regional disparities, regional policy depends largely on the political party in power. In the Third World, government intervention in territorial terms has generally revolved around attempts to maximize foreign revenues by gaining access to resources (e.g., in Brazil), although in many cases long-standing regional and ethnic rivalries may deeply shape this process. Thus, there is no single, "correct" way in which regional policies are formulated; their nature and extent depend heavily upon the specific conditions and problems faced by individual

countries, their position within the global system of states, and the internal political status of the nation involved.

Any number of government activities could be construed as *implicit* elements of regional policy. For example, the U.S. program of Social Security taxes employees and employers based on wages received and paid, and pays people who have retired from employment after they have paid Social Security taxes. This inter-generational transfer payment becomes a powerful determinant of regional fortunes when retirees move, with their Social Security income stream, to warmer, cheaper regions.

Uneven regional implications of government tax or tax-accounting regulation were implied in the section on macroeconomic policy, above. Allowing businesses to claim a large proportion of expenses on long-lived investments in order to offset taxable income in the current year makes new investments more attractive. Businesses contemplating significant new productive investments are more likely to consider relocation than businesses constrained to minor investment. While this helps the nation as a whole, it hastens inter-regional restructuring, as changing technology and changed regional characteristics modify the relative attractiveness of regions for particular business activities.

National government regulation of environmental pollution increases the business costs of location in congested regions (raising these costs nearer their real, social costs), and makes less congested regions appear more attractive. However, devolution of environmental regulation to lower levels of government would imply a policy of letting regions decide how much degradation to bear. These regional decisions will probably reflect disparities in regional wealth and education, rather than overall social costs. In such ways does environmental policy become implicit regional policy.

Business Climate: Code Word or Empty Words?

Studies of corporate location decisions often use the phrase "business climate" as a variable across potential locations. The phrase is a vague one, relating to a range of local conditions that can affect the ability of fixed capital investment to yield high returns. What are these conditions, aside from the measurable and therefore separate variables of wage rates, accessibility and transport costs, tax rates, and local availability of specialized suppliers and services?

Fixed constraints of any sort restrict profitability, because rapid reaction to changes and market opportunities is an important source of profit. Thus, strict regulation of the relationship between the firm and its workers is considered an element of a poor business climate. Most industrialized countries have national standards for a regular workweek, minimum wage, and on-the-job safety. But local areas and localized unions may develop additional constraints on layoffs or redundancies, on job classifications and duties, and union-membership requirements. In the U.S., some states prohibit labor unions from insisting that all workers in a unionized facility join (or pay membership fees to) the union. These states are termed "right-to-work" states, and the leverage and viability of labor unions is much lower in them. The presence or absence of right-to-work regulation in a state has been used by some researchers as the single variable to indicate business climate. Other constraints include regulation of environmental contaminants, pricing of output, or distribution channels. Financial companies are heavily regulated, and often choose operating locations that allow maximum flexibility of interest rates charged, financial reporting required, or distribution of services.

The regulatory flexibility afforded to private corporations in some national and regional locations increases the power of corporations to control their operations. It is also a measure of the relative power of business interests in the national or regional political setting. The sources of this political power include government leadership by business-supported parties or individuals, lack of a history of countervailing power (e.g., labor unions), or a scarcity of employment opportunities. Nations or regions undergoing a shift from a subsistence economy or an extractive or agricultural economy to an industrial

economy are likely to afford this flexibility to business interests, especially if their political leadership recognizes gains to be achieved from industrialization. Highly industrialized nations or regions are likely to develop demands for regulation and restriction of business flexibility. The tradeoff for business interests are the economic amenities and infrastructure of an area of intense development, versus the regulatory flexibility of developing locations.

While any company, individual or government, wants maximum flexibility, it also benefits from stability and predictability of its environment. Therefore, stability of government and labor relations is as important a part of business climate as the current quality of those relations. Political unrest is anathema to business investment, for many reasons. Tax and regulatory agreements that were assumed into an investment decision may be changed by a new government. Sovereign national governments may insist on departure or disinvestment of a foreign-based company. Economic instability can result in a government's requirement of new or higher fees or taxes, or the deterioration of government services on which the facility depends. Since past stability is the best guide to future stability, recent political, economic (including tax-rate), and labor histories of a place are heavily weighted components of business-climate evaluations.

Many organizations and trade journals attempt to rank locations (countries, states, metropolitan areas) according to "business climate" or other attributes. M. Ross Boyle compared several of these rankings of U.S. states. His comparison illustrated the extent to which the rankings depend on the variables measured. The variables selected by the organization or trade journal reflect the particular audience targeted. Corporate consulting firms emphasize cost-minimizing variables of greatest importance to branch-plant locations of large companies. Entrepreneurial organizations emphasize states' rates of enterprise formation and growth. Other organizations, more concerned with members' or readers' quality of life, include employment, wage, and "amenity" variables important to attracting and maintaining key employees. Obviously, the ranking of a state or region varies along such varied indices, and users should beware. Our conclusion is the essential conclusion of industrial geography: locations vary in their suitability for particular activities.

Making Use of Incentives

In Chapter 8, we saw that information is one of the biggest obstacles to location decision making. In no context is this more true than governmental incentives for industrial attraction and retention. Each of the types of incentives mentioned in the previous section may be offered by different utility companies, different levels of government, and different government agencies. Countries with federal government systems provide the most complexity. Let's suppose that a company has analyzed logistical considerations to narrow its search for a warehousing-and-distribution center to one area of the midwestern United States. To compare available incentives and subsidies, the company must contact the several state governments, many local governments, and several major utility (electricity and natural gas) companies that serve the area of interest. The complexity doesn't stop there, however. Recall the many dimensions of government regional policy: subsidy of labor, capital, transport, and energy inputs; level and type of taxation; regulation of labor relations, physical environmental practices, and competition. Different branches of government may oversee each dimension. Most states and localities have established a central ombudsman office for "one-stop-shopping" for incentives, which helps the time-pressed business manager.

Finally, each local government may offer different configurations of incentive packages, making comparison difficult.

A time-pressed and therefore distracted manager may accept first answers as final. However, the bare outlines of incentive programs do not reveal all the operational elements that make them more or less useful for the industrial operation. Think of all the variables involved in a state-sponsored job-training program; how can a decision maker rely on the kind of overview provided by Table 9.2.? In addition, governments are frequently willing to negotiate greater incentives for serious site-seekers that are willing to make numerical employment or invest-ment commitments – numbers, pay levels, longevity, local economic linkages. Besides the plethora of government agencies, other actors can be brought to the negotiating table. Local businesses, such as banks and utility companies, may enter the negoti-ation to subsidize some aspect of the investment and location process.

While anything and everything can and should be negotiated, it is important to recognize the public purpose with which the government agencies have been entrusted: leveraging public resources to improve local economic conditions. In devising a negotiating stance, companies must estimate their maximum possible impact on these conditions (from direct and indirect employment and taxes paid to prestige, to the development of the local industrial complex) and recognize their value to the region over a five- to ten-year period as the outer limits of expected incentives and subsidies.

CASE STUDY

The Billion-Dollar Attraction: McDonnell-Douglas's MD-12X

During 1990 and 1991, the Douglas Aircraft Corporation considered developing a new commercial aircraft, a wide-body jet to compete with the highly successful Boeing 747. Developing a new aircraft model is a very risky venture, in that the costs of development can approach or surpass the net worth of almost any company, and the returns depend on the decisions of a handful of large airlines. Development costs include years of employment of highly paid engineers and designers (in aeronautics, avionics, electronics, mechanicals, software) followed by massive capital outlays for plant and equip-ment. For the proposed new aircraft (given the name MD-12X), development costs were expected to exceed $4 billion, for a parent company (McDonnell-Douglas Corporation) that was $2.6 billion in debt, having lost $40 million in 1990 on sales of $14.6 billion worth of military and commercial aircraft, missiles, and related services. The company recognized that it could not take on the project alone, and began a search for an equity partner that would share the capital risk, the development costs, and the technological capability of the development project. A foreign partner would assist the ability to sell the new plane in foreign countries. Given the dominance of the Airbus consortium in Europe, Douglas Aircraft looked toward Asia for a partner.

These development costs represent tremendous economic benefits to the labor market and taxing jurisdictions affected by the development process. If the process and new model are successful, thousands of manufacturing, marketing, and engineering jobs follow: highly paid jobs that are secure for as long (or as short) as the model meets continued demand by airlines. States, counties, cities, and even countries routinely compete for major new facilities that provide such employment and tax revenue possibilities. Douglas Aircraft had insufficient space at its one, Long Beach, California facility to develop and build the new model there. The company's management was also dissatisfied with California's environmental protection regulations, and with the costs of taxes and wages. The company announced that it would be likely to invest elsewhere, and expected substantial competition for its $800 million plant and 7,400 eventual jobs. The company sought to use this competition in a novel way: to request that governments around potential sites propose how they would reduce the huge capital risk facing the company. In other words, in addition to a corporate partner in the development and manufacture of the new plane, Douglas sought a governmental partner. The governmental partner(s), however, would not share in potential profits, but would benefit if the jobs and eventual taxes materialized.

Scores of municipalities and states offered incentives. What kinds of incentives were relevant? What does an aircraft manufacturer need? Components are themselves very expensive, highly finished products, so their transport cost is not an issue. Availability of all transport modes is an issue: components arrive via air, truck, rail, and ship. The product clearly can be flown out. On-site runways are needed, along with sufficient space for huge buildings, hangars, and staging space: the company stipulated a 10,500 foot runway adjacent to 600 acres. Engineers and designers with the specialties mentioned above are not available everywhere, though they can be recruited to most metropolitan regions with sufficient pay. Interestingly, while the market location is completely irrelevant, a company as dependent on military contracts as is McDonnell-Douglas benefits from having its operations (even its commercial aircraft operations) within the jurisdictions of Congressmen and women with influence in military procurement.

During the summer of 1991, the company announced nine "finalist" locations (Figure 9.2): the major airports at Kansas City (Missouri), Shreveport (Louisiana), Tulsa (Oklahoma), Salt Lake City (Utah); a smaller, non-commercial airport near Fort Worth (Texas); former U.S. Air Force bases near Mesa (Arizona), Houston (Texas), and Mobile (Alabama); and land adjacent to an existing Air Force base near St. Louis, Missouri. Each location provided hundreds of acres of land adjacent to at least one runway (in most cases, the runway was currently capable of handling jumbo jets); most had rail access on site. The last five locations were in jurisdictions with powerful Congressional delegations, particularly influential over military programs and budgets.

What kinds of incentives did these locations offer? Douglas Aircraft preferred a particular package of government-built and -equipped facilities to a hotchpotch of bonds, tax breaks, and infrastructure improvements. Specifically, the company requested (in a 120-page request for proposals) that state

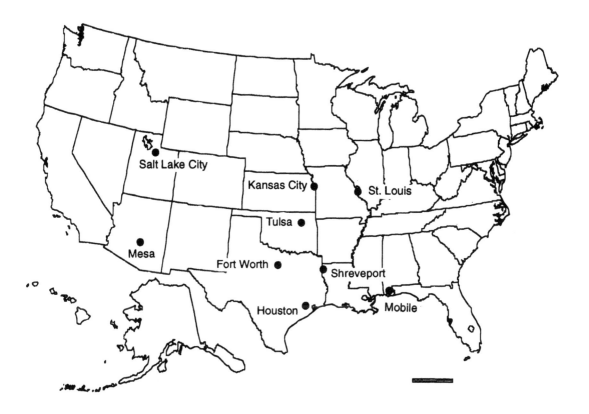

Figure 9.2 Potential sites for MD-12X aircraft production

and/or local government issue bonds to build the $800 million facility and provide up to $400 million in equipment, and then lease the facility to the company, with no actual lease payments made until the first planes are sold. Such a clear-cut proposal did not materialize, however, and states and municipalities pooled resources to offer site improvements,

After all the proposals and site inspections, the MD-12X project remained on hold. In November 1991, the company reached an agreement to sell up to 40 percent of the commercial aerospace division to Taiwan Aerospace Corporation (itself a public–private consortium) for up to $2 billion (rather than creating a special joint venture to produce specifically the proposed aircraft). Much of the aircraft's MD-12X body would be built in Taiwan. The agreement was tentative, however, and foundered during the next year. The company continued the search for an equity partner and postponed the development of the new aircraft, along with the designation of a production site.

Sources

Goodman, A. (1991) "Cities racing for jet plant, but it may be a long shot," *St. Louis Post-Dispatch*, September 1: E-1-ff.

Kandebo, S.W. (1992) "McDonnell Douglas–Taiwan pact: bad deal for the American economy," *Aviation Week & Space Technology*, January 13: 58–9.
Velocci, A.L. (1993) "Aerospace giant puts credibility on line," *Aviation Week & Space Technology*, April 19: 44–8.

Implications for Regional Policy

What can we learn from our categorization of regional policy instruments, our comparison of instruments and the industrial location process, and the ways in which businesses perceive and use location incentives? Regional policy is best formulated with a clear and prioritized statement of goals, so that policy makers can choose among possible instruments. A careful match is required among the primary policy objectives (including specification of the types of activities to be developed), the ways different policy instruments affect investment decisions, and the current characteristics of the locale.

Assess goals

Focused, coherent regional policy requires a set of prioritized goals. Relevant goals could entail rates of labor unemployment, average wages paid, the property or income base that can be taxed, development of new or complementary economic sectors, or congestion, pollution, and other attributes of over-development. Centralized policies generally aim to reduce regional disparities along any of these measures. Competitive or unilateral policies generally aim to maximize or minimize these measures. Whether centralized or competitive, regional policies will be most effective if they are designed to produce clearly defined (and prioritized) results, and if implicit government policies are not at odds with explicit policies.

With respect to employment and economic development, the greatest resources of nations and their component regions are the people, organizations, and physical investments already present. This is true because of their large size relative to the people, organizations, and investments that can be attracted in a given year, and because of locational inertia that binds these resources to the nation or region. Government actions must enhance rather than harm these resources. Therefore, industrial retention and productivity increases via governmental investment in physical, educational, communications, and technological infrastructure are more important regional policy goals than industrial attraction. This is especially true in regions where these key resources – people, organizations, and physical investments – are present in abundance. For example, a long-established industrial region with substantial labor unemployment is likely to have tremendous needs for productivity enhancement, worker retraining, and infrastructural improvement. A region that has little activity in manufacturing or service provision for national or international markets, because of little investment in those sectors or because of long disinvestment, may have relatively fewer internal resources for job growth or entrepreneurship. Industrial attraction appropriately plays a larger role in its regional policy.

The traditional emphasis on attracting manufacturing activity has largely been supplanted by a recognition that many service activities have locational flexibility, provide net income to the region, and increase the competitiveness of other regional activities. Retaining, improving, and attracting service industry investment relies on the same range of incentives, subsidies, and information as for manufacturing. However, labor is even more important to service activities – from skilled mechanics for an airline maintenance facility to dependable, articulate receptionists for a telemarketing operation.

Tools for industrial attraction

For the limited objective of industrial attraction, which incentives work? Empirical research has not yielded consistent results. The greatest inconsistency has revolved around the importance of tax rates and tax-based incentives. The largest number of studies in the United States has suggested that interstate and interregional differentials in tax rates play a small role in industrial location decisions, and make little difference in an area's rate of industrial investment. However, some studies have found tax differentials to be important, especially as "tie-breaking" considerations. A few conclusions emerge from these various studies and from an understanding of the objective requirements and subjective process of industrial location:

- Tax differentials are not very important considerations in the typical, quickly decided, capacity-driven, short-distance location decision of a medium-sized company. Expedience and logistical concerns matter much more.
- For decision-making processes in which tax differentials are weighed, different tax rates are more important for different industries (capital-intensive facilities being more sensitive to property-tax rates, labor-intensive facilities being sensitive to payroll taxes, and many manufacturing activities being sensitive to inventory taxes).
- Unusually high tax rates are a bigger disincentive than are unusually low rates an incentive.
- A jurisdiction needs to be most concerned about its various tax rates relative to jurisdictions that are similar in location, metropolitan status, industry mix, and occupational mix. Therefore, jurisdictions designing competitive or unilateral industrial retention and attraction policies need to recognize the jurisdictions with which they are "competing," and the basis for the competition. A large metropolitan region with a long history of industrialization does not need to match the tax or wage rate of a distant, remote, rural area: they are not in direct competition, because few industrial activities that thrive in one will thrive in the other.

Aside from tax differentials and tax-based incentives, attraction policies benefit from recognizing and ameliorating the expense of new industrial investment, especially when an ongoing operation is being branched or relocated. Assuming that local economic development officials target operations for which the local region is indeed a good location, incentives can focus on the barriers to relocation: regulatory, real estate, and other information; physical infrastructure; and employee recruitment. Negotiating and subsidizing these barriers will reduce the company's cost of the actual move or acquisition, allowing the region's good operating environment to weigh more heavily in the company's decision-making process.

As we have suggested repeatedly, labor concerns are of critical importance. While national and regional governments in capitalist economies have little direct, immediate control over wage levels, other important characteristics of the labor force and labor employment are subject to government influence. These include educational attainment, formally-gained skills, specialized training programs for individual employers, and employee-related taxes (payroll, unemployment insurance, even personal income taxes that reduce the purchasing power of workers' earnings).

It is difficult to over-emphasize the importance of information to industrial attraction, retention, and productivity-enhancement efforts. Government programs that are unknown are unused, and most businesses do not know about most government programs. Local authorities around the world have increased their use of ombudsman or "one-stop-shopping" offices, which can present, discuss, and negotiate all manner of programs. However, government programs are only the last fillip of information needed by companies confronted with an imminent physical investment decision. Information about land availability, public utility availability, zoning and development restrictions is notoriously difficult for a company to compile for several potential sites. Lack of familiarity with land availability and government agencies in remote areas is a source of locational inertia.

Timely and accurate provision of information is a reasonable use of the public's money and effort, because it should improve the quality of investment decisions made, increasing the efficiency of the general economy. As long as information is presented with minimal hyperbole and without wasteful, lavish expenditures, an increase in locational information available to investors is not a zero-sum game.

As we recognize the volatility of capital investment and of production, we see an additional, important role for governments: assisting those harmed by the volatility and mobility. Both regional and national governments (as well as supra-national governments such as the European Union) need to build a consensus for change and development by insuring that immobile populations are provided assistance to change and cope with change. For example, the mobility of fixed capital investment across regions and companies over the long term exceeds the mobility of some individual workers or small companies. Displaced workers and uncompetitive, smaller companies benefit from resources that increase their mobility across regions, products, occupations, and technologies. Relevant resources include financial loans, training, information concerning opportunities in other sectors or regions, and the kinds of technology dissemination programs discussed earlier in the chapter. This kind of assistance reduces the fears and fighting of change by those least able to adjust to or manage the change. Given the importance of change to industrial societies, adjustment assistance is more beneficial than fighting change.

THE POST-KEYNESIAN STATE AND URBAN POLICY

An especially important part of the role of contemporary governments is the way in which they regulate and finance urban areas, in which live the majority of the population of the industrialized world. Urban policy, as this collective set of policies is generally called, has changed dramatically in the last two decades, which have witnessed the rise of the "post-Keynesian" state. In the U.S. and the U.K., particularly, the rise of the post-Keynesian state rested upon shared conservative ideological affinities (e.g., a faith in the market, consumer sovereignty,

and competitive individualism), the use of tax cuts to stimulate industrial restructuring, deregulatory measures, and privatization. Deindustrialization, recession, slowing productivity growth, and deteriorating trade balances set the stage for the rise of increasingly powerful conservative coalitions. Unable to put a tolerable ceiling on inflation, unemployment, and interest rates in the 1970s, Keynesian interventionism became increasingly discredited, and with it, numerous liberal political parties and programs. The rise and fall of Keynesian policies are contemporary with Fordist production. The Keynesian state was largely legitimated by the benefits of Fordist production, particularly the provision of collective goods that depended upon a continuously rising productivity of labor. Thus, the transition from national to global markets, i.e., the internationalization of the U.S. and British economies within an increasingly competitive world system, threatened the viability of nationally based, state welfarism.

Keynesian urban policy in the U.S. paid attention to the problems of inner cities, as well as the rapidly growing suburban areas. Inner city programs included Urban Renewal in the 1950s, the Model Cities Program of the 1960s, and the Housing and Community Development Act of 1974. Suburban policy included federal subsidies for construction of all sorts of local public infrastructure. The last vestiges of Keynesian urban policy in the U.S. occurred under the Carter Administration, which implemented the Comprehensive Employment and Training Act in 1978 to offer cities fiscal aid, subsidies, and public service employment programs and instigated the Urban Development Action Grant (UDAG) program to subsidize the rehabilitation of older industrial and commercial spaces. During the Reagan Administration, UDAGs and the Comprehensive Employment Training Act (CETA) were both terminated, and federal subsidies to localities for mass transit education, water treatment, medical care, and public housing were sharply reduced. Repercussions are becoming acutely felt by many states, currently experiencing difficulties in balancing their budgets even while trimming services and raising taxes. In addition, the Reagan Administration initiated sharp reductions in federal income taxes and a rapid growth in military expenditures, which combined to produce historically unprecedented federal budget deficits. The assault on the welfare state had mixed results: while some domestic programs suffered sharp contractions, others, including Social Security, Medicaid, Medicare, Food Stamps, and Aid to Families with Dependent Children (AFDC) were politically popular enough to escape relatively unscathed.

In the U.K. by the late 1970s, welfarist prescriptions about urban decay had been greatly discredited. Government itself came to acknowledge the largely economic determinants behind inner city decline. The new urban agenda heralded the saving graces of an "enterprise culture," which alone would liberate marginalized populations from their demoralizing dependencies. The collapse of British Keynesian policy was apparent in the fall of the Labour government in 1979 and the ascendancy of Thatcherism, an event that can be interpreted as a critical conjuncture that witnessed the fracture of three main pillars upon which post-World War II social democratic politics were constituted – Fordism, Welfarism, and Keynesianism. In the conservative tenets of the 1980s, the need to substitute an enterprise culture for the dependence of welfarism was linked to a determination to deflate the swollen, anti-business public bureaucracy.

Several similarities between Reaganite and Thatcherite urban policies and their consequences should be emphasized. In both cases, a reduction of national government funding to urban areas was evident, raising the political leverage of private investors over local public authorities. In both the U.K. and the U.S., interurban and interregional competition sharply accelerated, forcing public authorities to partake in auctions to attract increasingly footloose capital. On both sides of the Atlantic, older industrial cities suffered while service-based urban areas flourished.

Tools of Post-Keynesian Policy

How was post-Keynesian urban policy implemented? The principal tools included enterprise zones, urban development corporations, urban subsidies, and public–private partnerships. Each is briefly examined here.

Enterprise Zones

Originally proposed as a means of unleashing market forces in distressed inner cities through the simple removal of government controls (an approach designed to produce "little Hong Kongs"), Enterprise Zones (EZs) evolved into more complex attempts to generate private investment through active public subsidies (income and capital gains tax breaks, depreciation allowances, labor training programs) in selected urban areas. EZs were implemented first in the U.K. beginning in 1981, and offer an explicit example of a British policy initiative deliberately imitated in the U.S. British EZs, begun as an experiment to provide inner cities with tax relief and freedom from planning restrictions, absorbed substantial public funds, making the cost per net new job generated prohibitively high for a government formally subscribing to a policy of economic nonintervention. Often the real beneficiaries of this public largesse were property owners rather than industrial firms, as, for example, when EZs proved to be particularly attractive to land-intensive distributive and warehouse activities.

EZs were enthusiastically embraced in the U.S. State governments, particularly in the South, were quick to pass EZ enabling legislation. The traditional reservations about EZs include arguments that they provide incentives to capital-intensive rather than labor-intensive firms, that the most likely beneficiaries would be large, multi-location firms with few local linkages, and that EZs would simply transfer jobs from firms and communities adjacent to them rather than generate new employment.

Urban Development Corporations

An early attempt by the Thatcher government to circumvent local authorities was the creation of Urban

Development Corporations to stimulate property investment and make sites available to developers. These quasi-public agencies were accorded effective control over development in their designated territories. Apart from compulsory powers to acquire, assemble, and dispose of land and other property, they can allocate subsidies by leveraging private investments. The outcome has been spectacular: the London Docklands has become the largest urban renewal project in Europe. British Urban Development Corporations are similar to the Urban Redevelopment Authorities present in the U.S. Both offer a direct channel to the private sector, and can respond flexibly and promptly. As agencies eager to deal with, and sympathetic to, business priorities, they can readily secure the confidence of developers and private financiers.

Post-Keynesian urban funding

A significant dimension of post-Keynesian urban policy was the change in the volume, composition, and regulations concerning national government subsidies extended to urban areas. In the U.S., federal funding programs for urban areas originated with the Urban Development Action Grants initiated by the Nixon Administration in 1974. In the 1980s, UDAGs earmarked for specific purposes were folded into block grants. This shift in funding strategies gave local political authorities considerably more control over the allocation of federal funds during a period in which interurban competition rose sharply. Legislation governing the allocation of funds to needy areas was abandoned in favor of criteria emphasizing economic growth, which facilitated the construction of downtown shopping malls, sports facilities, hotels, convention centers, and restaurant and entertainment complexes. To the considerable extent that block grants subsidized downtown developments without regulating them, they contributed to the surge of gentrification in central city areas. Opposition was generally muted as "anti-growth," and the costs of such projects in terms of residential displacement rarely entered into discourse about their contribution to a broadly inclusive "public

good." Post-Keynesian urban policy thus successively subordinated social needs to the prerequisites of real estate firms, bankers, the construction industry, and related interests.

Public–Private Partnerships

Underlying many urban initiatives in both nations was the institution of public–private partnerships. Involvement by the public sector allows for supervision and oversight of projects in which private developers would otherwise enjoy relative autonomy. Public promotion of private development is assumed to produce positive externalities that include improving a city's economic base and raising real estate values, thereby generating additional tax revenues that in turn permit greater resources for social needs. Moreover, the long-term gains of the new employment created for workers and the new or improved public services provided for consumers outweigh the ostensibly short-term public subsidies deployed. Corporate interests are attracted to cities that deploy powers of eminent domain, assemble and package land parcels, erect new infrastructures, and offer various subsidies. The reshaped built environment generated by such ventures instills a common civic pride while providing the basis for further boosterism when selling the city to potential investors and tourists. Private interests are held to contribute expertise, experience, financial resources, and technological and managerial efficiency frequently inaccessible to the public domain. Public sector involvement in such ventures often allows private interests to assume an aura of having made contributions to civic welfare, an important marketing asset otherwise unavailable to them. U.S. examples of such ventures include public authorities, and government-owned corporations that raise funds on private capital markets for public projects.

In the U.K. in the 1980s, the partnership concept was adopted enthusiastically by businesses and government. The latter frequently does so as a means of endorsing business participation in the formative stages of public planning. The private sector has a clear notion of its stake in the partnership arrangement. However, in many assessments of these programs, the opportunity costs of public sector involvement were rarely acknowledged, including funding that might otherwise have been extended to distressed neighborhoods. Often such "partnerships" amounted to little more than a thin excuse for privately-led – but publicly-subsidized – development.

Concluding Comments on Post-Keynesian Urban Policy

Broadly speaking, the post-Keynesian state has largely disassociated itself from explicit urban policy making. In certain cases, most notably enterprise zones, the Thatcher and Reagan Administrations resorted to explicitly interventionist strategies. But this tactic was the exception, not the rule, and by and large the advent of conservatism has heralded the withdrawal of government funding from many urban areas. This process has reconfigured the matrix of opportunities and constraints within which urban planning is practiced, and led to a broad based shift in the priorities of local governments, which are increasingly less concerned with issues of social redistribution, compensation for negative externalities, provision of public services, and so forth, and more enthralled with questions of economic competitiveness, attracting investment capital, and the production of a favorable "business climate." As planners have come to involve themselves more directly in economic development, market rationality and local competitiveness have replaced comprehensiveness and equity as the primary criteria by which planning projects are judged. Thus, long-term capital budgeting, master planning, and a concern with the environment have gradually given way to short-term concerns of job generation, looser regulations, and tax relief. Planning is hence concerned more with promoting development and less with regulating its aftermath. Although this strategy became widespread in the 1980s, it is not altogether new: long-standing connections exist between local governments and private capital in the form of "growth coalitions," alliances obscured by the pretense that privately-led, market-driven growth is inherently optimal for all parties concerned. In both the U.S. and the U.K., post-Keynesian urban policy sharply accelerated the competition among communities for jobs and public and private investment. Desperate for jobs, many localities vie with one another with ever-greater concessions to attract firms, including foreign companies, in an auction resembling a zero-sum game. The effects of such a competition are hardly beneficial to those with the least political clout.

SUMMARY

Because governments structure so much of our lives, they have tremendous influence over industrial location. They exercise this influence in the regulation of markets, the maintenance of macroeconomic conditions, the location of their own activities, their procurement policies, and in their deliberate attempts to manage the location of private investment. This chapter provided an overview of each of these routes by which governments influence activity location, with examples from the U.S. and the U.K. Governments vary not only across countries, but over time, as political circumstances change. We have made note of some of the political changes in the two countries over the past 30 years, with clear implications for the shift of economic power between public and private sectors, and among regions and urban areas within the countries.

The behavior of governments is the most obvious juncture among the political, the economic, and the geographic. This juncture is important for those who would understand industrial location, as well as for those who would try to forecast government behavior in order to make location decisions.

SUGGESTED READING

Boyle, M.R. (1989) "Grading the state business climate report cards," *The Survey of Regional Literature* 11: 2–7.

Fainstein, S. (1991) "Promoting economic development: urban planning in the United States and Great Britain," *Journal of the American Planning Association* 57: 22–33.

Hansen, N., Higgins, B., and Savoie, D. (1990) *Regional Policy in a Changing World*. New York and London: Plenum Press.

Kline, S.J. and Kash, D.E. (1992) "Technology policy: what should it do?," *IEEE Technology and Society Magazine* 11(2) (May, June).

Logan, J. and Molotch, H. (1987) *Urban Fortunes: The Political Economy of Place*. Berkeley: University of California Press.

Markusen, A., Hall, P., Campbell, S., and Deitrick, S. (1991) *The Rise of the Gunbelt: The Military Remapping of Industrial America*. New York and Oxford: Oxford University Press.

Milward, H.B. and Newman, H.H. (1989) "State incentive packages and the industrial location decision," *Economic Development Quarterly* 3 (3): 203–22.

Schmenner, R.W. (1982) *Making Business Location Decisions*. Englewood Cliffs, New Jersey: Prentice-Hall.

Wolman, H. (1986) "The Reagan urban policy and its impacts," *Urban Affairs Quarterly* 21: 311–35.

10

SECTOR-SPECIFIC CASE STUDIES

Many of the themes articulated in this book can be illustrated through the use of case studies of different industries. Uneven development, business cycles, rising and falling comparative advantages, industrial restructuring, a changing technological mix, and state policy are all evident in the differing historical geographies of several industries. While the specifics vary from sector to sector, the broad outlines are similar. This chapter offers case studies of four manufacturing sectors (textile and garment production, steel, automobiles, and semiconductor electronics) and two service sectors (financial services and telecommunications).

THE TEXTILE AND GARMENT INDUSTRIES

Though the textile and garment (or apparel) industries are often confused and the terms used interchangeably, in fact there are several distinct differences between them. The textile industry is engaged in the production of cloth, not clothing. However, roughly 50 percent of the output of the textile industry is used by the garment industry. In other words, there is a very strong forward linkage from the textile industry to garments, or conversely, a strong backward linkage from garments to textiles. The uses of textiles other than garments include household goods, such as furnishings and carpets, and industrial goods, such as upholstery and insulation. There are also significant differences in the production process and geographical location of each

of these industries: the garment industry tends to be considerably more labor intensive and less technologically sophisticated than the textile industry, and relatively more urbanized.

The textile industry is generally acknowledged to be the leading sector of the early Industrial Revolution. Textiles were the first commodity to be mass produced and mass consumed. The textile industry formed the motor of industrialization in early industrial Britain, continental Europe, Japan, and the U.S., and continues to do so in the newly industrializing countries of East Asia and elsewhere today. The reasons for this have largely to do with the characteristics of the textile industry, which will be explored in more depth below.

First, the textile industry, in comparison to many other sectors, is relatively labor intensive. This means that it uses relatively large amounts of labor per unit output, and that wages form the largest single cost (50–70 percent of total costs) faced by employers. The cost of labor is therefore by far the most important locational determinant in the historical geography of the textile industry. The industry has historically been a very low paying one: often, after years of employment in textiles, many workers earn only the minimum wage. Further, employers have frequently sought to cut wages in order to minimize costs during downturns of the business cycle. Not surprisingly, the textile industry gave birth to some of the first labor unions.

Second, the textile industry has traditionally been characterized by a relatively low degree of technological sophistication. Early textile mills frequently employed no more than ten to twelve young women at a time, and therefore did not rely heavily on economies of scale. Further, capital costs of production in the textile industry were relatively minimal. The cost of a spinning jenny or a cotton gin was minuscule compared to the enormous capital investments necessary for railroads or shipbuilding. Consequently, the textile industry has relatively few backward linkages and has low employment multipliers, and thus has not generated sizable numbers of jobs where it has been located.

Third, the textile industry is a relatively easy industry to enter and exit. Textile production involves low start-up costs and fixed investments. Because it is so easy to enter, the textile industry has historically been a very competitive one, and has generated low rates of profit. Further, because it places few demands on the infrastructure and can use a relatively unskilled supply of labor, the textile industry has called for relatively little government involvement. In contrast, much more capital-intensive industries such as steel, automobiles, or shipbuilding often required substantial government investments. The historical geography of textile and garment production reveals a changing series of comparative advantages in which the industry moved from place to place in a perpetual search for cheap labor. The industry began at the dawn of the Industrial Revolution in Britain in the late eighteenth century, in which the textile industry established some of the first factories, although other observers cited Flanders and Bruge in Belgium. British cities such as Manchester, York, Leeds, and Birmingham, which William Blake aptly labelled the "dark satanic mills," became the first industrial powerhouses of the world. In the nineteenth century, Britain alone produced half of the world's textiles. One particularly important backward linkage from the emerging textile industry was to agriculture, where the textile industry's growing demand for cotton and wool sped up the mechanization of agricultural production. Further, the growing demand for cotton stimulated the demand for cotton imports and hence, overseas production of cotton through the plantation system. Even today, when the textile industry first begins in many Third World countries, rural landlords are often among the first to initiate textile mills.

The textile industry was fundamental to British colonial policy. The British textile industry quickly saturated its domestic market and became increasingly export-oriented as British producers were forced to seek out new markets both in Europe and in its emerging colonies around the rest of the world. When Britain began to colonize South Asia, the British found, to their dismay, a thriving indigenous textile industry. They promptly destroyed the Indian handicraft textile production, and soon India became a net importer of British textile products (an issue Indian nationalists would raise in the twentieth century). Similarly, Britain allied itself with the American South during the U.S. Civil War of the 1860s in large part because it relied heavily on southern cotton exports.

In the early nineteenth century, the textile industry began to diffuse to the European mainland. In France, cities such as Lille and Lyon became important centers for the production of cloth and silk, respectively. In northern Italy, Milan and Florence became thriving centers of textile production. In Russia, among the last of the European nations to industrialize, St. Petersburg, long that nation's center of modernity, began as a textile center.

In North America, the textile industry took root among the first industrialized part of the continent, New England, particularly Massachusetts, Rhode Island, and Connecticut. Cities such as Lowell, Massachusetts, and Manchester, New Hampshire (site of the world's largest textile factories), were the primary foci of the textile industry. These industries employed large numbers of immigrants, particularly from Ireland, throughout the nineteenth century. In the 1860s, the textile industry was revolutionized when steam power was introduced, and the power loom began to replace the hand loom. In contrast to the textile industry, the garment industry became overwhelmingly concentrated in New York City. In 1900, New York City alone produced three-quarters of the total U.S. output of garments. In the lower east side of Manhattan, large numbers of Jewish immigrants were employed under very arduous working conditions, working long hours and with very low rates of pay.

In the 1930s through the 1950s – the end of one Kondratieff cycle and the beginning of another – the textile industry changed its geography again. New England, which had been industrialized around the textile industry for more than a century, lost its comparative advantage as large numbers of firms relocated to North and South Carolina and Georgia in search of cheaper labor (Figure 10.1). Internationally, Japan emerged as a leading textile producer after the Meiji Reformation of 1868; today, the Japanese textile industry is extremely capital intensive and one of the world's largest users of robots.

Following World War II, two important trends began to reshape the textile industry dramatically: internationalization and the introduction of synthetic materials. The internationalization of textile production began in the 1960s when Japan and the U.S. began to lose their comparative advantage in this industry. Since World War II, employment in the textile industry has declined in Europe, North America, and Japan. (However, a thriving garment industry still remains in cities such as New York and Los Angeles, frequently using immigrant labor from Third World countries.) Compare Figures 10.2 and 10.3. As the textile industry began to abandon places with high labor costs in the western industrialized world, it began to sprout up in a variety of Third World locations, in particular the famous "Four Tiger" nations of East Asia: South Korea, Taiwan, Hong Kong, and Singapore. Textiles were particularly important in the early industrialization of South Korea, while garment production was more significant to Hong Kong. In both cases, while these industries sought out much cheaper supplies of labor in East Asia, they became significantly more capital intensive, introducing, for example, the shuttleless loom and laser cutters as they relocated overseas.

The unstable geography of textile production is evident yet again today as the Four Tigers look over their shoulders at competition from nations with

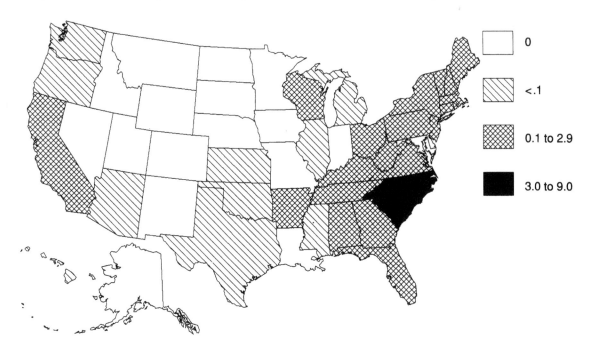

Figure 10.1 Textiles as percent of total employment, 1990

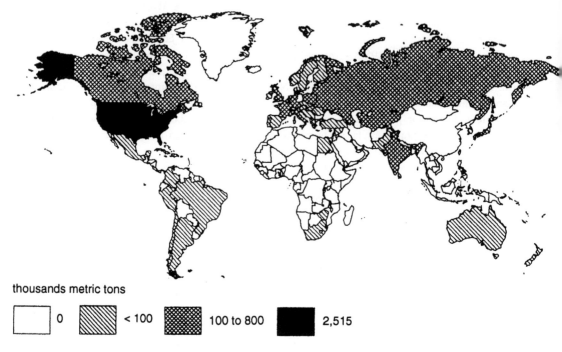

thousands metric tons

| | 0 | | < 100 | | 100 to 800 | | 2,515 |

Figure 10.2 Total textiles output by country, 1950

even lower labor costs, such as China, which has established a growing textile industry in many of its coastal special economic zones, such as Guangdong. Further, the textile industry has begun to migrate to new "tigers" such as Malaysia, Thailand, Pakistan, and even the Seychelles Islands. Typically in these Third World countries, the textile firms employ large numbers of women workers who work long hours in unsafe working conditions. Many garment workers who produce the clothes for export to the wealthy West suffer from tuberculosis and eye disease. Indeed, the textile and garment industries have long been one of the most exploitative of any sector.

In the post-World War II era, the international geography of textile production has become increasingly regulated. The prime example of this trend is the international Multi Fiber Agreement (MFA), signed in 1973, which establishes national textile and garment output limits for different nations through a series of quotas. The MFA has become an important factor in regulating the production and inter-

national trading patterns of the textile and garment industries which form 20 percent of world trade. However, some countries have attempted to circumvent MFA quotas by using false labelling systems.

The second important post-World War II trend that reshaped the textile and garment industry involved the growth of synthetic fibers such as rayon, nylon, dacron, and polyester. Because these synthetic fibers are essentially petroleum by-products, they are consequently susceptible to changes in the price of petroleum, their largest backward linkage. The production of synthetic materials tends to be more capital intensive than is the production of natural materials such as cotton, and therefore generally occurs through larger firms. Today, the world output of textiles and garments is divided roughly evenly between synthetic and natural materials.

The internationalization of production and the growth of synthetic fibers are reflected in a variety of multinational firms, such as Benetton, and the diffusion around the world of western brand names.

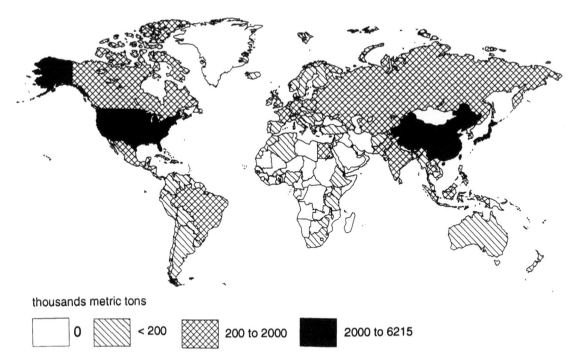

thousands metric tons

0	< 200	200 to 2000	2000 to 6215

Figure 10.3 Total textiles output by country, 1990

such as Levi Strauss and Wrangler. These trends in part reflect the nature of the demand for garments, which, in contrast to other commodities (such as many services), is income inelastic (that is, the demand does not rise dramatically as incomes increase over time). Thus, in order to insure that the demand for their product remains relatively high, garment firms are consciously involved in the manipulation of demand through heavy advertising and constant changes in style; thus, constant style changes are hardly a spontaneous expression of changing consumer demand, but reflect the deliberate intervention of firms in the creation and recreation of the demand for their product over time.

THE IRON AND STEEL INDUSTRY

Few industries are as critical to industrial society as that which produces iron and steel. There is virtually no economic sector that does not incorporate steel in one form or another; thus, the steel industry has enormous forward linkages, including automobiles, railroads, ships, aircraft, construction, machinery, wire, pipes, tools, and a variety of household goods.

The backward linkages of the steel industry consist primarily of iron ore, which is purified to form pig iron, coal, which is purified to form coke, and limestone. While the process of producing steel has changed over the last several centuries, the basics remain much the same: coke is heated in a blast furnace to roughly 3,000°F., and then iron, mixed with limestone to attract impurities, is heated to form molten steel, which is compressed into red-hot ingots often weighing several tons apiece. The ingots are shaped to form sheets and beams, cleaned with acid, tempered and allowed to harden, and finally galvanized with a coating of zinc to prevent corrosion.

There are several characteristics of the production of steel that are critical to its geography. First, the steel industry is highly capital intensive: steel production generally takes place in enormous plants that frequently run 24 hours per day (they are expensive to shut down and start up), and rely heavily on economies of scale. The steel industry has traditionally employed large numbers of workers and has been historically a well-paying, unionized sector with a strict job hierarchy. Second, the steel industry, unlike textiles and garments, involves very high start-up costs and significant quantities of investment capital; it is, therefore, a relatively difficult industry to enter, and for that reason exhibits an oligopolistic market structure. A third characteristic, derived from the others, is that the steel industry is relatively immobile over space, that is, its geography is slow to change and is characterized by significant inertia. This means that frequently, even when the profit-maximizing conditions for the production of steel have changed, steel companies may be reluctant to shut down plants in one area and open up new ones in another. Because steel is a heavy, bulky commodity, transport costs have played an important role in the location of steel plants. Fourth, the steel industry has traditionally relied on very heavy government intervention, frequently in the forms of a significant infrastructure and in the mediation of disputes with labor.

The historical geography of the steel industry began with small-scale firms in the British Midlands, many of which produced on a very modest scale. In the U.S., the first steel plants were found scattered throughout the Northeast. In both nations, the industry bore little resemblance to the large, energy intensive firms that we see today. Rather, it consisted of many small, competitive firms that primarily served local markets, producing steel using wood and charcoal fuel obtained from nearby forests. Steel production was an occupation dominated by highly skilled "puddlers" and ironmasters.

Tremendous changes in the production of steel and subsequent geography of the steel industry occurred in the late nineteenth- and early twentieth-century Kondratieff cycles, in which the largest forward linkage of the industry was the railroad sector. Steel production became increasingly oligopolistic as steel firms began to serve national markets rather than local ones, beginning in the 1880s. The U.S. surpassed

Britain as the world's largest steel producer, and steel became the mainstay of the national economy. It was during this period that famous "robber barons" such as Andrew Carnegie and John Rockefeller arose. Although several companies such as Kaiser, Republic, and Bethlehem Steel were important to this process, one firm stood out above all: the U.S. Steel Company, founded in 1901, which ultimately came to produce two-thirds of U.S. steel output around the turn of the century.

The steel industry was instrumental in the formation of the U.S. Manufacturing Belt, the states along the southern shores of the Great Lakes that acquired a comparative advantage in heavy industry in the late nineteenth century (Figure 10.4). This region was close to ample supplies of Appalachian coal, as well as iron ore from Michigan and northern Minnesota, which could be transported over the Great Lakes by water, revealing a classic case of Weberian transport cost minimization at work. Several cities within this region became famous centers of steel production, including Buffalo, New York; Youngstown and Cleveland, Ohio; Pittsburgh and Johnstown, Pennsylvania; Hamilton, Ontario; Chicago, Illinois; and Gary, Indiana.

In North America, manufacturing frequently faced a shortage rather than an abundance of labor, creating labor markets that attracted large numbers of immigrants to well-paying, frequently unionized jobs. As firms faced labor shortages, and correspondingly high prices for labor, they had a powerful incentive to create new technological innovations. In the steel industry, technological innovation was expressed through the introduction of the Bessemer process and the open hearth furnace, a new and highly productive way of producing steel. With the introduction of the Bessemer process, the attraction of iron ore deposits began to rise, that of coal began to decline, and steel firms faced new cost/revenue considerations in their siting decisions.

By the 1920s, the steel industry's largest market was not the railroad industry, but automobiles, and throughout the early twentieth century the two industries' fortunes rose and fell together. By the 1930s, the industry had become increasingly

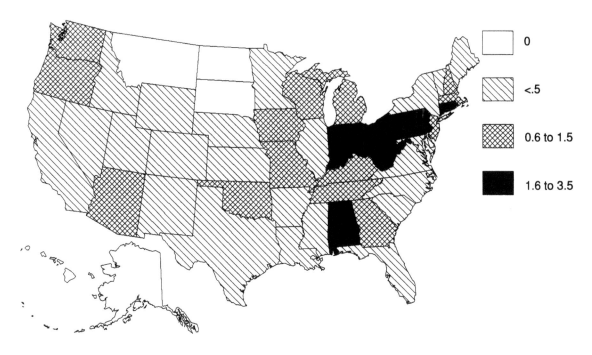

Figure 10.4 Iron and steel as percent of total employment, 1990

unionized through the United Steelworkers of America. Following World War II, however, a prolonged and gradual decline in the U.S. and British competitive position in steel began to occur, particularly as they lagged rather than led technologically. A significant technological change of this era was the electric furnace and the basic oxygen furnace, which replaced the open hearth furnace. The delay in the introduction of technological innovations among British and American steel firms partly reflected their oligopolistic market structure, in which large firms, not facing significant competition, did not encourage sufficient investments into research and development. One of the fundamental rules of capitalist production, however, is that firms must engage in technological change to maximize productivity and profitability in the long run. Only belatedly did American steel firms come to adopt new technologies such as the continuous casting methods and rolling mills. Because U.S. and British

steel companies did not adapt sufficiently quickly to the new environment of the late twentieth century, they suffered significantly from the growing influx of imports from other parts of the world. A prime victim of this process was the U.S. Steel Company, which witnessed its hegemony's gradual erosion.

Simultaneously, a new geography of steel production began to emerge. In the U.K., steel producing jobs evaporated and moved overseas. In the U.S., they relocated to the South and West of the United States (notably California, Texas, and Alabama), in contrast to the traditional concentration in the Manufacturing Belt. This new geography began to emerge as early as World War II, when the federal government constructed a number of steel plants, only to sell them at a discount to private companies. Later, when steel firms began to rely heavily on foreign ores and ocean transportation, they erected steel plants in coastal areas in states such as Maryland and California.

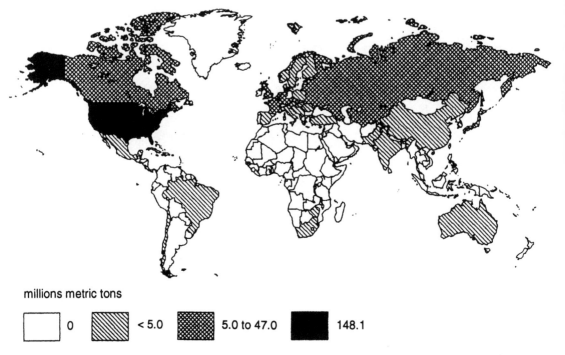

millions metric tons

☐ 0		▨ < 5.0		▨ 5.0 to 47.0		■ 148.1	

Figure 10.5 Total iron and steel output by country, 1950

British and U.S. steel firms also faced rising inter-national competition, particularly from companies based in Japan, Germany, Spain, South Korea, and Brazil. Firms in these nations were typically more technologically innovative and employed lower cost labor, and were frequently government-owned or subsidized. Thus, in the 1970s and 1980s, U.S. steel companies found their competitive position interna-tionally eroding steadily: in 1950 they produced roughly a half of the world's steel output, but by the 1980s this proportion dropped to roughly 12 percent (see Figures 10.5 and 10.6). To make matters worse, the 1970s saw a period of rapidly rising petroleum prices, which in turn both inflated the cost of producing steel and reduced the demand for steel (for example, with lighter automobiles). Simultaneously, new environmental protection legis-lation mandated the implementation of expensive anti-pollution devices such as smokestack scrubbers that in turn raised the price of steel. In the 1980s,

the rising value of the U.S. dollar and relatively high interest rates further discouraged steel production. Thus, beset by declining profits and angry share holders, steel firms in the U.S., Canada, and the U.K., and to a lesser extent in France, initiated a significant series of plant closures and layoffs. This process generated enormous difficulty for the communities that had their economic base swept out from underneath them, as well as unions, many of which saw their membership decline; for many unemployed steel workers with few other usable skills, deindustrialization meant prolonged economic hardship.

The response of the western steel industry to its declining competitive condition internationally has been varied. Some firms have called for protectionism, decrying the unfair competition from their subsidized competitors overseas. Other firms have diversified, purchasing assets in other sectors: the U.S. Steel Company, for example, purchased the Marathon Oil

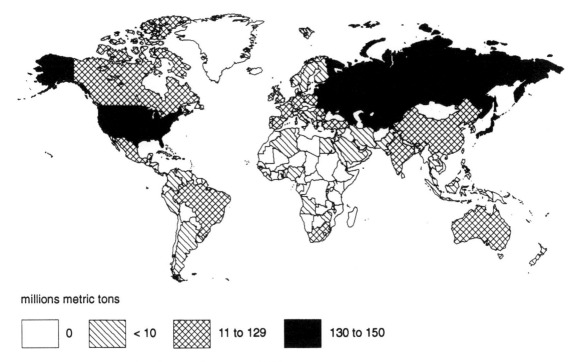

millions metric tons

0	< 10	11 to 129	130 to 150

Figure 10.6 Total iron and steel output by country, 1990

Corporation and became USX, and now derives more revenues from the output of petroleum than it does from steel. Other steel companies invested heavily in chemicals, real estate, and plastics. A third response consisted of the emergence of the so-called mini-mills, small, relatively capital-intensive, non-unionized firms that started as early as the 1960s to produce specialized output on a relatively small scale. In contrast to the large blast furnaces, mini-mills are relatively vertically disintegrated, and tend to serve relatively specialized markets. They tend to use scrap steel as a primary input, and often make relatively small quantities of steel bars, wire, and other household products. By 1990, they generated roughly 25 percent of U.S. steel output. The emergence of mini-mills reflects the extensive modernization of U.S. industrial facilities that occurred in the 1980s, and to a great extent has served to increase output, if not employment, in manufacturing.

THE AUTOMOBILE INDUSTRY

A third industry that played an important role in the formation of the economic landscapes of the industrial West is the automobile industry. Essentially, automobile production is a propulsive industry that consists of the assembly of some 15,000 individual parts. The industry has enormous backward linkages to a variety of other industrial sectors, including steel, plastic, lead, copper, chrome, glass, rubber, textiles, electronics, machine tools, and ball bearings. The production of automobiles also involves numerous inputs of services as well, including finance, legal services, advertising, and public relations. In many respects, the automobile industry resembles another significant heavy industrial sector, steel. Both industries tend to be very capital intensive and oligopolistic in their market structure, and both have traditionally employed large numbers of unionized laborers. Unlike the steel

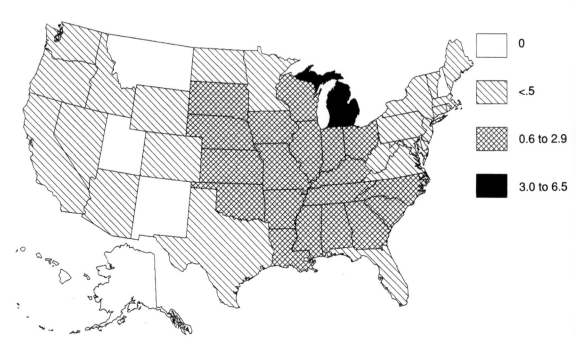

	0
	<.5
	0.6 to 2.9
	3.0 to 6.5

Figure 10.7 Automobiles and parts as percent of total state employment, 1990

industry, however, the automobile industry serves consumer demand directly (although sales are mediated through wholesalers), and thus factors such as advertising, interest rates, and brand loyalty figure prominently in shaping the demand for automobiles.

The origins of the historical geography of automobile production can be found in late nineteenth-century Europe. It was in Germany in the 1890s that the internal combustion engine was invented, leading to the emergence of the Mercedes in 1901, which incorporated French experiments with chassis design, in turn reflecting the ties to carriage manufacturers. However, early European producers focused primarily on the upper income luxury market, whereas their American counterparts targeted the middle class. By World War I, a significant European automobile industry had emerged, including firms such as Peugeot, Fiat, Triumph, and Volkswagen.

In the U.S., the automobile industry became concentrated heavily around the area of Detroit,

Michigan (Figure 10.7). Why did the industry cluster so heavily around that single location? The common answer to this issue is that Henry Ford, founder of the Ford automobile industry, was from the area. However, far from being the product of Ford's preferences, the Detroit region offered a significant series of advantages in the production of automobiles over other areas, that is, its comparative advantage in automobile production reflected a historical series of other industries that had previously located in the region. For example, the Detroit area had earlier been a center of wagon and buggy production, and numerous firms that could produce parts for the automobile industry were already located there. Second, in the Detroit region, numerous banks existed with excess investment capital left over from the evacuation of the lumber industry in the area. Third, the Detroit area offered cheap transportation over the Great Lakes and railroad lines, by which coal and iron inputs could be shipped cheaply.

Ford introduced a number of innovative techniques in the production of automobiles that had wide ranging impacts on industry in general, including the assembly line and interchangeable parts in 1909. Indeed, the Ford Motor Company initiated the entire regime of production frequently labeled "Fordist" production, one that relies heavily on economies of scale, mass markets, and homogeneous products (in contrast to the emerging "post-Fordist" forms of production increasingly evident in the late twentieth century). The earliest automobile producing centers in the Detroit area were typically very highly integrated, including inputs of coal and iron at one end and outputs of complete automobiles at the other; the Ford Company also owned its own shipping lines, its own transport systems, and its own mines from which coal and iron ore could be extracted. The emergence of the automobile industry in the Detroit area gave a further boost to the comparative advantage of manufacturing in the southern Great Lakes areas, increasing the demand for output in other industries such as glass, textiles, chemicals, and rubber.

One of the keys to the success of the Ford Motor Company was that it made the automobile affordable to the middle classes, for example, making a $300 car affordable for people who earned at the time $5 per day. The success of the Detroit based motor companies is seen in the growing ownership of automobiles from the 1920s onward: in 1920, there was one car for every thirteen Americans; in 1930, there was one for every five; and by 1980, one car for every three. Why did the demand for automobiles rise so heavily? Two significant answers to this question may be found. First, during a period of rising incomes, such as the boom of the 1920s and again in the 1950s, more people can afford automobiles. However, an additional answer is found in the changing value of time: as incomes rise, the opportunity cost of time also rises, making the automobile more attractive relative to its substitutes, such as mass transportation.

Probably no innovation has so deeply reshaped the fabric of industrial society as the automobile. This innovation has had enormous forward linkages that arise from the unprecedented mobility that it offers to large numbers of people. First, the automobile radically transformed urban labor markets, allowing commuting fields to extend much farther from their origin than previously. Second, as commuting fields extended, low-density urban sprawl became widespread, particularly through the widespread inhabitation of the single family home. Third, the emergence of the automobile changed the pattern of local and federal government expenditures: in the 1920s many municipal governments were already subsidizing the construction of urban parkways, and in the 1950s the federal government became heavily involved in the $56 billion interstate highway system, which radically transformed the urban geography of the United States. Fourth, the automobile changed retailing patterns significantly, frequently causing many downtown retailers to suffer declines in sales and simultaneously creating new landscapes in suburban areas, including institutions of suburbia such as shopping malls, drive-ins, and commercial ribbon developments. Fifth, the automobile had significant impacts on urban mass transit systems, many of which (such as trolley lines) suffered bankruptcy. Sixth, the movement of goods as well as people was transformed when trucking replaced railroads as the primary means of moving commodities between cities. Seventh, the automobile had significant environmental impacts, radically increasing the demand for petroleum (40 percent of U.S. energy consumption is devoted exclusively to the automobile), and contributing heavily to deteriorating air quality in the form of smog. Finally, the automobile had numerous important cultural and perceptual effects as it rewove the fabric of everyday life, changing the nature of personal contacts and neighborhoods, and, finally, the safety of 40,000 Americans who are killed each year in automobile accidents.

As the demand for automobiles rose, the industry became steadily more oligopolistic in its market structure. In 1914, there were more than 300 automobile producers, including such firms as Knight, Marian, Rambler, Stoddard, Studebaker, and a variety of models which have existed to this day. By 1923, the number of U.S. automobile producers had

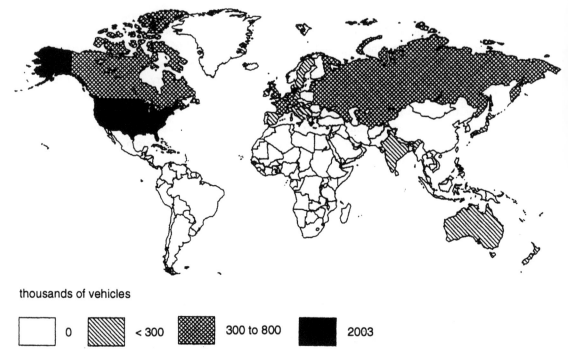

thousands of vehicles

| | 0 | | < 300 | | 300 to 800 | | 2003 |

Figure 10.8 Total automobile production, 1950

dropped to 108, and in the 1920s when General Motors introduced the front wheel drive, it surpassed Ford as the largest U.S. automobile producer. By 1927, the number of U.S. automobile producers had dropped to forty-four, and in the 1930s, when the Japanese introduced the automatic transmission, the number of automobile producers dropped yet further. By 1946, only nine automobile producers remained, and by 1970, 95 percent of U.S. automobile production was accounted for by the famous Big Three producers, Ford, General Motors, and Chrysler Corporation. The 1950s may perhaps be seen as the heyday of U.S. automobile production, in which many producers created enormous bulbous lead sleds: the 1959 Cadillac, for example, weighed more than 6,000 pounds and was loaded down with so much chrome that it achieved only nine miles per gallon. In the 1950s, it is estimated that almost one-fifth of U.S. employment was attributed directly or indirectly to the automobile industry.

By the 1960s, however, it became increasingly apparent that U.S. automobile producers faced significant international competition. Unlike their counterparts overseas, many U.S. automobile managers of firms did not reinvest their profits instead paying out high wages and high dividends to shareholders. Further, the large elephantine bureaucracies of U.S. automobile producers frequently favored short-run profits over long-run investments. As automobile producers elsewhere began to turn out automobiles in large quantities, the U.S. suffered a declining share of world automobile output: in 1965 the U.S. produced 72 percent of all of the automobiles in the world, but by 1980, this proportion had dropped to 30 percent. (Compare Figures 10.8 and 10.9.) A turning point in the international competitiveness of the U.S. automobile industry occurred in the 1970s, when the price of petroleum created significant energy crises. Unlike Japanese and European automobile producers, U.S. auto firms di

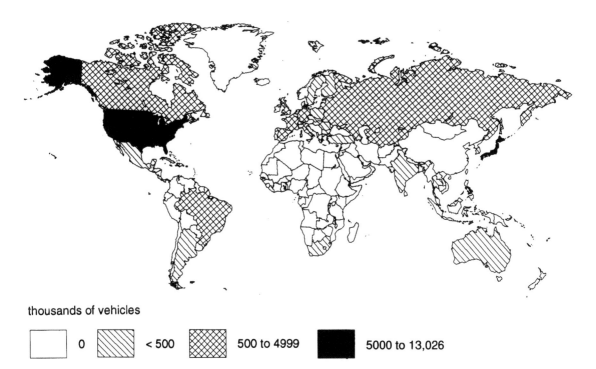

thousands of vehicles

| | 0 | | < 500 | | 500 to 4999 | | 5000 to 13,026 |

Figure 10.9 Total automobile production, 1990

not respond to the growing demand for small cars. It was this lack of innovativeness and responsiveness to the changing domestic demand for automobiles that produced a window of opportunity for many foreign producers of automobiles. In the process, Japan became the largest producer of automobiles in the world, making names such as Toyota, Nissan, Honda, Mazda, Subaru, Isuzu, and Mitsubishi common to households around the world. Similarly, German firms such as VW, Mercedes-Benz, BMW, Porsche, and Audi became well known. Further, countries such as South Korea also began to produce automobiles, in particular, the Hyundai. Thus, as with the steel and textile industries, the sunset industry of the West, in this case the automobile, became a sunrise industry, particularly for the growing industrial economies of East Asia.

There are several reasons for the success of foreign car producers. First, many of them are relatively more capital intensive, often employing robots and machinery unavailable to producers in the United States. Second, many foreign producers employ superior management techniques. Third, foreign producers, particularly those that produce automobile parts in Third World countries, have lower labor costs than those faced by American firms. Fourth, many foreign producers adjusted more rapidly to changes in the demand for different sizes and styles of automobiles and often produce automobiles that are superior in quality.

Besieged by imports, U.S. automobile producers found themselves faced with a growing overcapacity and a declining share of the market. As with steel production, the responses of U.S. automobile producers varied markedly. Many engaged in a series of plant closures and layoffs, further accelerating the deindustrialization of the U.S. in general, and in particular that of the Manufacturing Belt. Second, many automobile producers, particularly Chrysler and Ford, called for protectionism, decrying the allegedly subsidized competition that they faced overseas. Third, many automobile producers began to move plants or parts of plants overseas, particularly to Brazil and Mexico. Fourth, many U.S. firms invested in new technological innovations, modernizing the industry and making the production of U.S. automobiles more capital intensive. Fifth, many U.S. automobile producers engaged in a series of joint ventures and strategic alliances with foreign firms. For example, Chrysler and Mitsubishi engaged jointly in the production of the Colt, and GM and Toyota jointly created a new automobile factory, the Nummi plant in California. Along the same lines Chrysler began to make minivans for the Renault Corporation, and Ford purchased 20 percent of the stock of the Mazda Corporation.

One of the most dramatic tendencies in automobile production in the late twentieth century has been a spectacular growth of international subcontracting and the internationalization of automobile production. This process is readily evident in the trend toward international investment by automobile companies. U.S. companies (especially General Motors and Ford) have operated plants in Europe since the earliest days of the industry. However, these plants served the market of the given country, until the removal of barriers to trade among states of the European Common Market in 1957, and thus concentrated in the countries with the largest domestic markets (U.K., Germany, France). Investment in Europe now is motivated by a desire to have sufficient scale economies and sufficiently low costs to compete in the entire European market. The threat of increased import barriers on the part of the European Union or the United States has motivated large investments by Japanese automobile companies. In Europe, Japanese investment in automobile assembly has concentrated in the U.K. In the U.S., Japanese firms have concentrated new ("greenfield") plants (also called "transplants") in parts of the Midwest including Ohio, Kentucky, and Tennessee. This has helped to raise hopes of a reindustrialization of the Midwest. Given the low rate of job growth in the 1980s, many localities have engaged in a ferocious competition to attract these foreign firms. They operate with several advantages, including the state subsidies they have received, along with more flexible labor arrangements (with or without labor unions) and low expenditures on retiree benefits compared with the older automobile companies in the U.S. and the EU.

ELECTRONICS

A large, interrelated complex of industries has electronic technology as its basis. These industries include communications equipment, broadcast equipment and receivers, computing equipment, industrial and scientific instrumentation, and control systems for manufacturing, missiles, aircraft, and warfare. Driving the development of all these industries are the improvements in electronic components and computer software.

Electronic equipment dates from the early years of the twentieth century, with the development of radio transmission. Electronic technology got its biggest boost during and shortly after World War II. The exigencies of war provided resources and direction for the development of radar, improved communications systems, and standardized manufacturing techniques for vacuum-tube components. The invention of the solid-state transistor in 1947 led a new round of technical and commercial development, with great corporate and locational impact. Despite the end of the war, Cold War tensions kept military research expenditures high in the U.S. and the U.K. These public expenditures subsidized commercial electronics development during the 1950s.

The development, manufacture, and marketing of solid-state electronic components exemplified the product life cycle model quite well, with additional regularities that gave rise to models of industry cycles. The technological chasm between vacuum tubes and transistors reduced the competitive advantage of established electric companies in the northeastern U.S. While the new technology was in an early stage, scale economies were small, and critical technology was embodied in key people rather than in elaborate machinery. Hundreds of new semiconductor component companies have been started in each decade since

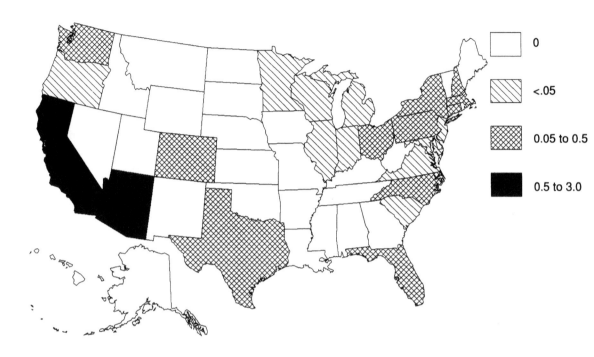

Figure 10.10 Semiconductors as percent of total employment, 1990

the transistor's invention, each company relying on key individuals with a new product. From the technological origins in Bell Telephone Laboratories in New Jersey, commercialization of semiconductor electronics spread to the electrical giants in New York, Pennsylvania, and Connecticut. However, these companies had huge investments in older technologies, and were slow to exploit the technical and commercial advantages of transistors and other solid-state electronics. The inventor of the transistor, William Shockley, established a company in Palo Alto, California. That company, Shockley Transistor, was not successful, but its partners went on to found many of the most successful companies in the industry. The benefits of agglomeration (technical workers, sector-specific venture capital, and the development of equipment suppliers that served small companies) encouraged the concentration of new, growing operations that gave the region the moniker "Silicon Valley," after the most important semiconducting material used in electronics.

Meanwhile, military and national intelligence contracts on the East Coast produced a concentration of electronic equipment companies in Cambridge, Massachusetts. For a host of public and private reasons (zoning, real estate development, security), the more successful of these companies established operations in western suburbs, along a new circumferential highway: Route 128. During the early 1960s, two other companies gained long-lasting success in the development and manufacture of semiconductor components and equipment based on these components. Motorola decided to establish a separate facility in Arizona, and Texas Instruments was founded and grew near Dallas (Figure 10.10).

By the late 1960s, a pattern was established that entailed steep, short-lived product life cycles. New products were introduced at very high prices, facing demand only from military and commercial markets who needed ever more powerful computers, control systems, or communications capability. As more companies began production of similar products, and as the unit cost fell in the originating company, the price fell and the product gained more widespread acceptance. Overlaying this neat progression of product cycles were steadily increasing capital requirements, as ever more powerful integrated circuits came to require increasingly sophisticated design and production equipment. Many small companies failed, and many large companies (especially companies with large stakes in electrical and other product lines) decided to leave the industry. In the U.S., with the dominance of older electrical companies in the Northeast and of specifically semiconductor companies in the West, this trend manifested itself in the increased importance of the West within the industry. No simple generalization suffices, however. IBM remained the world's largest producer and consumer of electronic components during the 1970s and 1980s, and maintained semiconductor fabrication facilities in the northeastern U.S.

The rest of the world was not stagnant during this period of U.S. technological, economic, and military dominance. Some U.S. companies pursued an international division of labor within the industry, establishing labor-intensive component assembly operations in low-wage locations in the Caribbean and East Asia. The location logic was simple, at first: rather than make very large capital investment in assembly equipment that might be obsolete with the next product cycle, companies made smaller (often foreign-government-subsidized) investments in labor-intensive operations, pulled far from supply sources or markets by low-wage labor and inexpensive air transport of the very small components. By the 1970s, Hong Kong and Taiwan had burgeoning electronic equipment industries, so that some of the assembled components found markets in Asia. Also by the 1970s, Japanese electric companies had invested large sums in the development of semiconductor technology and production capability. These large, well-capitalized companies had an important advantage over the smaller U.S. companies in an industry where competition drove prices lower, faster with each new product. Companies in England, France, Germany, and the Netherlands developed technical capability in semiconductor components,

as did Korea. Within each country, localized agglomerations remained important for the development and design functions, while the actual fabrication of the silicon chips diffused across each country. Some of the largest companies have been able to maintain development facilities outside of agglomerations, such as IBM's facility in Greenock, Scotland. Much of Japan's electronics industry is concentrated in the island of Kyushu, the so-called "Silicon Island."

In the popular consciousness the electronics component industry is a model for all "high technology" industries, exhibiting some key location changes discussed in this book: international product life cycles, international division of activities, industrial filtering within a country, industry agglomeration, and vertical disintegration of functions within one complex production system, the management of learning and experience curves. This perception, along with the rapid growth of the industry and its constituent companies and the fundamental nature of semiconductor-based integrated circuits in many industries and services, explains why so much attention has been paid to the sector.

FINANCIAL SERVICES

Financial services, in all their variety, are critical for the origins and growth of all companies. In addition, these services form an important employment base themselves. To understand their location and their influence on corporate location, we must disaggregate this broad sector:

1 **Commercial banks** hold most of their financial assets in loans to firms (i.e., as venture capital) to invest in production and real estate, as well as to individuals (e.g., for cars, boats, or college tuition). They also often offer a series of "retail banking" functions such as checking and savings accounts, credit cards, travelers' checks, and safe deposit boxes.
2 **Savings banks (thrifts; building societies)** are banks that primarily extend loans to buyers of homes (mortgages).

3 **Investment banks** typically are involved in buying and selling of securities (stocks and bonds) and related markets (e.g., futures, options, and commodities), as well as raising funds for corporate mergers and acquisitions. Both commercial and investment banks may also be involved in foreign exchange markets and provide investment advice for a fee.

4 **Insurance** consists of a series of interrelated sectors, including life, medical, casualty, automobile, and commercial and residential property insurance, and as such acts as a way of spreading risk among large groups of investors. Insurance companies typically have two sets of costs and revenues, including the premiums charged for their policies on the one hand and the returns to investment, largely property, on the other.

A critical function of most financial services is the provision of **credit**, which allows the costs of producing and purchasing goods to be separated in time and space. Banks play a particularly important role in this regard, serving as intermediaries between borrowers and savers, each of which has highly specialized needs and products in terms of risk, return, liquidity, and so forth. Banking transactions, therefore, can be conceived as a flow of money from savers to borrowers, on the assumption that the borrowers will earn a rate of return on the funds sufficient to pay the bank's interest and still profit from their efforts. Banks earn their profits from the difference in borrowing and lending rates.

The origins of the banking industry lay in Renaissance Italy. In the fifteenth and sixteenth centuries, goldsmiths, pawnbrokers, and money changers began to store precious metals for their clients, issuing deposit certificates in return that were payable upon demand. Since it was unlikely that all their depositors would demand their assets simultaneously, bankers could lend part of their deposits out to those who needed investment capital, for a profit. Gradually their functions expanded to include trade in bullion, arranging for transfers of funds among depositors, and reinvesting, either directly or through loans, the money left in their

safekeeping. With an expanding capitalist world market, the Italian merchant-bankers progressively monopolized the international exchange of money and credit, inventing the bill of exchange, essentially a contract governing later payment somewhere else, stocks, bonds, and the double entry accounting system. The emergence of deferred payment, particularly in the form of credit, was a prime force in the emergence of a money economy and in the formation of a capitalist nation-state.

Because banking is so important to regional and national economies, it is invariably a highly regulated system throughout the world. Banks play important roles in affecting nations' money supplies, which in turn strongly affect interest and inflation rates. In most nations, the money supply is dictated by a national bank, such as the Bank of England or the Bank of Tokyo in Japan; in the U.S., it is dictated through the Federal Reserve, actually a series of nine banks created by the federal government in 1917 to act as a form of a national bank. Government regulation of national banking systems, strengthened after the bank failures of the Great Depression, served the needs of bank owners and bank-dependent businesses and households through the post-World War II economic boom. However, in the 1970s, the once solid structure began to unravel.

As several trends converged, the 1970s mark a significant turning point in the evolution of western capitalist economies. Beginning in 1971, the Bretton Woods agreement, initiated by the U.S. after World War II to regulate the international financial system, collapsed. Shortly thereafter, led by the U.S., the world abandoned the gold standard by which the U.S. dollar had been pegged to $35 per ounce of gold. Second, the price of petroleum rose rapidly, in large part due to the efforts of the Organization of Petroleum Exporting Countries (OPEC) to limit its supply, forcing much of the western world into a serious recession and causing inflation rates to rise rapidly. Flush with "petrodollars" invested by OPEC nations, many western banks, in turn, recycled them in the form of Third World debt, giving rise to the global debt crisis. Third, many banks introduced extensive new telecommunications systems, largely

Table 10.1 Leading world debtors, 1990

Nation	Total debt ($ bill.)	Per capita GNP ($)	Ratio of debt service to exports
Brazil	120.0	2,020	26.7
Mexico	107.0	1,820	30.1
Argentina	60.0	2,370	45.3
Venezuela	35.0	3,230	22.4
Nigeria	31.0	370	10.0
Philippines	30.0	590	22.7
Yugoslavia	22.0	2,480	13.3
Morocco	22.0	620	23.4
Chile	21.0	1,310	21.1
Peru	19.0	1,430	12.5
Columbia	17.2	1,220	30.7
Ivory Coast	14.2	750	19.6
Ecuador	11.0	1,040	20.7
Bolivia	5.7	570	22.1
Costa Rica	4.8	1,590	12.1
Jamaica	4.5	960	25.8
Uruguay	4.5	2,180	24.4

as a means of accessing global markets (this issue will be addressed in more detail below). Finally, many western nations deregulated their financial sectors, lifting many government controls.

The Internationalization of Financial Services

Like many industries, finance became progressively globalized in the late twentieth century. Four sets of events played a particularly important role in this regard: the Third World debt crisis, fiscal policies of the U.S. government in the 1980s, the emergence of Japan, and the introduction of new telecommunications systems. Each of these will be briefly examined in turn, as will their geographic consequences.

In the 1970s, the Organization of Petroleum Exporting Countries (OPEC) constricted the flow of oil to the West, causing the price to rise steeply (from $3/barrel in 1970 to $30/barrel in 1980). One result was widespread recession and inflation in Europe, Japan, and the U.S. Another was a massive

influx of "petrodollars" to OPEC nations, many of which, particularly lightly populated Arabic nations, had limited domestic investment opportunities. OPEC countries, in turn, heavily invested hundreds of billions of dollars in western banks, particularly in London and New York. Flush with excess funds, these banks in turn began to lend extensively to those most in need of investment capital, i.e., the Third World, for whom the petroleum shocks had decimated a generation of development. The result of this "recycling" of petrodollars was an unprecedented explosion in foreign debt, amounting to more than $1.3 trillion by 1990. Thus, the debt crisis of the 1980s was intimately linked to the oil shocks of the 1970s. The largest debtors were primarily located in Latin America, including Brazil, Mexico, and Argentina (Table 10.1). As the prices of many Third World nations' exports (largely foods and raw materials) dropped in the 1980s, and as interests rates rose worldwide, their capacity to repay their debt declined accordingly. Many large "money center" banks, therefore, found the debt crisis working cruelly against them, and were forced to write off large portions (angering their shareholders) or sell it at sharply reduced prices on the secondary debt market. Thus, the debt crisis deeply intertwined the capital markets of the First and Third World.

A second factor contributing to the globalization of finance was a series of changes in U.S. federal government fiscal policy. Under the Reagan Administration (1980–8), which cut federal income taxes and increased military expenditures, the U.S. federal budget deficits climbed to record levels, frequently exceeding $150 billion annually. Because the federal government is a preferred client for large commercial banks – its chance of default is virtually zero – it could borrow enormous sums of capital from private sources. By 1987, federal borrowing absorbed 40 percent of all investment funds in the U.S., which increased interest rates in the U.S. Investors in other nations were attracted by these higher rates of return. Additional funds were therefore provided in large quantities by foreigners, particularly Japanese (and to a lesser extent European) banks, who purchased large quantities of Treasury Department securities, creating, in effect, a debt owed to another nation. Indeed, by the mid-1980s, the U.S. was the world's largest debtor nation, although as a percent of GNP this debt was low compared to many Third World nations. Because the U.S. economy exerts such a heavy influence on those of other nations, its policies in the 1980s served unintentionally to link various capital markets firmly together.

Third, the globalization of finance reflects the emergence of Japan as the world's premier creditor nation. Japan's remarkable post-World War II economic recovery has seen it generate enormous quantities of investment capital, with relatively few domestic investment opportunities. This pool of funds reflects the nation's high rate of savings, which frequently exceeds 20 percent of earnings, compared to 3 to 5 percent in the U.S. and much of Europe. By the 1980s, Japan had become the world's largest player in finance: the 10 largest banks in the world, led by Dai Ichi Kangyo, were all Japanese. In the securities markets, giants such as Nomura, Nikko,

and Daiwa traded large volumes of shares around the world. Tokyo became the world's largest stock market. Japan's powerful status was symbolic of the rapid growth of the East Asian economies and the shift in much of the world's economic activity from the Atlantic to the Pacific spheres.

A fourth process contributing to the globalization of financial markets was the introduction by financial firms of an extensive network of national and international communications systems. Because they are so information-intensive in nature, banks and securities firms have turned to leased telephone networks as well as satellites and fiber optics systems in the formation of global capital markets. Electronic funds transfer systems form the nerve center of the international financial economy, allowing banks to move capital around at a moment's notice, arbitraging interest rate differentials, taking advantage of favorable exchange rates, and avoiding political unrest. In the securities markets, global telecommunications systems have also facilitated the emergence of the 24-hour trading day, linking stock markets through computerized trading programs (Figure 10.11). Subject to the process of digitization, therefore, information and capital become two sides of the same coin. The same process, of course, also heightens the susceptibility of these networks to disruptions (e.g., computer viruses and international transmissions of stock market disruptions).

Among the most profound geographic repercussions of the globalization of finance has been the growth of "world cities," particularly London, New York, and Tokyo, each of which seems to be more closely attuned to the rhythms of the global economy than the nation-state in which it is located. London, for example, boomed under the impetus of the Euromarket in the 1980s, and has become detached from the rest of Britain. Similarly, New York rebounded from the crisis of the mid-1970s with a massive influx of petrodollars and new investment funds (i.e., pension and mutual funds) that sustained a prolonged bull market on Wall Street. Tokyo, the epicenter of the gargantuan Japanese financial market, has become the world's largest center of finance capital, with one-third of the world's stocks

12 11 10 9 8 7 6 5 4 3 2 1 0 1 2 3 4 5 6 7 8 9 10 11

Hours difference from Greenwich MeanTime

Figure 10.11 Eight-hour trading days of leading world financial centers

by volume and twelve of its largest banks by assets. Given the extensive backward linkages of these firms, the effects are considerable. In each metropolitan area, a large agglomeration of banks and related firms generates well-paying jobs; in each, soaring incomes for a wealthy stratum of traders and professionals has sent real estate prices soaring, unleashing rounds of gentrification and a corresponding impoverishment for disadvantaged populations.

A related geographic manifestation of the new, hypermobile capital markets has been the growth of offshore banking, a reflection of the shift from traditional banking services (loans and deposits) to lucrative nontraditional functions, including debt repackaging, foreign exchange transactions, and cash management. The growth of offshore banking, usually in response to favorable tax laws, has stimulated banking in such places as Panama, Bahrain, and the Cayman Islands.

U.S. Banking: Geography of Growth and Decline

Banking shares with many other service activities a market and labor orientation with respect to location. However, the banking sector is heavily regulated by government in every country, including government ownership of banks in many countries. This intensive government involvement in the sector clearly affects the location and behavior in the sector.

The U.S. has approximately 12,000 U.S. commercial banks (i.e., depository institutions chartered as banks and regulated by state and/or federal regulatory agencies), holding $3.5 trillion in financial assets. In 1989, the 44 commercial banks with assets above $10 billion (a mere 0.3 percent of the total number of bank companies) held 38 percent of total commercial bank assets. Between

1991 and 1992 there was a spate of mergers among the largest U.S. bank companies, yielding this list of largest banks: Citicorp, with over $200 billion in assets; Bank of America and Security Pacific Bank with just under $200 billion in assets; Chemical Bank and Manufacturers Hanover Trust with $130 billion in assets; NCNB and C&S-Sovran (now NationsBank) with over $110 billion in assets; and Chase Manhattan with nearly $100 billion in assets. Thus, the nation's five largest banks hold between one-fifth and one-quarter of total commercial bank assets. The five largest banks in Canada and in France hold about 80 percent of the commercial bank assets in each of those countries, and in the U.K. the proportion is 70 percent. So even with the mega-mergers, the U.S. banking industry is not very concentrated along international parameters. State regulation is a strong reason for that relative lack of concentration, at least when concentration is measured at the national scale. However, the level and trend of bank concentration are much stronger at the local scale.

Banks' increasing reliance on elaborate computer systems has increased scale economies. Studies suggest that during the 1980s, large banks (assets in the $1–5 billion range) were better able to reduce costs by availing themselves of regulatory changes, systems changes, and product changes than were medium-sized banks. There are other sources of potential scale economies in large, multiple branch banks: diversification of credit and liquidity risk; marketing and name-recognition; and raising deposits and reserve funds. Computerization did not slow the growth of finance and bank employment during the 1970s and 1980s. However, consultants have predicted the loss of 300,000 jobs in U.S. commercial banks (18–20 percent of the current work force), over the 1990s. Jobs of all types will be lost, though clerical occupations (tellers, data entry, and paper handlers) face greater proportional reductions. The reasons for employment decline include mergers, followed by consolidation and elimination of branches and duplicative support services; reduced expenditures on information-handling systems and staffs; and increased contracting out so that systems

become more fully utilized. The mergers of very large banks announced during 1991 alone caused the end of nearly 30,000 jobs.

Despite the local nature of key retail and small-business banking markets, the increased importance of other bank services and the interstate purchase of banks increase the possibility that certain financial centers export banking services, thereby gaining income and employment from sales beyond their local markets. However, local variation in bank-related employment reflects more than inter-regional imports and exports of bank services: local demand for bank services varies in the region-specific, cyclical nature of credit demand. The combination of localized markets for financial services and inter-regional bank holding companies implies a geographic as well as organizational hierarchy. What are the likely determinants of the balance of operations at various levels of the hierarchy, and of their location? Large banks outside of New York that are acquiring far-flung bank companies may bring net employment growth to their home regions, wherever those home regions are located.

The location of these regional centers is of substantial importance, and cannot be assumed from current patterns (Figure 10.12). Not every regional center is destined to become a center for newly integrated financial services; the rise of Charlotte (North Carolina), Atlanta (Georgia), and Columbus (Ohio) as bank headquarters cities reflects peculiarities of regional economic fortunes, state banking regulation, bank corporate strategy, and particular bank executives. Aside from these unique considerations, the benefits of agglomeration and of scale suggest that relatively few, dispersed, larger metropolitan areas will benefit from consolidated financial service operations.

At the local-regional scale, rationalization and scale increases will make banking a growth or an economic-base sector for relatively few places. For most local economies, employment in depository institutions will continue to decline for several more years, followed by relative stability. State and local governments will try to attract banks' non-market-oriented operations – bank company

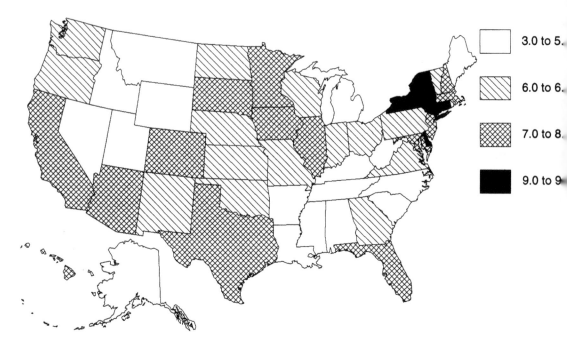

Figure 10.12 Banking as percent of total employment, 1990

headquarters and back-office support operations – with an even greater intensity, as their size and growth stands in increasing opposition to branch operations. The locational factors to which headquarters are most sensitive include highly trained and multiply skilled labor force, availability of business and professional services, and elaborate communication infrastructures. The locational factors to which support operations are most sensitive include reliable and low-wage labor force and elaborate communication infrastructures. While focused attempts to attract these operations may be ill-placed, the kinds of public and private investments useful in such an attempt are also useful for a wide range of expanding economic activities.

The Importance of Regulation on Banking Change and Location

In response to the array of high-interest options available to small savers in the inflationary 1970s, the U.S. Congress deregulated the interest rates offered by savings and commercial banks gradually over 1980–5. To help savings banks pay higher interest, their reliance on long-term, low-interest residential mortgages was reduced; they were allowed to lend to businesses, to commercial real estate owners and developers, and to increase their consumer loans. In sum, thrift institutions were allowed (and the market compelled them) to engage in riskier lending. That did not matter to small, retail depositors, whose deposits were insured (up to $100,000 per account) by a federally chartered agency. So long as deposits are insured, however, depositors do not judge the security of banks' assets. In addition, as loan losses reduce the net worth of the bank company, its owners become more highly leveraged, facing risk on ever-smaller amounts of net worth, and increasing the bidding for lending and deposits in hopes of increasing profits that can be added to net worth. The resultant moral hazard is for increasingly risky behavior not punished in the marketplace for deposits or loans. Thrift institution

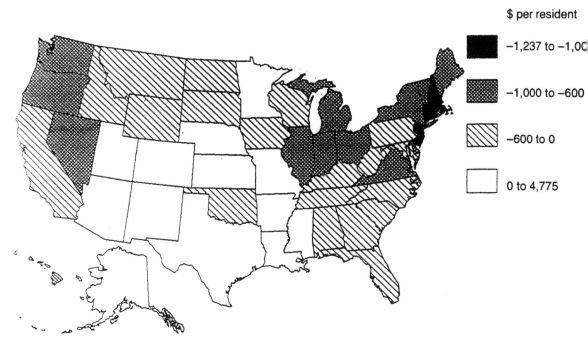

Figure 10.13 Per capita costs and benefits of the U.S. savings and loan bailout

faced grave problems during the 1980s, reflecting their limited scope, managements' limited experiences, and the swiftness of major regulatory change.

At this juncture, the U.S. government entered the scene as depositors called upon it to save the banking industry. The notoriously expensive S&L bailout began to unfold as the FSLIC began to use taxpayer dollars to shore up threatened deposits extended primarily by commercial lenders. As more and more S&Ls became insolvent, the total cost of bailing them out rose exponentially. The U.S. Congress passed the Financial Institutions Reform, Recovery and Enforcement Act of 1989, which promised $157 billion to save the thrift industry but offered initial funds of only $50 billion. The bailout would be managed through the Resolution Trust Corporation (RTC), which would sell off insolvent S&Ls and their assets at subsidized prices to wealthy investors.

A disproportionate share of troubled S&Ls was located in the southwestern U.S., particularly Texas. Long a region of rapid, unconstrained growth, the

Southwest enjoyed spectacular gains in real estate values in the late 1970s and early 1980s, particularly given rising global petroleum prices. Gorged by junk bonds and unfettered by deregulation, many S&Ls from around the U.S. invested heavily in commercial property in cities such as Houston, Dallas, Phoenix, and dozens of other Sunbelt metropolises. When the price of oil dropped markedly in the 1980s and when the dimensions of real estate overinvestment and speculation became increasingly obvious, the same region suffered the most acutely. The S&L bailout thus had an uneven geography (Figure 10.13); on a per capita basis, northeastern and midwestern states lost the most, suffering up to $1,237 per resident (in Connecticut), while southwestern states gained heavily (up to $4,775 per resident in Texas).

Compared with savings banks, the problems of commercial banks were mitigated by the broader scope of their operations (reducing, for example, the interest-rate risk that squeezed the thrift sector).

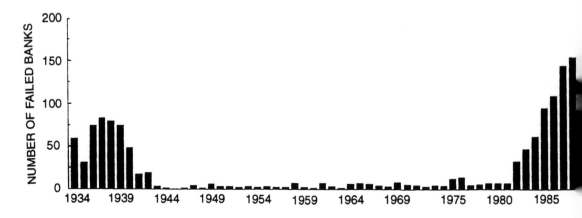

Figure 10.14 U.S. commercial bank failures, 1934–90

Much of the attention paid to the growing crisis of U.S. commercial banks has focused on the critical role of Third World debt, which affects primarily the large "money center" banks headquartered overwhelmingly in New York. As banks throughout the U.S. became increasingly involved in commercial real estate, however, the growing crisis in U.S. banking extended progressively throughout the banking hierarchy (Figure 10.14). The rate of commercial bank failures rose during the late 1980s, and this pattern varied by state. In total, about 1.5 percent of extant U.S. banks failed in each of the three years. However, the distribution is quite skewed. The preponderance of outright bank failures were in energy-producing states whose financial situations declined mightily in the late 1980s — falling too far too fast for mergers to prevent the fall of banks with substantial nonperforming assets and declining collateral protection. In 1989, 84 percent of the failed banks were in the four states of Texas, Oklahoma, Louisiana, and Colorado. In addition to an economic base weighted toward energy development, these states had very restrictive banking regulations, preventing branch banks, preventing banks from crossing county lines, discouraging operations of out-of-state banks. There is an apparent danger in severe restrictions on

statewide or interstate operations, when combined with poor economic conditions, to yield high numbers of failed banks.

TELECOMMUNICATIONS

Until the advent of electronic communications in the late nineteenth century, transportation and communications were synonymous and all information was necessarily conveyed through the physical movement of bodies in time and space (at a maximum rate of roughly 40 miles per hour). Electronic communications, however, ruptured this unity, permitting "out of body" experiences. With electrification, the distances between transactions expanded infinitely, while the time necessary to transmit them shrank to zero. Electronic communications thus allowed for a dramatic reconfiguration of the spatial and temporal matrices of opportunities and constraints in which everyday life unfolded. The introduction of electric communications was an integral part of the historical construction of modernity and the commodification of time and space. The history of electronic communications is intimately intertwined with the emergence and geography of industrial capitalism and is punctuated by the introduction of several innovations of far-reaching importance. Morse's invention of the telegraph in the 1830s and the development of the first commercial lines in the 1840s had widespread effects, particularly with regard to the uneven circulation of information among highly specialized commercial circuits, enhancing the status of some cities and eroding that of others. Electrical lighting and photography in the late nineteenth century inaugurated new rhythms of daily life and new ways of experiencing time and space. In the same vein, by the 1920s, radio became the public's most common source of news, entertainment, and commercial information.

Alexander Bell's patent of the telephone in 1876 created what is still by far the most commonly employed form of two-way telecommunications. The telephone had important repercussions for the structure of urban areas, accelerating the explosion of suburbia and downtown service complexes. The telephone allowed the formation of "communities without propinquity," or groups of people sharing common interests but not common places and contributing to the dissolution of urban neighborhoods. From the beginning, however, the effects of this technology reverberated differentially across class and space, disproportionately enhancing the power of some social groups and disempowering others. The telephone did not simply allow communication over longer distances, it threatened class relations by extending the boundaries of who could speak with whom, altering modes of friendship, romance, business, and other formal and informal networks.

Today, it is difficult to exaggerate the role of electronic communications in the everyday life of people who inhabit the industrial world. Their perceptions of the world – and of themselves – are deeply shaped by the information that floods the electronic airways. News, entertainment, and advertising – all deeply ideological texts – are shaped by their electronic "wrapping." By providing much of the corpus of taken-for-granted knowledge that individuals draw in their everyday lives, electronic media form a critical network of assets and constraints that structure the world of lived experience.

Within geography, the impacts of telecommunications have generally been approached from the perspective of time–space convergence. The numerous telecommunications systems introduced during the late twentieth century have accelerated the "annihilation of space by time" through an acceleration of the turnover time of capital. The cultural and ideological appropriation of this transformation is manifested in the form of "time–space compression," the rapid integration of vast amounts of information from far-flung locales into the rhythms of everyday life. Time–space convergence or compression reflects capitalism's incessant tendency to produce new economic and social forms, new technologies, new ideologies, new political practices, and new geographies, a phenomenon that has occurred repeatedly in the history of capitalism. During the period in which modernist culture became hegemonic, for example, the introduction of railroads and the telegraph initiated new ways of experiencing time and space that found their ways into the fabric of everyday life as they were internalized, generally without critical reflection, by the masses of late eighteenth- and early nineteenth-century Europe and North America.

It should be apparent that the electronic mediation of information does not simply expand the time–space parameters of social interaction, it qualitatively changes the nature of communication itself. Just as written texts erect a certain distance between author and reader, allowing time for contemplation and repeated receptions of messages, so too do electronic communications alter the ways in which information is collected, represented, transmitted, and received. Unlike written texts or speech, electronic data, images, messages, advertisements, and texts can be splintered, shifted, and recombined in an infinite number of nonlinear ways. The fact that information can and has become "disembodied," therefore, has enormous implications for the time–space stretching of social relations.

Finally, telecommunications have also allowed for the steady decentralization of "back offices," which perform tedious clerical and data entry functions. Traditionally clustered near headquarters in central business districts or in adjacent suburbs, back offices have become increasingly mobile and capable of seeking out pools of cheap, semi-skilled labor around the world. For example, U.S. airlines and insurance firms have relocated large numbers of back office jobs to the Caribbean and to Ireland. In the formation of global office networks characterized by international linkages, back offices reveal many of the archetypal aspects of post-Fordist production.

SUMMARY

This chapter has emphasized the way that the historical development of each of the six sectors included shifts in the distribution of productive capacity, both among and within countries. The shifting patterns of production reflected changes in national and international regulation, in production, transport, and communications technology, as well as the spread of industrialization and, especially, industrial capitalism. These changes affected the materials- versus labor- versus market-orientation of the sectors, the importance of scale economies in production and marketing, and the facilities' benefit from agglomeration. By studying industrial location at this aggregate level, rather than the individual-organization level of Chapter 8, we are drawn to the general tendencies of industrial location. We recognize the importance of context and environment to organizational decisions. Nonetheless, earlier chapters have made it clear that individual companies and governments have some discretion in their actions. The study of industrial location requires that we shift back and forth among these levels of analysis.

SUGGESTED READING

Ballance, P. and Sinclair, S. (1983) *Collapse and Survival: Industry Strategies in a Changing World*. London: George Allen & Unwin.

Dicken, P. (1992) *Global Shift: The Internationalization of Economic Activity*. 2nd edition. New York: Guilford Press. Chapter 8.

Edwards, G. (1982) "Four sectors: textiles, man-made fibres, shipbuilding, aircraft," in J. Pinder (ed.) *National Industrial Strategies and the World Economy*. London: Croom Helm.

Florida, R. and Kenney, M. (1992) "Restructuring in place: Japanese investment, production organization, and the geography of steel," *Economic Geography* 68: 146–73.

Hogan, W. (1991) *Global Steel in the 1990s: Growth or Decline*. Lexington, Mass.: Lexington Books.

Holly, B. (1987) "Regulation, competition, and technology: the restructuring of the U.S. Holmes Commercial Banking System," *Environment and Planning A* 19: 633–52.

Holmes, J. (1991) "The continental integration of the North American automobile industry: from the Auto Pact to the FTA," *Environment and Planning A* 23: 177–92.

Langdale, J. (1989) "The geography of international business telecommunications: the role of leased networks," *Annals of the Association of American Geographers* 79: 501–22.

Main, A., Florida, R. and Kenney, M. (1988) "The new geography of automobile production: Japanese transplants in North America," *Economic Geography* 64: 352–73.

Mayer, M. (1990) *The Greatest-Ever Bank Robbery: The Collapse of the Savings and Loan Industry*. New York: Scribner's.

Moss, M. (1987) "Telecommunications, world cities, and urban policy," *Urban Studies* 24: 534–46.

Schoenberger, E. (1988) "From Fordism to flexible accumulation: technology, competitive strategies and international location," *Environment and Planning D: Society and Space* 6: 245–62.

Scott, A. (1987) "The semiconductor industry in South East Asia: organization, location and the international division of labor," *Regional Studies* 21: 143–60.

Scott, A. and Storper, M. (1987) "High technology industry and regional development: a theoretical critique and reconstruction," *International Social Science Journal* 112: 215–32.

Thrift, N. (1987) "The fixers: the urban geography of international commercial capital," in J. Henderson and M. Castells (eds.) *Global Restructuring and Territorial Development*. Beverly Hills: Sage.

Tiffany, P. (1988) *The Decline of American Steel: How Management, Labor and Government went Wrong*. New York: Oxford University Press.

Walter, I. (1988) *Global Competition in Financial Services: Market Structure, Protection, and Trade Liberalization*. Cambridge, Mass.: Ballinger.

Warf, B. (1989) "Telecommunications and the globalization of financial services," *Professional Geographer* 31: 257–71.

11

CONNECTING THE PIECES

We have approached the study of industrial location from several directions, or through several lenses: microeconomic, macroeconomic, and social-institutional. We have seen the location of a discrete facility as a problem of economic optimization. We have put the decision in the context of the solution to a set of other problems: capacity shortfalls, technological obsolescence, recognition of new market opportunities. In this context, location of a facility is not an end in itself, but merely one possible way of resolving a set of roadblocks on the way to organizational survival or growth. We have noted all the variables that affect the outcome of a production investment, and the consequent necessity to decompose the location decision into a set of smaller logistical decisions and competitive considerations. We have enumerated ways in which governments attempt to steer these investment decisions in ways that the governments deem beneficial. Finally, we have looked at *industrial location as a system* of related activities (economic, organizational, technological, political, and social) that have changed as technologies, social institutions, and public policies have changed. These institutions and policies are also affected by industrial change and industrial location change. These interactions matter to each of us as we attempt to raise families, earn our living, and pay our taxes. In a set of recapitulations below, we attempt to show the complementarities of these approaches to a complex subject.

THE CORPORATE PERSPECTIVE

The small-business owner seldom makes explicit location decisions. *New-firm* formation, or entrepreneurship, is a highly localized phenomenon, dependent upon the individual's credit rating, access to capital, client contacts, and (often) access to alternative sources of wage or salary income. These requirements tie the entrepreneur to a specific local region. *Stable small businesses* generally rely on localized markets and market information, and maintain "arm's-length" (contractual, non-ownership) linkages with input suppliers. Small-business decisions about procurement, expansions, contractions, and marketing seldom include explicit considerations of locations, as locational inertia is a dominant force. Considerations of location change, when required by expansion constraints, cost pressures, or market relocation, are critical and difficult: critical because of the large investment as a proportion of the organization's size, and difficult because of the number of variables to be analyzed.

Larger organizations make location decisions, often explicitly, every time a new investment is considered. Ordering specialized equipment for an existing facility, contemplating the acquisition of an available facility, gearing up to enter a new product or regional market – all these decisions entail fixed investment and thus, locational commitment. Whether the organization operates for profit or for public benefit, its objective is its own current and continued survival. To survive, it must provide its clients with the products or services they need, at a price that is more attractive than the clients' options (which might include other suppliers, other products, or doing without the good or service altogether). Where are the clients? What are the costs involved in producing the good or service? Which clients have which alternatives? Do the answers to these questions vary across potential locations, or over time? We are left with too many questions to answer simultaneously. We generally retreat to a series of questions, such as:

1 What is to be produced?
2 What is the current and planned production technology?
3 Which market is to be targeted, described by price level, location, and nature of competition?
4 What activities should our organization undertake in order to satisfy that target market? Design, market research, production design and engineering, component production, final assembly or customization, marketing, after-sales service?
5 What other organizations can provide aspects of this process that we do not want to provide internally? Where are they? Are there costs or benefits that result from our organization being far from or near to each of these suppliers?
6 What is the size of the operation that our organization will operate? What material, employee, service, and capital requirements does the operation have?
7 Where are these materials, employees, services, and capital available, with what proximity to the target market? At what cost can these items be procured? At this point, some of these considerations can be considered from a logistical perspective, in that their cost or revenue depends on distance from certain points of reference.
8 What is the nature of competition, or, more generally, of alternatives for the target market? How

might this change over the life of the investment be contemplated?
9 Given the alternatives facing the clients, the consequent returns facing our organization, and the opportunity costs of any fixed investment to be made, does it seem that these inputs can be procured and operated at a high enough rate of return to justify the capital expense?

Earlier in the book, we have noted the vast amounts of information about markets, technologies, suppliers, competitors, and their future characteristics, all required to answer these questions. In addition, however, good decisions result from good information about the organization itself. What functions can it perform well? What functions can other organizations perform better, and thus might be purchased? What are the strengths and knowledge of its employees? Are the employees or their key knowledge (about clients, process technology, the particular production in a given facility) mobile, or do they need to be employed *in situ*? Can the elements of the production process be separated and managed in separate locations? Or, again, should the organization cease providing some function for itself?

These questions about the organization itself are critical elements in the generation of **corporate** and **business strategies**: in what lines of business is the organization engaged, and how does it want to position itself in the market for that line of business? Each organization is unique in its mix of capabilities, employees, facilities, and clients. Therefore, we should expect different organizations – even those within the same industry – to answer these questions differently. To make this idea explicit with respect to facility location decisions, recall Figure 8.2 (see p.160). Close competitors in the same industry are unlikely to have exactly the same distribution of market centers (C_1 through C_3) and existing production locations (P_1 and P_2) as the organization graphed in that figure. Therefore, the logistically optimal location for an additional production facility would differ for the organizations. However, even with the same distribution of likely markets, available labor forces, tax differentials, and material inputs, competitors may invest in different locations. The differences reflect:

- differences in the organizations' dependence on external supply versus internal production, an element of corporate strategy
- differences in the degree of market responsiveness desired by the organizations, an element of business strategy
- differences in the organizations' abilities to expand on existing sites, a function of each organization's individual history and of the circumstances around different production facilities; or
- differences in the proportion of highly skilled labor needed within the two organizations' facilities – as a result of the different business strategies of the organizations, the different technologies developed by the organizations, the different scales of production in the two facilities to be located, or the capability of one organization to transfer technical labor as needed among its facilities while another organization has only one existing facility from which to obtain technical people when needed.

These differences among companies help explain the heterogeneity of investment locations within the same industry. From the corporate perspective, these differences are good, insofar as they present each competitor with a limited advantage over others. One organization may excel at large orders, another at customized orders; one at servicing one geographic market, another at serving a remote market. Despite the importance of perfect competition to economic theory, actual companies seek some form of monopolistic competition. The existing locations of each company's operations provide one source of monopolistic competition, as each company is best suited to serve a particular set of markets, or to implement a particular technology (because of its mix of locally available factors). From the perspective of the non-profit organization, these distinctions are inevitable, and help organizations define themselves. The

differences also make it difficult to comprehend, and impossible to predict, industrial location decisions, as we will discuss below.

THE SYSTEM-WIDE PERSPECTIVE

While it is very difficult for the individual organization to reach a good decision about the deployment of fixed investment, it is even more difficult to make sense of the total pattern of industrial investment. The reduction in the importance of transportation logistics in industrial location has made the task even more difficult. There is little wonder that people talk of "footloose" industry, investment waiting to occur anywhere a corporate manager, public decision maker, or influential politician wants it to occur. To the extent that fixed investment is footloose – at least, before it is put into place – this book is useless.

Let us look at this proposition, however. Is there no logic to industrial location decisions besides individual whim, power relationships within companies, and interplay between business and government? Are there any concerns or types of analysis common to most managers, before they commit millions of dollars into an expansion, acquisition, or new facility? Are there any observable tendencies resulting from shared concerns on the part of decision makers? We have certainly proposed some concerns and methods of analysis in this book, which apply even when transportation issues are of little importance. Let us review the guides we have developed so far.

Distinguishing Places and Activities

A second theme in this book, after the emphasis on industrial location decisions as investment decisions, is the conceptual method of distinguishing places according to characteristics relevant for the location of different kinds of activities. Now that we have studied this subject matter, what can we say about these distinguishing features? First, note that the geographic scale of our "places" varies with the characteristics that concern us. We can study industrial location at the international scale, the sub-national regional scale, the scale of the metropolitan area or local labor market, and the scale of individual sites within a metropolitan area. At each scale, different characteristics are paramount: investment regulations and macroeconomic conditions at the international scale; regional histories of industrialization, market development, and labor relations at the sub-national scale; wage rates, labor availability, tax rates at the metropolitan or local labor market scale; and access to public services such as highways, railways, sanitary sewers, and building codes at the site-specific scale. We can also say with certainty that these place characteristics change over time.

Furthermore, the changes are affected greatly by the nature and pattern of industrial investments, which influence macroeconomic conditions, labor relations, industrial suppliers, industrial markets, consumer markets, wage rates, and tax rates. While this implies a circular system, it is neither closed nor uni-directional. For example, a history of continuous industrial investment in a metropolitan region should increase the region's network of industrial suppliers, skilled workers, specialized services, and consumer markets, while reducing the rate of property and sales taxes required to raise public monies. Each of these tendencies increases the attractiveness of the metropolitan, industrial region for further investment. On the other hand, wages and salaries are likely to be high, reflecting the availability of jobs and the demand on the region's land.

However, the attraction to such an area, as well as the discouragement from the area, vary across activities. What kinds of activities benefit from such a set of regional characteristics? We have seen different ways of distinguishing investments: by industry, by function, by type of organization. A weight-losing, materials-processing industry would not benefit especially from such a regional location; neither would a simple painting or assembly operation of small parts. The local availability of skilled and technical labor, as well as the range of supporting services available in a thriving metropolitan area, is of greater importance to a small,

entrepreneurial company than to a large, multi-product, multi-location company that can transfer human resources, information, and services internally. The characteristics of a given activity change over time, as well, in ways synopsized in Chapter 6.

Interestingly enough, historical development has brought a diffusion of most industries across the regions and countries of the world. Steel making, textiles, electronics, and machinery are no longer the exclusive industries of a small set of countries or regions. While this perhaps indicates a reduced relevance of industry-by-industry matching for countries or large regions, there are still differences in the way that these industries are undertaken in different places. These differences, in the functions pursued, the size of company, or the capital or labor intensity of production, leave some utility in the understanding of location via place and activity characteristics. Materials orientation, labor orientation, agglomerative tendencies, and competitive reactions still affect different organizations in different ways.

Different Approaches to Different Questions

How we make use of these insights depends on the nature of our questions. To understand the *general tendencies of industrial development*, to gain a sense of where we are headed in what seem to be especially turbulent times, we need a broad, historical overview of industrialization. Such is the purpose of Chapter 7. Capitalist industrialization, the form of economic and social organization that dominates the world at the end of the twentieth century, has some general tendencies in the development of places. Its basic characteristics – the production of commodities for sale, dependent on the coordination of materials, capital, and labor, and the constant attempt to increase profit by reducing costs and by seeking monopoly returns – encourage the expansion of markets, uneven development of places, the specialization of places in certain types of production, and redeployment of investment in new industries and new places. By adding detail to the scenario in the form of a particular sector or a particular set of places, we can gain some general understanding of the past of a place and some insight into its future. The detail includes the economic specifics of resource endowment, local markets, transport connections, wage rates, tax rates, industrial agglomerations, as well as the social characteristics that underlie the wage and tax rates and influence their change over time. High-wage, resource-poor places develop a particular mix of activities; demographically growing, low-wage, resource-filled places develop another mix. The presence of the one place influences the behavior of companies and governments in the other, because late twentieth-century capitalism is nothing if not globally interdependent.

If our purpose is to assess the *impact of a change in government policy*, such as taxation or labor regulation, we need a careful specification of the change, along with the current characteristics of the region affected. The direction of the effect is usually straightforward, at least in the first round of effects. A policy change that will make employment more expensive for the employer will encourage the substitution of capital for labor, and will discourage subsequent investment by labor-intensive activities. However, the degree of discouragement varies by activity, according to:

- the other reasons for the activity's location (tied to local resources or markets?);
- he relative cost of labor in other regions;
- the degree of technical substitution that is possible; and
- the nature of competition in the industry (can the higher cost of employment be passed on to buyers of the activity's product?).

Second-round effects complicate the picture. If the increased cost of labor reflects costs of improving the labor force (through better education, training, or health care), the results will not be obvious immediately. Eventually, the creation of a different labor force will encourage investments of different kinds. Similar results may be obtained by fiscal policy that increases taxes but improves infrastructure. These are economic paths for studying the effects of policy changes. The social effects could be just as important, such as changes in family patterns and demographic change stemming from increased wages, better schooling, or changed work hours.

If our purpose is *analyzing a particular location decision*, or even more challenging, predicting of a particular decision, we need a great deal more, and more organization-specific, information. The heterogeneity of actual organizations makes this task much more difficult than the task of understanding general trends. The most important information pertains to the distribution of the organization's markets, suppliers, and existing facilities. This allows us to make some simple, logistical calculations, based on a normative model of cost minimization. The distribution of competitors' facilities also matters, however, as well as the organization's subjective assessment of its competitors' reactions (see Chapter 3). If we cannot know its subjective assessments, knowledge of its business strategy relative to those of its competitors will help us understand the organization's decision. Its strategy will influence the degree of the competition and the strength of the competitive reaction. Corporate and business strategies also affect the supplier and market linkages and the technology used. You can see that compiling this kind of information is quite painstaking, especially since eliciting responses to direct questions about strategy and competition is difficult.

The ideas presented in this book support a range of questions, and a range of approaches to those questions. The issue is complex, spanning the investment process of the individual company, the logistics of procurement, hiring, and marketing, and the development of regional and national economies. The phenomena are important, affecting corporate profitability, the availability of goods and services where they are demanded, the ability of people to earn money to sustain themselves and their families, and the interaction and evolution of identifiable places at whatever scale. It is reasonable for these issues to spawn different approaches. You should now be able to discern what analytical approach best suits the questions of greatest importance to you.

INDEX

Page numbers in italic print refer to figures and tables.

ASSESSMENT OF THE FATE OF SURROGATES FOR ENTERIC PATHOGENS RESULTING FROM THE SURCHARGING OF COMBINED SEWER SYSTEMS

Iosif Marios Scoullos

ASSESSMENT OF THE FATE OF SURROGATES FOR ENTERIC PATHOGENS RESULTING FROM THE SURCHARGING OF COMBINED SEWER SYSTEMS

DISSERTATION

Submitted in fulfillment of the requirements of
the Board for Doctorates of Delft University of Technology
and
of the Academic Board of the IHE Delft
Institute for Water Education
for
the Degree of DOCTOR
to be defended in public on
Wednesday 13, May 2020, at 12:30 hours
in Delft, the Netherlands

by

Iosif Marios SCOULLOS
Chemical Engineer, National Technical University of Athens, Greece
Master of Philosophy in Technology Policy, University of Cambridge, UK
born in Athens, Greece

This dissertation has been approved by the
promotor: Prof. dr. D. Brdjanovic and
copromotors: Dr. C.M. Lopez Vazquez and Dr. J. van de Vossenberg

Composition of the doctoral committee:

Rector Magnificus TU Delft	Chairman
Rector IHE Delft	Vice-Chairman
Prof. dr. D. Brdjanovic	IHE Delft / TU Delft, promotor
Dr. C.M. Lopez Vazquez	IHE Delft, copromotor
Dr. J. van de Vossenberg	IHE Delft, copromotor

Independent members:

Prof. dr. F. Clemens	TU Delft
Prof. dr. A. Mynett	TU Delft / IHE Delft
Prof. dr. C.L. Moe	Emory University, USA
Prof. dr. A.M. de Roda Husman	Utrecht University
Prof. dr. C. Zevenbergen	TU Delft / IHE Delft, reserve member

This research was conducted under the auspices of the Graduate School for
Socio-Economic and Natural Sciences of the Environment (SENSE).

This thesis was accomplished thanks to a Scholarship from the Academy of Athens, Vasiliki
Bekiari-Vekri Bequest and a Research Grant from the Alexander S. Onassis Public Benefit Foundation.

CRC Press/Balkema is an imprint of the Taylor & Francis Group, an informa business

Published by:
CRC Press/Balkema
Schipholweg 107C, 2316 XC, Leiden, the Netherlands
Pub.NL@taylorandfrancis.com
www.crcpress.com – www.taylorandfrancis.com
ISBN 978-0-367-55692-1

καὶ ἐὰν εἰδῶ τὰ μυστήρια πάντα καὶ πᾶσαν τὴν γνῶσιν, ἀγάπην δὲ μὴ ἔχω, οὐδέν εἰμι

To my parents

ACKNOWLEDGMENTS

The completion of this thesis has been an exciting and wonderful journey that I travelled together with many people to whom I am extremely grateful. The gradual deepening into mechanisms taking place in nature, for the benefit of society and the environment, is perhaps the most rewarding experience for the time, resources and energy invested.

First of all, my research and this PhD thesis would not be possible without the valuable Scholarship from the Academy of Athens from the Vasiliki Bekiari-Vekri Bequest and a Research Grant from the Alexander S. Onassis Public Benefit Foundation. I am deeply grateful for this opportunity to deepen my knowledge and to strengthen my commitment in finding practical solutions for a sustainable future.

I would like to thank my promotor, Professor Dr. Damir Brdjanovic for giving me the opportunity to do a PhD and for his overall supervision, guidance and support since the first day I arrived in Delft. I am very thankful for his important expertise and contribution in formulating this very interesting topic and during all the stages towards the completion of my thesis. Thank you very much! Second, I would like to thank my mentor Dr. Carlos Lopez Vazquez for his day to day supervision, guidance and countless hours of meetings. Carlos, you know in detail all the stages of my PhD research, all the challenges I met, how I managed to overcome them and how I have changed. Thank you for everything! I would like to equally thank Dr. Jack van de Vossenberg for mentoring me in the second half of my PhD. Thank you, Jack, for your invaluable ideas and support! In difficult times you helped me see things in a positive way and you helped me believe in my skills. I would also like to thank Dr. Michael Hammond, who mentored me in the first half of my research. Thank you for your support, your ideas and your encouragement. Sabita, my amazing MSc student, I would like to add you in this list just after my mentors, because although you were my student I learned a lot from you. Your contribution in the research, your self-motivation, your willingness to learn and your smile even in times of desperation motivated me too and I am grateful that I had the opportunity to work with you.

I would like to thank Prof. Maria Kennedy for her important input as member of my proposal defence committee and all the reviewers of my PhD proposal, my papers and my thesis. I would like to thank Dr. Zoran Vojinovic, Dr. Arlex Sanchez Torres, Dr. Solomon Seyoum and Dr. Eldon Raj Rene from IHE, Dr. James Shucksmith and Dr. Matteo Rubinato from the University of Sheffield, Prof. Marcos von Sperling from UFMG and Dr. Arjan van Dijk from RIVM for their precious suggestions, advice and insights.

Enormous thanks are due to the laboratory staff of IHE without the precious daily support of whom the lab work would not be possible. I would like to warmly thank Fred Kruis,

Frank Wiegman, Lyzette Robbemont, Peter Heerings, Ferdi Battes and Berend Lolkema. Special thanks to Peter, who spent many hours patiently answering my never ending microbiology-related questions. Furthermore, I am very thankful to Yuli, Mary, Fiona, Mona, Suzette, Yvonne, Pancho and Bruno for helpful advice and discussions related to the experimental methodology.

For their wise advice and support related to PhD issues I would like to extend my thanks to Prof. Ken Irvine, the PhD coordinators, and Jolanda Boots, Floor Felix, Bianca Wassenaar and Anique Karsten. Also, I am grateful to all PhD Fellows' Association Board (PAB) members and very good friends: Chris, Shahnoor, Mohan, Gonzalo, Nirajan, Motasem, Alex, Pedi, Sang Yeob, Musaed, Shakeel and Joel during my time in PAB and all the other PAB members. Special thanks to Angelica, my PhD Buddy.

Next, I would like to thank both for their friendship, their support and their scientific insights all my friends from IHE: My first friends in Delft, Chiara, Joy and Arda, my office mates, Nirajan, Motasem, Taha, Chol, Mirjana, Nang, Suniti, Anastasiia, Ramita, Feishu, Mink, and many more. A special place belongs of course to my fantastic group of friends that I met in IHE, Chris, Lea, Samayita, Shrutika, Teju, Viviana, Angelo, Gabriele, Mirko and extended group members Manu, Alessandra, Nirakar, Paolo, Ricardo, Georg. Special thanks also to Paulo, Mauricio, Mohaned, Laurens, Maria, Poolad, Jaka, Erika, Martha, Luana, Eva, Thaine, Natalia, Adele, Xiaoxia, Gemma, YuJin, Ilka, Ashwini and Ovidiu. I would like to thank all academic and non-academic staff members, PhD fellows and MSc students that I met in IHE and TU Delft and have been part of my daily life.

Among the above mentioned friends, I would like to dedicate a few more lines to some without whom my life wouldn't be the same. Chris, you deserve my gratitude for everything good that you taught me, especially to research before I reject any idea. Mohan, you are a continuous source of inspiration. You taught me to not burn easily and to not care about negative moments. Teju, thank you for all the philosophical discussions and time well spent together. Chiara, thank you for your trust and your friendship. Arda, thank you for your friendship and for teaching me to set priorities. Joy, thank you for being there for me and sharing your knowledge and positive energy. Suniti, thank you for teaching me to appreciate honesty and for reminding me to not think too much. Taha, thank you for your optimism and your wise words. Motasem, thank you for everything, you have been a constant pillar in the office and in my PhD life. Nirajan, thank you for everything, it was an honour to meet you. Mary, thank you for your valuable support and advice in lab and in life. Last, but not least, Shahnoor, a big thank you for your friendship, your kind smile, your trust, and your wise advice in every step of my PhD.

I would like to extend my thanks to my friends that I met in Delft outside of IHE and supported me too in this journey. First of all, I am thankful to have met Judith, Sasheeka, Kelvin, Rahul, Gynett, Geraldine, Kim and Marlon, José and Ella, Fr. Eli, Leroy, David, Rafael, Praveen, Jinto and Annu, Angelina, Larissa, Stanley, Marie and Koen, Calixto and many more. Sasheeka, I can't be thankful enough for your friendship, your support and your positive energy. Judith, thank you for being always there, a true friend. Kelvin, thank you for your friendship and for our interesting discussions. I'm also very thankful for having met Maria, Gemma, Nada, Weiwei, Camilo, Michelle, Roberta, Priya and Rev. Taco. Thank you for your friendship and precious lessons learned. Special thanks are due to Rev. Waltraut and Fr. Verbakel who have welcomed me and guided me in times of crisis and despair. And of course, last but not least, my friends from Greece and Cyprus who managed to stay in touch despite the distance: Ioanna, Lefteris, Virginia, Akis, Chrisa, John and Panagiotis.

Of course I owe the most to my family for bringing me up and giving me the opportunity to learn and learn from the wise. I thank my mother Marina for her love and support and my father Michael for instilling in me the sense of scientific curiosity that someone needs to start a PhD and the power of persistence that someone needs to finish a PhD. I thank my aunt Yvonne and my aunt Christiane with my uncle John, my cousins Marina, Constantine and Dimitris with their families, and all my relatives in England. Also I'm grateful to my grandparents and all my other relatives of blessed memory.

And not to forget the billions of microorganisms that sacrificed their tiny lives for science.

As a last remark, my thought goes to the millions of people affected by floods worldwide, unprotected and unprepared against the increasing consequences of climate change. May we never forget that we are all part of the same world!

SUMMARY

In the last decade (2009-2019), flooding has caused the death of over 48,000 people, and affected over 697 million people globally. This is expected to increase as a result of climate change, increased populations and urbanisation. Floods can cause infections due to the release of water-borne pathogenic microorganisms from surcharged combined sewers and other sources of faecal contamination. Pathogens can occur after the surcharge of sewer networks on surfaces commonly found in the urban environment such as concrete, asphalt, pavement blocks and grass.

The goal of this research was to contribute to a better understanding of how the concentration of water-borne pathogens on different urban and recreational surfaces is affected by different environmental conditions after urban floods. Also, in order to assess the concentration of water-borne indicator organisms on flood-prone urban surfaces it is necessary to identify the most reliable method for the recovery of these organisms.

The inactivation of faecal indicator bacteria *Escherichia coli* in the water phase was studied in an open stirred reactor, under controlled exposure to simulated sunlight, mimicking the effect of different latitudes and seasons, and different concentrations of total suspended solids (TSS) corresponding to different levels of dilution and runoff. While attachment of bacteria on the solid particles did not take place, the inactivation rate coefficient, k (d^{-1}), was found to depend on light intensity, I (W.m^{-2}), and duration of exposure to sunlight, T (h d^{-1}), in a linear way ($k=k_D+0.03 \cdot I$ and $k=k_D+0.65 \cdot T$, respectively) and on the concentration of TSS (mg.L^{-1}), in an inversely proportional exponential way ($k=k_D+14.57e^{-0.02 \cdot [TSS]}$). The first order inactivation rate coefficient in dark conditions, $k_D=0.37$ d^{-1}, represents the effect of stresses other than light.

Four different sampling methods were compared for retrieving samples from concrete, asphalt, pavement blocks and grass: swabbing, direct agar contact, stamping and adhesive tape-lifting. The surfaces were inoculated with known amounts of *E. coli*. A glass surface was used as control. Contact plating had the highest log recovery ratio, 96.1% on glass, for concentrations up to 10^3 CFU.100 cm^{-2} of *E. coli*, but this method has a limited range of bacterial numbers because it is not possible to dilute or concentrate the samples. Swabbing was the most reliable technique because it could be used for a wide range of concentrations with high recovery ratios of up to 96.2% for 10^5 CFU.100 cm^{-2} of *E. coli*. Comparatively, the indirect methods of stamping and tape had no additional advantages.

Further experiments using the swabbing technique revealed that the water accumulated on rougher surfaces affected the swabbing recovery ratios when the samples got diluted. Swabbing any amount of sample higher than what the swab heads could absorb (0.15 mL) reduced the recovery ratio. Furthermore, swabbing was more efficient without the use of a detergent (Tween 80) in wetting solution and eluent. After the recession of an artificial flood, swabbing and contact plating were confirmed as reliable methods to sample and enumerate the presence of *E. coli* on different urban surfaces, with the log recoveries ranging from 21.0% to 59.0%, depending not only on the sampling method, but also on the actual amount of bacteria on the surfaces.

The inactivation of faecal indicator *E. coli* was studied under controlled exposure to simulated sunlight on a range of different surfaces found in urban environments: gravel, sand, asphalt, pavement blocks, concrete, playground rubber tiles and grass, using glass as control. The surfaces were inoculated with artificial flooding water containing 10^5 $CFU.mL^{-1}$ of *E. coli* and sampled periodically using the sterile cotton swab technique, after lowering the water level. The results show that inactivation in dark conditions was not statistically significant for any surface, suggesting that chemical composition and pH (varying from 6.5 to 9.2) did not affect significantly the inactivation rates in the short term. The highest light-induced inactivation rates for *E. coli* after the floodwater recess, observed on rubber (>3.46 h^{-1}) and asphalt (2.7 h^{-1}), were attributed to thermal stress and loss of surface moisture.

The inactivation of *E. coli*, *Bacillus subtilis* spores, and bacteriophage MS2, that all are surrogates for different groups of pathogens, was studied under controlled exposure to simulated sunlight. Concrete, asphalt and pavement blocks were inoculated with artificial floodwater containing these organisms. The research took into account the pH of the water that is exposed to these surfaces and its role on the survival of the organisms. The results showed that inactivation in dark conditions was not statistically significant for any organism and surface, suggesting that pH alone (varying from 7.0 to 9.6) did not affect significantly the inactivation rates in the short term. The highest light inactivation was seen on *E. coli* on a concrete surface (pH 9.6) with an inactivation rate of 1.85 h^{-1}. MS2 phage had the highest light inactivation on asphalt with a rate of 1.3 h^{-1}. No inactivation of *B. subtilis* spores was observed on any of the surfaces on both light and dark conditions. In general, the light inactivation on all surfaces followed the ascending order of *B. subtilis* spores < MS2 phage < *E. coli* except on asphalt, where the light inactivation of *E. coli* and MS2 was found to be same and that of *B. subtilis* spores was the least.

This study suggests that given the sunlight conditions after an urban flood, the concentration of indicator organisms, TSS and the type of flooded surfaces it is possible to estimate the fate of selected water-borne pathogens. The observations and results presented in this study, in combination with Quantitative Microbial Risk Assessment (QMRA) and mapping of urban surfaces, can be used to develop policy-making tools for rapid implementation of appropriate measures to mitigate public health risks after flooding.

SAMENVATTING

In het afgelopen decennium (2009-2019) hebben overstromingen geleid tot de dood van meer dan 48.000 mensen, en hebben ze wereldwijd de levens van meer dan 697 miljoen mensen beïnvloed. Het is de verwachting, dat deze aantallen nog zullen toenemen als gevolg van klimaatverandering, toenemende bevolkingsgroei en verstedelijking. Overstromingen kunnen infecties veroorzaken door het vrijkomen van uit water afkomstige pathogene (ziekteverwekkende) micro-organismen uit overgelopen riolen en andere bronnen van fecale besmetting. Pathogenen kunnen worden achtergelaten nadat rioolwater oppervlakten heeft overstroomd in de stedelijke omgeving, zoals beton, asfalt, plaveisel en gras.

Het doel van het onderzoek was het bijdragen aan een betere verklaring van hoe de concentratie van uit water afkomstige pathogenen op verschillende stedelijke en recreatieve oppervlakken beïnvloed wordt door verschillende omgevingsfactoren na stedelijke overstromingen. Daarom diende een betrouwbare methode ontwikkeld te worden om uit water afkomstige indicator-organismen te isoleren van de overstroomde stedelijke oppervlakken.

De inactivatie van fecale indicatorbacterie *Escherichia coli* in de waterfase werd onderzocht in een open geroerde reactor, waarmee het effect van verschillende breedtegraden en seizoenen nagebootst konden worden. Dat werd gedaan met gecontroleerde blootstelling aan kunstmatig zonlicht. Er werd getest bij verschillende concentraties van totale hoeveelheid deeltjes in suspensie (total suspended solids, TSS) die overeenkomen met verschillende niveaus van verdunning en afvoer. Er vond geen hechting plaats van bacteriën aan de vaste deeltjes, en het werd duidelijk dat de snelheids-coëfficiënt voor inactivatie, k (d^{-1}), lineair afhankelijk was van lichtsterkte, I (W.m^{-2}), en aan de tijdsduur van blootstelling aan zonlicht, T (h d^{-1}), waarbij respectievelijk gold $k=k_D+0.03 \cdot I$ en $k=k_D+0.65 \cdot T$. De invloed van de concentratie van TSS (mg.L^{-1}) op de inactivatie coefficient was omgekeerd evenredig exponentieel ($k=k_D+14.57e^{-0.02 \cdot [TSS]}$). Het effect van andere factoren dan licht, weergegeven met een eerste orde snelheids-coëfficiënt voor inactivatie onder donkere omstandigheden, was k_D=0.37 d^{-1}.

Vier verschillende bemonsteringsmethodes voor het verzamelen van monsters van beton, asfalt, straattegels en gras werden met elkaar vergeleken: het afnemen van swabs met steriele wattenstaafjes, direct agar contact (contactplaten), afstempelen en het gebruik van zelfklevende tape. De test-oppervlakten werden geïnoculeerd met vastgestelde hoeveelheden *E. coli*. Een glazen oppervlak werd gebruikt als controle. Contactplaten gaven de hoogste terugwinning van de geïnoculeerde bacteriën, 96.1% op glas, voor concentraties tot 10^3 CFU.100 cm^{-2} *E. coli*. Deze methode heeft echter een beperkt meetbereik voor mogelijke concentraties van bacteriën, omdat het niet mogelijk is om de monsters te verdunnen of te concentreren. Het afnemen van swabs bleek daarna de meest betrouwbare techniek te zijn, omdat het gebruikt kon worden over een groot concentratie-bereik met hoge efficiëntie, tot 96.2% voor 10^5 CFU.100 cm^{-2} *E. coli*. In vergelijking met de andere twee methoden hadden de indirecte stempelmethode en het gebruik van tape geen toegevoegde waarde.

Verdere experimenten met de swab techniek toonden aan dat water op ruwere oppervlakten de efficiëntie van de swab techniek nadelig beïnvloedde zodra de monsters werden verdund. En zodra de swabs werden gebruikt voor een groter volume dan het wattenstaafje kon absorberen (0.15 mL) verminderde de terugwinning. Toevoeging een detergens (Tween 80) in de oplossingen om het wattenstaafje van de swab te bevochtigen en om de bacteriën van de swab te elueren, bleek minder efficiënt dan zonder de toevoeging. Na het verdwijnen van een kunstmatige overstroming bleken de swab methode en contactplaten opnieuw de meest betrouwbare methodes om de aanwezigheid van *E. coli* op verschillende stedelijke oppervlakten te analyseren. De log terugwinning varieerde van 21.0% tot 59.0%, en was afhankelijk van de bemonsteringsmethode en de oorspronkelijke hoeveelheid bacteriën op de oppervlakten.

Bij gecontroleerde blootstelling aan kunstmatig zonlicht werd de inactivatie van de fecale indicator *E. coli* onderzocht op een reeks verschillende oppervlakten uit stedelijke omgevingen: grind, zand, asfalt, straattegels, beton, en rubberen tegels die gebruikt worden op speelplaatsen, met gras als controle. De oppervlakten werden geïnoculeerd door kunstmatige overstromingen met water wat 10^5 CFU.mL^{-1} *E. coli* bevatte, en werden na het verlagen van het waterniveau regelmatig bemonsterd met de steriele watten-staafjestechniek. Uit de resultaten bleek dat inactivatie in het donker statistisch gezien niet significant was, voor alle organismen en oppervlaktes, met de aanname dat chemische samenstelling en pH (variërend van 6.5 tot 9.2) de inactivatiesnelheid op korte termijn niet noemenswaardig beïnvloedden. Door licht geïnduceerde snelheden van inactivatie, na het verdwijnen van het overstromingswater, waren echter wel significant, en de snelste inactivatie was voor *E. coli*, op rubber (>3.46 h^{-1}) en asfalt (2.7 h^{-1}). Deze afnames werden toegeschreven aan temperatuurstress en uitdroging van het oppervlak.

De inactivaties van *E. coli*, *Bacillus subtilis* sporen, en bacteriofaag MS2, tevens surrogaten voor verschillende groepen pathogenen, werden onderzocht onder gecontroleerde blootstelling aan gestimuleerd zonlicht. Beton, asfalt en straattegels werden geïnoculeerd met kunstmatig overstromingswater met deze organismen. Het onderzoek hield rekening met de pH van het water dat blootgesteld werd aan deze oppervlakten en de rol van de pH van het water wat betreft de overleving van de organismen. De resultaten toonden opnieuw aan dat donkere inactivatie niet significant was voor alle organismen en oppervlakten, er van uitgaande dat pH alleen (variërend van 7.0 tot 9.6) de inactivatiesnelheid op korte termijn niet noemenswaardig beïnvloedde. De hoogste licht-geïnduceerde inactivatie was zichtbaar bij *E. coli* op een oppervlakte van beton (pH 9.6) met een inactivatiesnelheid van 1.85 h^{-1}. Faag MS2 had de hoogste licht-geïnduceerde inactivatie op asfalt met een snelheid van 1.3 h^{-1}. Inactivatie van *B. subtilis* sporen werd niet waargenomen, zowel onder lichte als onder donkere omstandigheden. Over het algemeen was op alle oppervlakten de volgorde van de licht-geïnduceerde inactivatie van langzaam naar snel: *B. subtilis* sporen < MS2 phage < *E. coli*. De uitzondering was asfalt, waar licht-geïnduceerde inactivatie van *E. coli* en MS2 hetzelfde was en waar de inactivatie van *B. subtilis* sporen het traagst was.

Het onderzoek in dit proefschrift laat zien dat het mogelijk is om na een stedelijke overstroming, met behulp van de zonlichtomstandigheden, de concentratie van indicator-organismen, TSS en de typen overstroomde oppervlakten, in te schatten wat gebeurt met een aantal categorieën uit water afkomstige pathogenen. De waarnemingen en resultaten gepresenteerd in dit onderzoek zouden, gecombineerd met Quantitative Microbial Risk Assessment (QMRA) en het in kaart brengen van stedelijke oppervlakten, gebruikt kunnen worden om beleid te ontwikkelen voor een snelle uitvoering van geschikte maatregelen om na een overstroming volksgezondheidsrisico's te beperken.

ΠΕΡΙΛΗΨΗ

Την τελευταία δεκαετία (2009-2019), πλημμυρικά φαινόμενα προκάλεσαν τουλάχιστον 48.000 θανάτους και επηρέασαν πάνω από 697 εκατομμύρια ανθρώπους παγκοσμίως. Οι αριθμοί αυτοί αναμένεται να αυξηθούν ως συνέπεια της κλιματικής αλλαγής, της αύξησης του πληθυσμού και της αστικοποίησης. Οι πλημμύρες δύνανται να προκαλέσουν λοιμώξεις λόγω της απελευθέρωσης παθογόνων μικροοργανισμών που μεταδίδονται μέσω του νερού από υπερφορτωμένα παντορροϊκά αποχετευτικά δίκτυα και άλλες πηγές κοπρανώδους μολύνσεως. Παθογόνοι μικροοργανισμοί μπορούν να προκύψουν σε επιφάνειες που απαντώνται συχνά στο αστικό περιβάλλον, όπως σκυρόδεμα, άσφαλτος, πλάκες πεζοδρομίου και γρασίδι.

Σκοπός της παρούσας έρευνας ήταν η βαθύτερη κατανόηση της επίδρασης διαφορετικών περιβαλλοντικών συνθηκών στη συγκέντρωση παθογόνων μικροοργανισμών που μεταδίδονται μέσω του νερού σε διαφορετικές αστικές επιφάνειες και επιφάνειες αναψυχής ύστερα από αστικές πλημμύρες. Επίσης, για την αξιολόγηση της συγκέντρωσης οργανισμών-δεικτών που μεταδίδονται μέσω του νερού σε αστικές επιφάνειες εκτεθειμένες σε πλημμύρες είναι απαραίτητη η ταυτοποίηση της πιο αξιόπιστης μεθόδου για την ανάκτηση αυτών των οργανισμών.

Η μελέτη της φυσικής αδρανοποίησης του βακτηρίου-δείκτη κοπρανώδους μολύνσεως *Escherichia coli* στην υδατική φάση έλαβε χώρα σε ανοιχτό αναδευόμενο αντιδραστήρα, υπό ελεγχώμενη έκθεση σε προσομοιωμένη ηλιακή ακτινοβολία, μιμούμενη την επίδραση διαφορετικών γεωγραφικών πλατών και εποχών, και με διαφορετικές συγκεντρώσεις ολικών αιωρούμενων στερεών (TSS), αντιστοιχούντων σε διαφορετικά επίπεδα αραίωσης και απορροών μετά από πλημμύρες. Ο συντελεστής ρυθμού αδρανοποίησης, k (d^{-1}), βρέθηκε να αυξάνεται γραμμικά με την αύξηση της έντασης της φωτεινής ακτινοβολίας, I ($W.m^{-2}$), και της διάρκειας έκθεσης στο φως, T ($h\ d^{-1}$), ($k=k_D+0.03 \cdot I$ και $k=k_D+0.65 \cdot T$, αντιστοίχως) και με τη μείωση της συγκέντρωσης των TSS ($mg.L^{-1}$), με εκθετική σχέση ($k=k_D+14.57e^{-0.02 \cdot [TSS]}$), χωρίς να έχει λάβει χώρα σύνδεση των βακτηρίων στα στερεά σωματίδια. Ο συντελεστής ρυθμού αδρανοποίησης πρώτου βαθμού σε σκοτεινές συνθήκες, $k_D=0.37\ d^{-1}$, εκφράζει την επίδραση καταπονήσεων εκτός του φωτός.

Πραγματοποιήθηκε σύγκριση τεσσάρων διαφορετικών μεθόδων δειγματοληψίας για την ανάκτηση μικροβιακών δειγμάτων από σκυρόδεμα, άσφαλτο, πλάκες πεζοδρομίου και γρασίδι: λήψη επιχρισμάτων, άμεση επαφή άγαρ, συμπίεση και χρήση κολλητικής ταινίας. Οι επιφάνειες εμβολιάστηκαν με δεδομένες συγκεντρώσεις *E. coli*, ενώ γυάλινη επιφάνεια χρησιμοποιήθηκε ως επιφάνεια ελέγχου. Η μέθοδος άμεσης επαφής άγαρ παρουσίασε το υψηλότερο λογαριθμικό ποσοστό ανάκτησης, 96,1% στο γυαλί, για συγκεντρώσεις *E. coli* έως 10^3 $CFU.100\ cm^{-2}$, αλλά αυτή η μέθοδος ανιχνεύει περιορισμένο εύρος βακτηριακών συγκεντρώσεων επειδή δεν είναι δυνατή η αραίωση ή συμπύκνωση των δειγμάτων. Η λήψη επιχρισμάτων ήταν η πιο αξιόπιστη μέθοδος επειδή ανιχνεύει μεγάλο εύρος συγκεντρώσεων με υψηλά λογαριθμικά ποσοστά ανάκτησης, έως 96,2% για συγκεντρώσεις *E. coli* 10^5 CFU.100 cm^{-2}. Οι έμμεσες μέθοδοι με συμπίεση και χρήση κολλητικής ταινίας δεν εμφάνισαν συγκριτικά πλεονεκτήματα.

Περαιτέρω πειράματα με λήψη επιχρισμάτων έδειξαν πως νερό συσσωρευμένο σε τραχείες επιφάνειες επηρέασε τα ποσοστά ανάκτησης λόγω της αραίωσης των δειγμάτων. Η λήψη επιχρισμάτων δειγμάτων μεγαλύτερου όγκου απ' ό,τι μπορεί να απορροφήσει η κεφαλή (0,15 mL) παρουσίασε μειωμένα ποσοστά ανάκτησης. Επίσης, η λήψη επιχρισμάτων ήταν πιο αποδοτική χωρίς τη χρήση απορρυπαντικού (Tween 80) στα διαλύματα διαβροχής και εκλούσεως. Μετά την ύφεση τεχνητής πλημμύρας, η χρήση επιχρισμάτων και η άμεση επαφή άγαρ επαληθεύτηκαν ως αξιόπιστες μέθοδοι δειγματοληψίας και απαρίθμησης *E. coli* σε διαφορετικές αστικές επιφάνειες, με λογαριθμικά ποσοστά ανάκτισης μεταξύ 21,0% και 59,0%, αναλόγως όχι μόνο της μεθόδου δειγματοληψίας, αλλά και της συγκέντρωσης βακτηρίων στις επιφάνειες.

Η αδρανοποίηση του βακτηρίου-δείκτη κοπρανώδους μολύνσεως *E. coli* μελετήθηκε υπό ελεγχόμενη έκθεση σε προσομοιωμένη ηλιακή ακτινοβολία σε διαφορετικές επιφάνειες που απαντώνται στο αστικό περιβάλλον: χαλίκι, άμμο, άσφαλτο, πλάκες πεζοδρομίου, σκυρόδεμα, ελαστικές πλάκες παιδικής χαράς και γρασίδι, με τη χρήση γυαλιού ως επιφάνειας ελέγχου. Οι επιφάνειες εμβολιάστηκαν με τεχνητά πλημμυρικά ύδατα περιέχοντα *E. coli* σε συγκέντρωση 10^5 CFU.mL^{-1}. Περιοδική δειγματοληψία έλαβε χώρα με τη λήψη επιχρισμάτων μετά τη μείωση του επιπέδου του νερού. Τα αποτελέσματα έδειξαν πως η αδρανοποίηση σε σκοτεινές συνθήκες δεν ήταν στατιστικώς σημαντική για καμία επιφάνεια και συνάγεται ότι η χημική σύνθεση και το pH (κυμαινόμενο από 6,5 έως 9,2) δεν επηρέασαν σημαντικά το ρυθμό αδρανοποίησης βραχυπρόθεσμα. Οι υψηλότεροι ρυθμοί αδρανοποίησης *E. coli* υπό την επίδραση φωτός μετά την ύφεση των πλημμυρικών υδάτων, οι οποίοι παρατηρήθηκαν στις ελαστικές πλάκες (>3,46 h^{-1}) και την άσφαλτο (2,7 h^{-1}), αποδόθηκαν στο θερμικό σοκ και στην απώλεια της υγρασίας των επιφανειών.

Η αδρανοποίηση *E. coli*, σπόρων *Bacillus subtilis* και βακτηριοφάγων MS2, οι οποίοι όλοι είναι υποκατάστατα διαφορετικών ομάδων παθογόνων, μελετήθηκε υπό ελεγχόμενη έκθεση σε προσομοιωμένη ηλιακή ακτινοβολία. Σκυρόδεμα, άσφαλτος και πλάκες πεζοδρομίου εμβολιάστηκαν με τεχνητά πλημμυρικά ύδατα περιέχοντα αυτούς τους οργανισμούς. Η έρευνα έλαβε υπ' όψιν το pH του νερού στις διαφορετικές επιφάνειες και το ρόλο του pH στην αδρανοποίηση των οργανισμών. Τα αποτελέσματα έδειξαν πως η αδρανοποίηση σε σκοτεινές συνθήκες δεν ήταν στατιστικώς σημαντική για κανένα οργανισμό και καμία επιφάνεια και συνάγεται ότι το pH (κυμαινόμενο από 7,0 έως 9,6) από μόνο του δεν επηρέασε σημαντικά το ρυθμό αδρανοποίησης βραχυπρόθεσμα. Ο υψηλότερος ρυθμός αδρανοποίησης (1,85 h^{-1}) υπό την επίδραση φωτός παρατηρήθηκε στο *E. coli* σε επιφάνεια σκυροδέματος (pH 9,6). Ο βακτηριοφάγος MS2 παρουσίασε τον υψηλότερο ρυθμό αδρανοποίησης υπό την επίδραση φωτός στην άσφαλτο (1,3 h^{-1}). Δεν παρατηρήθηκε αδρανοποίηση των σπόρων *B. subtilis* σε καμία από τις επιφάνειες, τόσο σε σκοτεινές όσο και σε φωτεινές συνθήκες. Γενικά, η αδρανοποίηση υπό την επίδραση φωτός ακολούθησε την αύξουσα σειρά σπόροι *B. subtilis* < βακτηριοφάγοι MS2 < *E. coli* σε όλες τις επιφάνειες εκτός από την άσφαλτο, όπου το *E. coli* και ο MS2 αδρανοποιήθηκαν με ίδιο ρυθμό.

Από την έρευνα προκύπτει πως γνωρίζοντας τις συνθήκες ηλιακής ακτινοβολίας ύστερα από αστικές πλημμύρες, τη συγκέντρωση οργανισμών-δεικτών και TSS, και το είδος των πλημμυρισμένων επιφανειών, είναι δυνατόν να εκτιμηθεί η τύχη επιλεγμένων παθογόνων που μεταδίδονται μέσω του νερού. Οι παρατηρήσεις και τα αποτελέσματα που παρουσιάζονται στην παρούσα μελέτη, σε συνδυασμό με Ποσοτική Εκτίμηση Μικροβιολογικής Επικινδυνότητας (QMRA) και χαρτογράφηση των αστικών επιφανειών, μπορούν να χρησιμοποιηθούν για την ανάπτυξη εργαλείων χάραξης πολιτικής για την ταχεία υλοποίηση κατάλληλων μέτρων για την άμβλυνση κινδύνων δημόσιας υγείας ύστερα από πλημμύρες.

CONTENTS

1

INTRODUCTION

The present thesis is a contribution in addressing one of the most acute health problems of our days, namely the spreading of water borne diseases through floods, by studying the inactivation of indicator organisms in characteristic background surfaces of various urban environments. Considering the close relationship between floods and climate change, this work contributes also to the adaptation to climate change.

1.1 URBAN FLOODS AND PUBLIC HEALTH

Floods worldwide pose a range of threats to human life, health and livelihoods. In the last ten years (2009-2019), flooding has caused the death of more than 48,000 people, and affected over 697 million people. In 2010 alone, reported flood disasters killed over 8,000 people directly (EM-DAT, 2019). The occurrence of extreme rainfalls, their impacts, and phenomena like urban flooding are predicted to become more frequent and intense because of climate change, rapid urbanization, high population density and due to the failure and aging of the sewage infrastructure (Arnell and Gosling, 2016; ten Veldhuis et al., 2010). For instance, it has been projected that in Europe, 250,000-400,000 additional people will be affected per year by river flooding by the 2080s, if no effective measures are taken, which is more than double the numbers in the period 1961–1990 (Menne and Murray, 2013). In many parts of the world and in Africa in particular, most flood-related fatalities occur mainly due to enhanced vulnerability to disasters and lack of, or poor, flood management schemes. In developing countries the majority of flood-related deaths are caused by diarrhoea and other water-borne diseases, or from drowning and snake bites (Jonkman and Kelman, 2005; Jha et al., 2012).

Floods can be classified according to their (i) cause (high rainfall, tidal extremes, storm waves, lack of drains or drains blocked by waste and debris, land use changes like urban expansion or engineering works upstream, waste dumping, structural failure, etc.), (ii) nature (regularity, speed of onset, velocity and depth of water, spatial and temporal scale, etc.) and (iii) health outcomes (Ahern et al., 2005). The health impacts of floods (physical, chemical or biological) can be either short-term, with disease symptoms or mortality appearing during or immediately after the flood, or long-term, that usually appear later as a result of the damages caused to infrastructure (Alderman et al., 2012).

The deaths which are the most easily monitored and connected with floods are those related to physical health effects, which occur from drowning (two-thirds of direct deaths) or trauma in flash and coastal floods. Nonfatal injuries are, together with the exacerbation of chronic illness, the main cause of mortality among affected populations shortly after the flood. However, the identification and attribution of deaths to a single cause is often particularly difficult because of coexistence of many hazards. After hurricane Katrina/Rita in Greater New Orleans, 7,500 incidents of nonfatal "injuries" were recorded among residents and relief workers. It is believed that many more were affected. In general, little information is available on the occurrence of nonfatal injuries during floods because they are not always reported or related to floods. Although the international Emergency Events Database (EM-DAT) records such injuries, this data is much less robust than reports of fatalities (Ahern et al., 2005; Alderman et al., 2012). If handling of "injuries" and the data related to them is difficult in a developed country like the USA, it is easily understood that the situation in developing countries under flood crisis is in many cases totally out of control.

In terms of chemical health effects, the causal relationships between floods, pollution from pesticides and other agricultural and industrial wastes (dioxin, heavy metals, etc.) released during flooding, contamination from sewage and health outcomes are still inconclusive. Apart from diarrhoea, exposure to such contaminants is linked to cancer, cardiovascular, gastrointestinal, kidney, liver and neurological diseases (Euripidou and Murray, 2004). Chronic illness and related conditions such as cardiovascular disease, cancer, chronic lung diseases and diabetes can also be worsened by disasters, increasing a person's vulnerability to synergetic adverse health outcomes following floods (Alderman et al., 2012).

The research presented in this work focuses on water-borne pathogens. However, it is worth mentioning that, in parallel, vector-borne diseases also increase during floods because of increased exposure to vectors (mosquitoes), which breed in or close to stagnant or slow-moving, usually polluted waters and rodents transmitting pathogens, changes in their habitat and compromise in vector control programs during floods (Ahern et al., 2005; Alderman et al., 2012). There have been developments in mapping the current and potential future distribution of important disease vector species (Liu-Helmersson et al., 2019). Australia, Europe and North America are projected to have the largest increases in human exposure to vector mosquitoes due to climate change (Monaghan et al., 2018). For instance, in Europe the spread of vector mosquitoes may extend from the southern part eastward and northward due to climate change (Fischer et al., 2011; Roiz et al., 2011; Caminade et al., 2012).

Although flood-related mortality has been studied in both developed and developing countries, evidence about the effects of floods in public health is limited. Detailed data is limited because rigorous epidemiological studies of flooding are difficult to carry out, especially in developing countries where most of the affected populations live, and because it is difficult to quantify the true burden of ill health due to flood events, especially when most cases are not adequately investigated, classified and monitored (Alderman et al., 2012; Jha et al., 2012)

1.2 WATER-BORNE DISEASES RELATED TO FLOODS

Depending on the source of contamination, urban flood water quality varies. Flood water flowing as a result of high surface runoff in heavy rainfall may contain suspended and dissolved particles, as well as faeces from animals and birds (de Man et al., 2014). In addition, several urban surfaces such as streets and pathways get contaminated with human faeces from combined sewer overflows (CSOs) during extreme events (ten Veldhuis et al., 2010). As a result, numerous human and animal pathogens present in faeces may also be present in urban flood water. Although immediately after floods outbreaks rarely occur, despite the high risks of transmission of communicable diseases

(WHO, 2019), flooding events are statistically associated with disease, and one of the potential sources of this link is combined sewer overflows (CSOs) (Curriero et al., 2001).

Evidence about risks from climate change with respect to infectious diseases is still limited (Semenza and Menne, 2009; Randelph and Rogers, 2010; Semenza et al, 2012; Kovats et al., 2014) and the surveillance of health effects of disasters remains inadequate (Fewtrell and Bartram, 2001; Kovats et al., 2014). Case studies from Africa and Asia show that floods and other events linked to climate change may exacerbate the risk of water-borne infectious diseases (Cissé, 2019). Diarrhoeal diseases have been found to increase during higher than average rainfall and associated flooding in Kenya (Okaka and Odhiambo, 2018). In high-income countries like Norway, the United States, former Czechoslovakia, the risk of faecal-oral diseases is mainly associated with cases of residing in flooded dwellings and contact with flood waters (Ahern et al., 2005). The risks are high when infrastructure, water supply systems and drinking water facilities are seriously damaged and when people have to leave their dwellings. Contamination from the handling of bodies of diseased people and animals can also be a risk if correct precautions are not taken. In general, the extent of transmission of diseases and the risk of epidemics after floods depend on population density and displacement, as well as the extent to which the natural environment has been altered or disrupted. Common diseases resulting from water contamination include cholera, diarrheal disease, hepatitis A and E, leptospirosis, parasitic diseases, rotavirus, shigellosis and typhoid fever. Deaths related with many of these diseases can occur during a relatively long period following the reported flood and are not always recorded in disaster databases (Alderman et al., 2012; Jha et al., 2012).

Pathogens can gain access to the human body through the gastrointestinal tract, the respiratory tract, or the skin, through wounds and abrasions (Bitton, 1999). Direct contact and ingestion of contaminated water (either flood water or contaminated tap water) is very common during severe flood events. During disasters, electricity cut offs often take place. Most of the drinking water supply systems are pressurized and even in the gravity systems, the pipes are pressurized. However, when there is no electricity, the pumps are not functioning, which results into intrusion of contaminated ground and flood water in the water supply systems, which are often not appropriately flushed and disinfected before being used again. Inappropriate cross-connections between sewage and water supply networks can also take place (Laine et al., 2011). The faecal-oral transmission cycle of pathogens is presented in Fig. 1.1 and the most common flood-borne pathogens and related human diseases transmitted by water are presented in Table 1.1.

Wound infections, dermatitis, conjunctivitis and ear, nose and throat infections are the most usual water-borne diseases related to floods (Alderman et al., 2012). During and after floods there is a higher risk for gastrointestinal diseases like diarrheal disease, cholera and norovirus-based gastroenteritis. The main reasons for this are poor hygiene and sanitation conditions, inadequate provision of clean drinking water, over-crowding and resettlement, contact with flood water and consumption of crops grown on soil which was contaminated during floods with wastewater from municipal and livestock operations (Wakuma Abaya et al., 2009). For some diseases such as hepatitis A and E it is known that they are transmitted through the faecal-oral route by ingestion of contaminated food or water. Person-to-person transmission is rare. Outbreaks of hepatitis E after floods, which often affect large populations, is frequent in areas where the virus is endemic and the spread of the disease is mainly attributed to the contamination of water sources (Watson et al., 2007; Alderman et al., 2012).

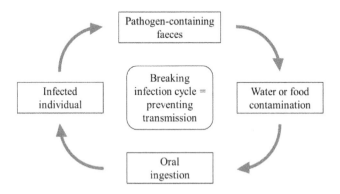

Figure 1.1 Faecal-oral transmission cycle of pathogens (Strande et al., 2014).

A common type of infectious diseases occurring after floods including upper respiratory tract infections with flu-like symptoms (throat infections, coughs and general symptoms, earache, skin rashes, inflammatory dermatoses and infectious skin conditions) common among inhabitants and construction workers are attributed to damp buildings (Noe et al., 2007; Carroll et al., 2010), due to growth of indoor mould, although their role in this process is not well understood. Conditions found in urban environments and surfaces, such as those of flooded buildings of various types, particularly under dark and poorly aerated conditions can support the existence and growth of mould, bacteria, protozoa and algae. Microbial contaminants grow, persist and produce toxins in these places. In addition, indoor humidity and temperature affect the transmission of aerolised respiratory viruses (Taylor et al., 2011). Different species of mould can impact public health either through direct respiratory infection, generation of a harmful immune response or severe reactions when mycotoxins are ingested. Measurements of the change in the ratio of indoor microbial levels before and after floods show that the 2009 flood in Taiwan resulted in an important increase in the concentrations of indoor fungi, some of which, e.g. *Aspergillus versicolor* are associated with negative health outcomes (Hsu et al., 2011; Alderman et al., 2012).

Table 1.1. Potential flood-related pathogens and associated human diseases in parentheses below the respective pathogens (Pond, 2013; Taylor et al., 2011; Gloyna, 1971).

Bacteria	Protozoa	Viruses	Fungi	Helminths
Aeromonas spp.	*Acanthamoeba* spp.	Adenovirus	*Candida albicans*	*Ascaris*
Campylobacter spp.	*Cryptosporidium*	Coxsackievirus	*Candida parapsilosis*	*Fasciolopsis buski* (fasciolopsiasis)
Enterococcus spp.	*Entamoeba* spp. (amoebic dysentery)	Echovirus (aseptic meningitis)	*Torulopsis glabrata*	*Schistosoma* spp. (schistosomiasis)
Escherichia coli (gastroenteritis)	*Giardia duodenalis*	Enterovirus		
Helicobacter pylori	*Giardia lamblia*	Hepatitis A virus		
Legionella spp. (legionnaires disease)	*Naegleria fowleri*	Norovirus		
Leptospira spp. (leptospirosis)		Parvovirus		
		Rotavirus		
Listeria monocytogenes (listeriosis)				
Mycobacterium spp.				
Pseudomonas spp.				
Salmonella spp. (typhoid fever)				
Shigella spp. (shigellosis)				
Vibrio spp. (cholera)				
Yersinia enterolitica (enteric yersiniosis)				

Leptospirosis, an acute febrile illness, is contracted through direct contact of human skin with areas contaminated with the urine of infected rodents. It is the only flood-related water-borne disease proven to be epidemic by the World Health Organization. Outbreaks of leptospirosis after flood events have been observed globally in urban and rural areas, especially in Latin America and Asia and with higher risks in highly populated areas with suboptimal drainage, low-lying areas and small island states. An example was the outbreak of leptospirosis in the Philippines after the devastating cyclones of 2009, with 2,158 confirmed cases and 167 deaths reported. Timely diagnosis and appropriate antibiotic therapy are necessary to avoid the progression of disease, system failure and fatalities (Gaynor et al., 2007; Jha et al., 2012).

One of the major sources of pathogens related to water-borne diseases, along with open defecation and contamination of water supply sources with faecal sludge, is urban wastewater. Wastewater is a mixture of natural and man-made organic and inorganic substances discharged into the sewers from households (domestic/sanitary wastewater, also called sewage), from industries (industrial/trade wastewater), from roads and roofs, as well as groundwater infiltrated into sewers. The quantity and quality of wastewater in sewers depends on the time of day and the season of the year, the infiltration and surface run-off, as well as the per capita water usage in an area, local habits and diet. In average, 99.9% of the volume of wastewater is water. The remaining part (0.1%) consists of faeces, food particles, grease, oils, soap, salts, metals, detergents, plastics, sand and grit, which can be converted by wastewater treatment into a manageable sludge, while leaving only a small portion in the final effluent (0.003% dry solids) disposed of in the environment (Gray, 1989). Wastewater needs to be treated for a series of reasons. First of all, the inactivation of pathogens is crucial for the protection of public health and hygiene. Also, the removal (and recovery) of water and waste materials present in the wastewaters can prevent the appearance of adverse ecological effects. The disposal of treated wastewater should also happen without nuisance or offence to the society and the environment. Therefore, it is important to recycle and recover nutrients and other valuable components of wastewater. The purpose of modern treatment plants is to achieve all the above mentioned goals in an economically feasible and sustainable way, according to legal environmental and public health standards (Gray, 1989) and in accordance to the sustainable development goals (SDGs) (UN, 2015). In most of the developing countries sewage is still discharged "raw", without any treatment or after often inadequate pre-treatment, while 2.4 billion people have no access to improved sanitation facilities (UNICEF and WHO, 2015), despite the progress made in the efforts to achieve the Millennium Development Goals (MDGs) and the ongoing ones to achieve the relevant SDGs.

Sewer networks can be either combined or separate. Combined sewers collect sanitary drainage from households, industrial sewage and surface runoff waters in a single pipe (or channel) system. In this case, during wet weather conditions, if the amount of wastewater exceeds the hydraulic capacity of the wastewater treatment plant (WWTP), wastewater (as a mixture of urban/industrial drainage and precipitation runoff) is either stored temporarily within the catchment or is directly discharged into the receiving water as CSO without any treatment, bypassing the WWTP (Rauch et al., 2002). Although urban runoff dilutes the pollutants included in the household and industrial drainage, CSOs contain significant loads of pathogens, next to various pollutants, as toxic substances and persistent materials (heavy metals, polycyclic aromatic hydrocarbons (PAHs), etc.). In addition, runoff water often contains pollutants and pathogens that have been accumulated on catchment surfaces and/or small pools during dry weather and washed off during rainfall events (Rauch et al., 2002). A study has shown that the probability of developing gastrointestinal diseases from incidentally ingesting water near CSOs has a range of 0.14-0.70 over the course of a year for people coming in contact with it, associated with the presence of faecal indicators *Streptococcus* and *Enterococcus* (Donovan et al., 2008).

A significant volume of the flow is also likely to emerge on urban surfaces such as streets, pavement tiles, soil and playgrounds, through surface discharge (surcharge) of wastewater out of manholes. Thus, urban and recreational surfaces can present a potential reservoir/source of pathogens (Shah et al., 2011). Urban areas have been identified as one of the main sources of faecal indicator organisms, with the highest values occurring for high-flow conditions during or after rainfall (Kay et al., 2008). Flood water mixed with sediments exhibited an increased abundance of putative pathogens after floods caused by Hurricane Harvey in Houston, USA, in both residential areas and public parks (Yu et al., 2018). After the Elbe river flood in 2002 in Germany, high numbers of pathogenic bacteria were detected in the mud, streets, playgrounds, and in the basement of flooded houses (Abraham and Wenderoth, 2005). In Toronto, Ontario, Canada, faecal indicator bacteria (FIB) have been reported in beach sand and in sand at playgrounds and sandboxes of urban settings and they were found to be detectable for several months (Staley et al., 2016). Concentrations of pathogens are often measured in samples from sewer flooding incidents. For instance, *E. coli*, intestinal enterococci and *Campylobacter* were found in samples from three sewer flooding incidents in the Netherlands (ten Veldhuis et al., 2010). Presence of pathogens like norovirus, rotavirus, enterovirus, *Giardia* oocyst, *Cryptosporidium*, *Campylobacter*, and *Salmonella* was reported in the canal and recreational lakes of Amsterdam, contaminated from CSOs (Schets et al., 2008). These are the most common pathogens causing gastrointestinal problems, even in the developed world (Wit et al., 2001). The estimated risk of infection in canals and recreational lakes in Amsterdam, polluted by combined sewers, raw sewage from houseboats and dog and bird faeces per exposure event has been measured around 0.0002%-0.007% with *Cryptosporidium* and 0.04%-0.2% with *Giardia* for occupational divers professionally exposed to canal water (Schets et al., 2008). Furthermore, possible associations between the occurrence of gastroenteritis and environmental strains of human enteric viruses in sewage in Tunisia was found using molecular detection and characterisation techniques (Sdiri-Loulizi et al., 2010). Sales-Ortells and Medema (2015) found a high concentration of *Campylobacter* at the Bellamyplein water square in Rotterdam, The Netherlands (cover image), serving with the dual function of recreational area in dry seasons and of storm water detention during extreme events. Rain simulation events with drinking water showed that the mean *Campylobacter* disease risk for children playing in the water square is $2.5 \cdot 10^{-3}$ in the presence of animal faeces contamination and $4.5 \cdot 10^{-2}$ per person per event in the presence of human faecal contamination, higher than the Dutch national incidence of *Campylobacter* disease. Also, a strong correlation between cholera outbreaks and flooding was reported in the urban area of Dhaka, Bangladesh (Reiner et al., 2012).

An important knowledge gap in the vast majority of cases is related to the lack of data about the time and conditions of the inactivation of pathogens in the environment, particularly on urban surfaces. This has triggered a major part of the research presented in this work.

1.3 INDICATOR AND SURROGATE ORGANISMS

Since the detection of individual pathogenic agents in wastewater is difficult, costly and time-consuming, indicator organisms are used to indicate the presence of faecal contamination and the potential presence of pathogens. Indicators are used to define the limits of pathogen concentrations above which a particular environment is susceptible to increased risk and contamination. Surrogates are organisms used as substitutes or models, imitating the growth and persistence of pathogens of interest in processes like disinfection, processing or other procedures (Busta et al., 2003).

Traditionally, microbial quality of water resources was assessed with the use of FIB only, mainly faecal coliforms and enterococci (Ashbolt et al., 2001). The indicator microorganism for which bacteriological analyses of water are most commonly performed is *E. coli*. *E. coli* can be found in the intestinal tract of humans, has been studied thoroughly and can grow easily under laboratory conditions (Gloyna, 1971). Prototype strains of *E. coli* (ATCC 25922, used in this study) and *Enterococcus faecalis* (UMRL 1053) are usually selected as representative FIBs (Fujioka and Yoneyama, 2002).

However, after the documentation of some outbreaks of cryptosporidiosis in waters which did meet the standards based on bacterial indicators (Ashbolt et al., 2001), it is assumed that *E. coli* is not a suitable indicator for environmentally stable human viral, spore and oocyst forming pathogens (Nieminski et al., 2010; Harwood et al., 2005; Bonadonna et al., 2002; Medema et al., 1997). Aerobic spores such as the *B. subtilis* spores used in Chapter 5 of this study have been proposed as suitable surrogates of *Cryptosporidium* because of their higher persistence to environmental extremes (Stelma, 2018; Bradford et al., 2016; Headd et al., 2016; Mazoua and Chauveheid, 2005; Rice et al., 1996). Furthermore, *B. subtilis* spores have been recommended as a surrogate for *Cryptosporidium* because they are safe, relatively cheap, can be measured easily and with high reproducibility and the recovery of spores from spiked water samples is acceptable. Also, they have similar shape with oocysts, although a bit smaller (US EPA, 2010).

Human enteric sewage-borne viruses, such as poliovirus, echovirus and coxsackievirus have been detected even in marine waters, causing diarrheal diseases among swimmers (Fujioka and Yoneyama, 2002). Faecal bacteriophages are not suitable as indicators of the presence of human enteric viruses because they are also present in animal faeces, and

9

because human enteric viruses have been detected also in water samples that do not contain bacteriophages (Brookes et al., 2004). However, bacteriophages can be used as surrogates of the behaviour of human enteric viruses, because of their similar size and morphology, and the low cost, the ease and the higher speed of detection compared to human enteric virus assays (Brookes et al., 2004; Dias et al., 2018). There is evidence that bacteriophage fate and transport in ambient waters may resemble that of viral pathogens more closely than FIB, suggesting that the former may be suitable surrogates under some environmental conditions (McMinn et al., 2017). Most studies until now have focused on coliphage systems and a range of bacterial hosts. The ideal host bacteria would be bacteria originating from human faeces only and lysed by phages that do not replicate in another host or the environment, such as *Bacteroides fragilis*, although the latter have low phage numbers for general use (Brookes et al., 2004). The MS2 bacteriophages used in this study (Chapter 5) are RNA coliphages infecting *E. coli* by attaching to the F-pili plasmids of the bacterial cells. They have been used in disinfection studies to model properties of human pathogens, especially noroviruses (Dawson et al., 2005).

It is important to note that although monitoring of microbial quality of flood water and of surface and recreational waters (e.g. in beaches and lakes) and the assessment of related public health risks from exposure to these waters is performed routinely on the basis of local and national guidelines (de Man et al., 2014; Mark et al., 2018; Sales-Ortells and Medema, 2015; ten Veldhuis et al., 2010), standard procedures and guidelines for FIB monitoring on urban and recreational surfaces (e.g. beach sand) are almost inexistent (Staley et al., 2016).

Finally, it is critical to have knowledge of the specific system of the site studied instead of relying on universal rules. In any case, the sampling must be coupled with deep understanding of the biological, physical and chemical factors which influence the processes of transport, distribution and inactivation of the pathogens (Brookes et al., 2004).

In this study, the inactivation of *E. coli,* the most common FIB, *B. subtilis* spores, chosen as a non-pathogenic surrogate for *C. parvum* oocysts and *Giardia* cysts, and *E. coli* bacteriophage MS2 as a surrogate for viral contamination was studied.

1.4 FACTORS AFFECTING THE SURVIVAL OF PATHOGENS

The spread of pathogenic microorganisms by contaminated flood waters is a serious global problem connected with the spread of diseases. The initial presence or absence of pathogens in flood waters and on surfaces after the recession of the waters does not provide a reliable direct indication of the risks to human health. The inactivation of the pathogens appear to play a crucial role in defining public health risks since most of the outbreaks of diseases occur within different time intervals after flooding events (Alderman et al., 2012; Du et al., 2010; Taylor et al., 2011).

1.4.1 Survival conditions in water

The inactivation of pathogens depends mainly on temperature, solar radiation and predation, and to a smaller extent on salinity and pressure. In order to understand the fate of pathogens and predict public health risks and safe conditions, it is necessary to link the timescales of hydrodynamic events with the timescales of pathogen inactivation (Brookes et al., 2004).

In a typical aqueous environment, enteric pathogens are not able to replicate, with the exception of some bacteria. Viruses and parasitic protozoa do not replicate because they rely on a specific animal or human host for replication. Infected hosts, with or without symptoms of illness, excrete these pathogens in large numbers. After entering the aqueous environment the pathogens are slowly being inactivated. Pathogen concentrations can increase when the water flow rate increases, due to the resuspension of settled pathogens. Sedimentation of the protozoa during periods of low flow rates is stronger than settling of viruses because protozoa are much larger than viruses. This means that during rainfall the protozoa that are settled in the sewer system get mobilised in addition to those coming from the direct discharge of raw wastewater (Schijven and de Roda Husman, 2005).

Cryptosporidium parvum is capable of surviving for long periods of time in water. The inactivation of *C. parvum* in natural river water follows first-order kinetics with rates ranging from 0.013 to 0.039 $\log_{10}$ per day. Inactivation in natural river water has been found higher than in synthetic hard water (Weber and Rutala, 2001). Retention time and contamination by water flow were considered to be the major factors influencing these rates (Brookes et al., 2004).

Suspended matter in water allows the encasement (adsorption/absorption) of microorganisms, offering protection from environmental stressors, such as grazing, adverse hydrodynamic conditions and irradiance, by offering shading (refraction, reflection and/or scattering). It can also offer increased access to nutrients of the particles or enhance settling (Walters et al., 2014). Faecal indicator bacteria have shown affinity for fine particle attachment, which increases their survival and deposition rate. Around 30-55% of faecal indicator organisms can be found attached to particles in storm water runoffs (Characklis et al., 2005). The fraction of microbes associated with settleable particles varies according to the microorganism types. The percentage is around 40% for faecal coliforms, *E. coli* and enterococci, 65% for *Clostridium perfringens* spores and 13% for total coliphages (Krometis et al., 2007). Bacterial mortality due to predation or environmental exposure has been shown to be reduced when cells attach to particles, as particles settle faster (Fries et al., 2008).

Also it is important to note that microorganisms can survive in nutrient-rich sediments for long periods of time. The ability of microorganisms to survive in aquatic sediments implies that faecal coliforms detected in the water column may not always indicate recent contamination but may be the result of sediment resuspension (Davies et al., 1995).

The present section elaborates on the inactivation of indicator organisms in the water column, without any sediment, but only with suspended solids in order to assess light attenuation and inactivation. The level of attachment to the suspended solids is therefore also tested.

1.4.2 Survival conditions on different urban surfaces

Water-borne outbreaks are an acute aftermath of flood disasters, mainly as a result of contaminated drinking water supply and poor access to waste management and healthcare services (Brown and Murray, 2013), but little is known about the concentration of pathogens on different surfaces after the recession of floodwater (Hellberg and Chu, 2016). Several studies have shown that water-borne pathogens from human and animal sources are present on flooded surfaces (Andrade et al., 2018; Rui et al., 2018; Sales-Ortells and Medema, 2015; ten Veldhuis et al., 2010) in concentrations higher than prior to flooding (Eccles et al., 2017). For example, a five-fold increase in the levels of *E. coli* was observed in the Salinas Valley, California, USA during months with a high water flow and frequent flooding downstream (Cooley et al., 2007).

Despite the obvious importance of the issue, most studies until now were based on case-specific events, making it very difficult to have a broader understanding of the mechanisms of inactivation of pathogens. Research on rainwater harvesting showed that the microbial communities collected from different roofing materials such as concrete tiles, acrylic-surfaced bituminous membrane, grass, steel, and asphalt fiberglass shingle were not the same as in ambient rainwater and varied depending on the material used (Bae et al., 2019). The effect of pH is presented separately below, but it is important to note that the age of concrete surfaces affects their pH. Carbonation, the process of hydroxide anions of calcium hydroxide being replaced over time by carbonate anions results in the formation of calcium carbonate and decrease of pH (Pade and Guimaraes, 2007).

Literature about survival of pathogens on surfaces other than agricultural soils or food industry and hospital environments is very limited. For example, a suspension of *C. parvum* oocysts on glass slides and air-dried at room temperature showed a 97% loss of viability after 2 hours and full inactivation after 4 hours (Robertson et al., 1992). More prolonged survival, up to 72 hours, has been reported when *Cryptosporidium* in a diarrheal stool was applied on a wooden surface (Weber and Rutala, 2001). In order to obtain a more realistic account of the overall health risks, it is necessary to assess the critical factors influencing the viability of pathogenic microorganisms in the urban environment, such as different environmental conditions and the properties of different surfaces. The scope of this thesis is exactly to focus on this research gap. Chapters 3, 4 and 5 elaborate on this topic.

1.4.3 Effect of temperature

The survival of pathogens such as *E. coli*, that may grow in natural waters, outside the animal host, is affected by temperature fluctuations in two ways: first by the growth rate, and secondly, by the rate of inactivation. When temperature is increased, growth is favoured and thermal and light inactivation as well. This double effect explains why water temperature variation between 12 °C and 40 °C does not seem to be important in the inactivation of bacteria, although inactivation does happen at higher temperatures (Wegelin et al., 1994). In contrast, the inactivation of viruses has been reported to increase steadily over a range from 20 °C to 50 °C (Reed, 2004). At the same time, there is more complication due to the lower solubility of oxygen at higher temperatures; this leads to less than expected light inactivation in increased temperatures (Giannakis et al., 2014).

The survival of pathogens that do not grow in natural waters when present outside of the animal host, such as *C. parvum*, is reduced as temperature increases, by a higher/faster inactivation. The effect of temperature on *Cryptosporidium* infectivity and/or viability can be described by a first-order inactivation function:

$$C(t) = C_0 e^{-kt} \tag{1.1}$$

$$k = 10^{-2.68} 10^{0.058T} \tag{1.2}$$

where t is time in days, C is the concentration of viable oocysts, k is the inactivation rate (dimensionless) and T is the temperature in Celsius degrees (Brookes et al., 2004).

The ability of oocysts to initiate an infection is related to finite carbohydrate energy reserves in the form of amylopectin, which is metabolised in direct response to the temperature. Inactivation at higher temperatures is a result of increased oocyst metabolic activity, with a close relationship between infectivity and the level of ATP in the oocyst (Peng et al., 2008).

Data suggests that temperature affects the rate of *C. parvum* inactivation regardless of whether the oocysts are freely suspended or bound within a matrix of faecal or organic material (Brookes et al., 2004). However, in faeces and soil the degradation can be more rapid because of microbial antagonism and/or predation (Brookes et al., 2004).

In the case of viruses, inactivation is observed at temperatures higher than 50 °C (Bertrand, 2012). According to WHO (2004), most of the virus proteins get destroyed when heated above 60 °C, except the lyophilized ones, which can endure up to 80 °C.

The behaviour of microorganisms in soil is much more complex than in water, due to the simultaneous presence of gaseous, water and solid phases, numerous chemicals, microorganisms, macrofauna and plants. Various physical properties which can vary over small distances are also important factors affecting the behaviours of microorganisms. For these reasons the inactivation rates of pathogens determined for different soils show a wide variation with respect to temperature (Peng et al., 2008).

1.4.4 Effect of pH

Most microorganisms show stability and optimal growth near neutral pH (7.0) (Sinclair et al., 2012). *E. coli* cannot grow at alkaline pH and it gets inactivated above pH 9.3 (Parhad and Rao, 1974). High pH induces bacterial inactivation either by enhancing the generation of reactive forms of oxygen (hydrogen peroxide, superoxide and hydroxyl radicle) that are lethal to bacteria or by reducing their ability to resist the sunlight effects (Curtis et al., 1992) or disruption of the internal pH by weak bases like ammonia.

To describe the effect of pH on the maximum specific growth rate $k_{max}(pH)$ (h^{-1}) the Cardinal pH Model equation (CPM, Eq.2, Rosso et al., 1995) can be used (Eq. 1.3).

$$k_{max}(pH) = k_{opt} \cdot \gamma_{pH}(pH)$$
$$= k_{opt} \cdot \frac{(pH - pH_{min}) \cdot (pH - pH_{max})}{(pH - pH_{min}) \cdot (pH - pH_{max}) - (pH - pH_{opt})^2} \qquad (1.3)$$

where $\gamma_{pH}(pH)$ (-) is the reduction of the maximum specific growth rate due to a deviation from the optimal pH (pH_{opt}) at which the maximum specific growth rate is equal to the optimal growth rate k_{opt} (h^{-1}). pH_{min} is the minimum pH below which growth is not possible and pH_{max} is the maximum pH beyond which growth is not possible.

For most enteric viruses, the optimal pH is between 3 and 9 (Melnick et al., 1980).

1.4.5 Effect of light

Sunlight has been identified as one of the predominant factors leading to inactivation of enteric microorganisms in surface waters (Schultz-Fademrecht et al., 2008; Burkhardt et al., 2000; Davies-Colley et al., 1999; Garvey et al., 1998). Inactivation rates of bacteria and bacteriophages under solar radiation are more than 10 times higher compared to a dark environment. The sensitivity of microorganisms, including viruses, to electromagnetic radiation depends on the wavelength and on the species of organisms and is described by biological weighting functions (Silverman and Nelson, 2016). The relative susceptibility to light, is higher for enterococci, followed by faecal coliforms, *E. coli*, somatic coliphages and F-RNA phages have the lowest inactivation (Sinton et al., 2002).

Although bacteria have greatest sensitivity to short wavelengths of light (Reed, 2004), the short wavelength light is faster attenuated in the water column having a smaller fraction of the incident radiation (Silverman and Nelson, 2016). Therefore, the lethal effect of sunlight on bacteria is more important in wavelengths higher than 320 nm (Oppezzo, 2012). These wavelengths can cause indirect (photo-chemical, particularly photo-oxidative) damage, when absorbed by photosensitiser macromolecules within microbial cells (endogenous, such as porphyrins, bilirubin or chlorophyll) or in the surrounding water (exogenous, such as dissolved organic matter or humic acids) (Maraccini et al., 2016).

Cryptosporidium oocysts are susceptible to solar inactivation and UV-B wavelengths have been experimentally identified as the most germicidal ones affecting the oocysts' infectivity in environmental waters. UV-B radiation is detrimental also to a wide range of organisms including bacteria, fungi, plants and annelids (King et al., 2008). In general, inactivation is more acute at shorter wavelengths, but the greatest inactivation occurs at full sunlight. Although the UV-B component of sunlight may directly damage biomolecules, such as DNA, UV and visible light more often cause indirect damage, when they are absorbed by photosensitiser molecules inside or around microbial cells (Reed, 2004). These molecules are raised by light into an excited state and can subsequently react with cellular biomolecules or with molecular oxygen (Reed, 2004). When reacting with oxygen, various reactive oxygen species (ROS) are produced, such as singlet oxygen, superoxide, hydroxyl radicals and hydrogen peroxide. ROS react with DNA, proteins and cell membrane components, leading to the inactivation of cells (Reed, 2004). Therefore, high levels of dissolved oxygen (DO) in water lead to increased light inactivation rates, up to 4-8 times higher compared to deoxygenated water, which shows that photo-oxidation is the principal reason for the rapid decrease in bacterial counts in water of low turbidity (Reed, 2004).

The wide range of wavelengths inactivating enterococci and F-RNA phages means that the effect of solar radiation on these microorganisms is photo-oxidative damage. On the other hand, the inactivation of faecal coliforms and somatic coliphages by a more specific range of wavelengths of UV-B shows that the effect of solar radiation on these microorganisms is photo-biological damage (Sinton et al., 2002; King et al., 2008). *E. coli* strains are slightly more resistant to death by solar radiation than other bacteria (*P. aerugenosa, S. flexneri, S. typhi,* and *S. enteritidis*). *E. coli* strains therefore serve as indicators to assess the effects of solar radiation on enteric bacteria. Past experiments showed a 3-log reduction of *E. coli* after 75 minutes of exposure under test conditions. However, experiments showed that *Str. faecalis* is slightly more resistant than *E. coli* (Wegelin et al., 1994).

The concentration of bacterial cells also affects the sensitivity to solar radiation. Very high concentrations show lower sensitivity compared to low or moderate density, because of shielding from one cell to another (Giannakis et al., 2014). Ultraviolet (UV) radiation represents around 3-4% of incoming solar (<2,800 nm) radiation. The hourly average incoming shortwave radiation around midday, at mid-latitude, during summer is 1,000 W.m^{-2}. This corresponds to around 30 W.m^{-2} of UV radiation and to an hourly UV dose of 360,000 mJ.cm^{-2}, which is much greater than typical water treatment dosages (20-120 mJ.cm^{-2}) (Brookes et al., 2004).

Oocyst inactivation due to UV exposure follows an exponential inactivation function similar to temperature:

$$C(t) = C_0 e^{-k_{UV}H} \qquad (1.4)$$

$$H = I_{UV}t \qquad (1.5)$$

where $C(t)$ is the concentration of viable oocysts at time t, H is UV dosage (intensity integrated over time), I_{UV} is light intensity, k_{UV} is the inactivation coefficient of UV light, which is approximately equal to 1.2 (Brookes et al., 2004).

Particulate matter in water often results in lower inactivation rates of microorganisms due to shielding of harmful UV rays, but the degree to which particle association can impact light inactivation of enteric microorganisms in surface waters is not yet clear (Walters et al., 2014). The vertical distribution of UV light is described by an exponential function equivalent to Beer's law:

$$I_{UV(z)} = I_{UV(surface)}e^{-\mu z} \qquad (1.6)$$

where μ is the attenuation coefficient of UV light, which is closely related to dissolved organic carbon (DOC) and has been measured at values between 0.1 and 40 m^{-1}. Given the wide range of μ values, the penetration depth of UV light (the depth reached by only 1% of surface irradiation) can range considerably (e.g. between 0.1 and 46 m) (Brookes et al., 2004).

Therefore, the impact of UV light on oocyst viability is governed by a double exponential function, which means that there is a sharp transitional depth in the water column between the area where UV light has a dramatic effect and the area where its effect is insignificant. This transitional depth varies a lot from one reservoir to another, depending on DOC (Craik et al., 2001; Brookes et al., 2004).

The underwater light field is influenced by the properties of the incoming irradiance and the optical properties of the aquatic medium, which can be divided into inherent properties and apparent properties (Baffico, 2013). Inherent properties, such as the absorption coefficient, the scattering coefficient etc., depend on the nature and composition of suspended particles and dissolved compounds. The apparent properties depend on the medium and on the geometrical structure of light and one example of such properties is the angular distribution of the underwater radiation (Baffico, 2013).

The presence of particulate matter in water reduces considerably the transmission of sunlight, either by shading (refraction, reflection or scattering of light) or by encasement. This in turn results in lower inactivation rates of the microorganisms. Many studies report a direct correlation of particle size and/or concentration on UV disinfection efficiency of secondary wastewater effluents, but more research needs to be done to understand to which degree they are correlated (Walters et al., 2013).

One of the most important parameters in characterising the attenuation (scattering and absorption) of UV light within solids is the attenuation coefficient. It represents the decline in light as a function of depth as a result of both scattering and absorption and it is defined by the following equation:

$$K = \frac{\log\left(\frac{E_1}{E_2}\right)}{z_1 - z_2} \tag{1.7}$$

where K is the attenuation coefficient (μm^{-1}), E_1 and E_2 is the scalar irradiance at depth one and depth two, respectively ($mW.cm^{-2}$) and z_1 and z_2 is the depth at point one and point two within the particle, respectively (μm). The attenuation coefficient equals the absorption coefficient when the medium strictly absorbs. Once the absorption coefficient is determined for the particle solids, light intensity can be calculated for any given depth into the solid (Loge et al., 1999).

1.5 SAMPLING METHODS

Most methods for recovery of microorganisms from surfaces found in the literature are designed for smooth and non-porous surfaces, such as stainless steel surfaces used in the food industry (Garayoa et al., 2207; Maunula et al., 2017; Moore and Griffith, 2007), and in healthcare environments (Rawlinson et al., 2019; Claro et al., 2015). In the food industry, methods and techniques to recover microorganisms from surfaces have been developed, namely non-microbiological surface sampling, including visual assessment, ATP bioluminescence, protein and other assays, and microbiological surface sampling, i.e. indirect methods such as swabbing, sponges and wipes, direct methods such as contact plates, dip slides, petrifilm, rollers and agar sausages, and molecular methods (Griffith, 2016). However, parameters such as the diversity of experimental conditions or samples hinder the choice of the best method (Griffith, 2016). Hospital environments in Europe are usually sampled only in response to outbreaks, but guidelines on monitoring do not provide with microbiological protocols (Otter et al., 2015). Sampling becomes even more complicated for urban surfaces such as concrete, pavement and asphalt that are rough and porous and for which there is no standardised procedure in the literature.

Table 1.2. Surface sampling methods, their typical applications, their main advantages and limitations and indicative references.

Methods		Typical applications	Advantages	Limitations	References
Direct agar contact (printing)	Contact agar plates (RODAC)	Small, smooth, nonporous, flat or gently curved surfaces	Simple, no processing needed	Sampling area limitations, not possible to count high concentrations, colonies and aggregations	Lemmen et al. 2001 ; Tamburini et al., 2015
	Dip slides / Petrifilm				
	Roller				Lutz et al., 2013
	Agar syringe/sausages				
Pouring agar (Direct Surface Agar Plating, DSAP)		Nonporous	High recovery (no transport issues)	Fixed surfaces, contamination, coalescence of colonies, incubation temperature	
Indirect contact	Stamp	Polyurethane foam cylinders	Autoclavable, "dilution" of high concentrations by serial impressions		Tresner and Hayes, 1970
	Adhesive tape/sheet	Mainly to examine samples by microscopy		Difficulty to recover from tape	Urzì et al, 2001; Yamaguchi et al., 2002
Swabbing	Rayon tipped swab	Smooth, nonporous, or crevices and corners	Good for small areas, corners, crevices	Sampling area limitations	Moore and Griffith, 2007; Ismaïl et al., 2013; Lahou and Uyttendaele, 2014
	FLOQSwab				(Finazzi et al., 2016)
Friction/ scrubbing	Sponges / Wipes	Smooth, nonporous	Easier to cover larger areas.	Need of handles or sterile gloves for each sample.	
	Brushes				Ismaïl et al., 2015
Vacuum methods	Vacuum filter sock (HEPA)	Porous or nonporous, smooth or rough	Multiple surface types	Large volume of dust makes detection more difficult, need for power source, desiccation	
	Microvacuum cassette				
	Wet vacuum	Nonporous, smooth or rough	Direct collection, less desiccation		
Bulk sample	Immersion/Rinsing	Porous or nonporous, smooth or rough	Multiple surface types	Need of sterile tools, no standard method for processing, difficult interpretation of results	Boehm et al., 2009; Taylor et al., 2013; Martínez-Bastidas et al., 2014; Staley et al., 2016
	Scraping, grinding				Ismaïl et al., 2015
	Chopping grass				Cao et al., 2018
	Soil Sprinkling				
Radiometry		Bacteria tagged with radioisotopes	Sensitive and accurate with most surfaces	Not possible in the field	

Most methods referring to porous and rough surfaces include either the collection of bulk samples of the material with methods such as scrapping and grinding or collection of water samples after immersion or rinsing. For example, Taylor et al. (2013) acquired bacterial samples from a 0.1% Tween solution after immersing and stirring in brick, plaster and wood coupons. Several recovery methods and techniques exist for bacteria and fungi contained in soil or rock. Non-destructive methods such as washing, scraping, printing off of surface microorganisms, or *in situ* tests of physiological activity by applying fluorogenic substrate analogues for the study of monuments and destructive methods like rock-coring or other method for aseptic collection of samples, can be followed by fragmentation to release the microorganisms to study natural rock deterioration by microorganisms (Hirsch et al., 1995). *In situ* microscopy can also contribute to demonstrating the presence of microorganisms on stone, but this does not allow precise identification of the organisms, for which molecular biology techniques such as high throughput sequencing and metagenomics are necessary (Mihajlovski et al., 2015).

Swabbing is used for the recovery of microbial cells, especially from hard-to-reach and irregular surfaces, but although the procedure and data evaluation are standardised and inexpensive, this method is also prone to significant variations due to multiple factors such as the size and roughness of the area swabbed, the material of the swab head, the wetting solution and extraction protocol, operator contributions, ambient temperature and humidity, as well as the morphological and physiological properties of the microorganisms (Digel et al., 2018).

According to Moore and Griffith (2007), the effectiveness of the swabbing technique depends on the percentage of bacteria that the swabs can collect from the sampled surface, on how many of these bacteria can be removed from the swabs and on the cultivation of these bacteria. The authors studied separately the parameters that affect the removal of bacteria from the surface (swab type, swab absorbency, wetting solutions affecting viability, de-clumping and detachment, wet or dry surface) by applying agar on the surfaces before and after swabbing and the parameters that affect the release of bacteria from the swab (swab type, wetting solutions) by directly inoculating swabs with a pipette. They found that swabs coated with a brush-textured nylon tip in combination with non-growth enhancing wetting solution allowed the highest recovery from wet surfaces, but noted that this combination may not always be ideal.

For any method, a variety of recovery ratios may be obtained. In an optimum method, the tool, the reagents and the equipment must produce the same end result even if they are used by different operators under the same sampling conditions (Ismaïl et al., 2013).

1.6 PROBLEM STATEMENT

The spread of diseases during and after flood events is largely connected with the spread of pathogenic microorganisms from CSOs and surcharged combined sewers, especially in poorly managed systems. Even in developed countries when recreational activities take place in the storm water detention and retention basins, there is a high risk for inhabitants, especially children to be infected. The literature review has indicated that more research needs to be undertaken in areas concerning the inactivation of pathogenic organisms after urban floods because the presence or absence of pathogens immediately after floods does not provide a direct indication of risk to human health. Some of the pathogens are inactivated by the influence of sunlight, while others are protected, depending on the surface type and its characteristics. The different parts of the system, their interactions and their relationships are illustrated in Fig. 1.2.

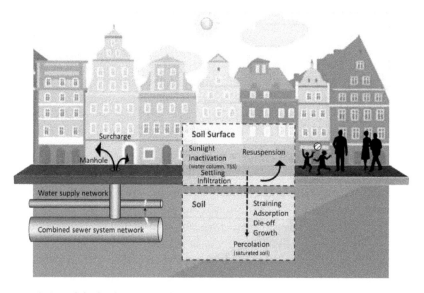

Figure 1.2. Simplified schematic of boundaries of research focus of this thesis.

Although sunlight-mediated inactivation is considered as the main cause of disinfection in waste stabilisation ponds, with mechanisms depending on the characteristics of the sunlight and its capacity to penetrate through water, as well as pH, temperature, nutrients available, salinity and DO (Dias et al., 2017), research on the survival of pathogens and indicator organisms on different urban surfaces is limited. Most studies are based on case-specific events, making it very difficult to have a broader understanding of all the complex processes taking place and to quantify the associated public health risks in the hours and days following the flooding events and the recession of flood waters.

In order to obtain a more realistic assessment of the overall health risks and to determine when access into flooded areas is safe, it is necessary to understand the critical factors influencing the inactivation of pathogens on different urban surfaces, for which literature is very limited. Laboratory experiments on different surfaces under controlled conditions of intensity and duration of sunlight and concentration of suspended solids can improve the understanding of the inactivation of indicator organisms. These results can be further applied to provide predictive information for effective public health management.

1.7 RESEARCH OBJECTIVES

1.7.1 Overall aim

The aim of this research is to contribute to the reduction of public health risks related to floods and caused by water-borne pathogens resulting from the surcharge of combined sewer systems. The main environmental factors that affect the inactivation of indicator organisms on different urban surfaces will be assessed. Since floods are increasingly connected to climate change, this research should be considered also as a contribution to climate change adaptation.

1.7.2 Research questions

The key research questions which occur based on the literature review are the following:

1. What is the most reliable method for collecting samples of faecal indicator bacteria from different urban surfaces after the recession of flood water?

2. What is the effect of solar radiation and total suspended solids on the inactivation of indicator organisms that originate from the surcharging of combined sewer systems?

3. What are the effects of solar radiation on the inactivation of different indicator organisms (originated from the surcharging of combined sewer systems) on different urban surfaces?

1.7.3 Research hypotheses

The following research hypotheses correspond to the research questions set above:

1. The most reliable method for collecting samples of faecal indicator bacteria from urban surfaces is swabbing.

2. Increased exposure to solar radiation increases inactivation of surrogate microorganisms for enteric pathogens in water. Therefore, increased suspended

solid concentrations reduce the inactivation rates due to limited penetration of light in the water column.

3. The pH of different surfaces affects solar inactivation of surrogate microorganisms for enteric pathogens on these surfaces.

1.8 RESEARCH APPROACH

In order to achieve the overall aim and to address the research questions, this thesis uses simulated laboratory experiments with certain indicator organisms to study processes that attempt to mimic natural phenomena taking place in the water column during floods and on urban surfaces after the recession of flood water. The thesis has been structured in six chapters (Fig. 1.2), including an introduction, four research phases and an outlook.

Chapter 1 is a general introduction on the background of the study, including a literature review relevant to the problem statement of this research.

In the first research phase (Chapter 2) the effect of artificial solar radiation on the inactivation of indicator bacteria *E. coli* was studied in a water column to mimic a water column occurring during a flood. This phase took place in a relatively deep vessel/unit to assess the effect of different total suspended solids' concentrations on the light attenuation and on the inactivation of the indicator organisms in the water column, corresponding to the phase when urban surfaces are still flooded.

In the second research phase (Chapter 3) the most reliable method was identified for sampling water-borne indicator *E. coli* from flood-prone urban surfaces where pathogens can occur after the surcharge of sewer networks: concrete, asphalt, pavement blocks and grass. Different sampling methods were compared: swabbing, direct agar contact, stamping and adhesive tape-lifting. The swabbing method was selected as the most appropriate one for the collection and enumeration of *E. coli* in samples from different surfaces in the next research phase.

In the third research phase (Chapter 4) the inactivation of indicator bacteria *E. coli* on different urban surfaces was studied. These were typical surfaces which are found in urban and sub-urban open areas and storm water detention and retention basins: asphalt, concrete, pavement blocks, gravel, sand, rubber tiles and grass, with glass as control. The experiments took place on the surfaces after the recession of artificial flood waters, mimicking the post-flood phase when the surfaces start drying after urban floods and people start using them for transport, recreational purposes etc. This allowed to better understand the effects of pH and other surface characteristics on bacterial inactivation.

The fourth research phase (Chapter 5) studied the inactivation of *E. coli*, *B. subtilis* spores, and bacteriophages MS2 under controlled exposure to simulated sunlight and to real sunlight. Concrete, asphalt and pavement blocks, and glass as control, were inoculated with artificial flood water of 1cm height to control the temperature and to keep the surfaces saturated, compared to the previous phase when drying was studied.

The outlook (Chapter 6) summarises and draws conclusions on the knowledge gained from the above studies. It also gives recommendations and prospects for future research.

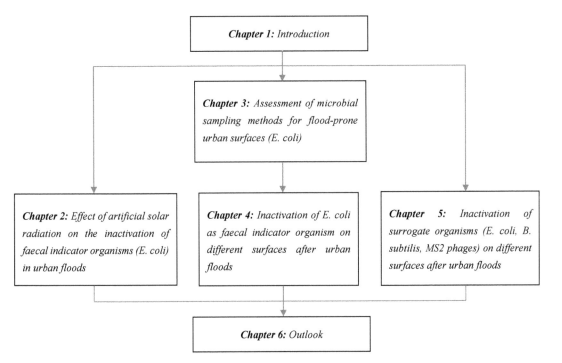

Figure 1.3 Structure of this PhD thesis.

2

EFFECT OF ARTIFICIAL SOLAR RADIATION ON FAECAL INDICATOR BACTERIA AFTER URBAN FLOODS

This research contributes to a better understanding of how the concentration of water-borne pathogens in contaminated shallow water bodies is affected by different environmental conditions. The inactivation of faecal indicator bacteria *E. coli* was studied in an open stirred reactor, under controlled exposure to simulated sunlight, mimicking the effect of different latitudes and seasons, and different concentrations of total suspended solids (TSS) corresponding to different levels of dilution and runoff. While attachment of bacteria on the solid particles did not take place, the inactivation rate coefficient, k (d^{-1}), was found to depend on light intensity, I ($W.m^{-2}$), and duration of exposure to sunlight, T ($h.d^{-1}$), in a linear way ($k=k_D+0.03 \cdot I$ and $k=k_D+0.65 \cdot T$, respectively) and on the concentration of TSS ($mg.L^{-1}$), in an inversely proportional exponential way ($k=k_D+14.57e^{-0.02 \cdot [TSS]}$). The first order inactivation rate coefficient in dark conditions, $k_D=0.37$ d^{-1}, represents the effect of stresses other than light. This study suggests that given the sunlight conditions during an urban flood, and the concentration of indicator organisms and TSS, the above equations can give an estimate of the fate of selected pathogens, allowing rapid implementation of appropriate measures to mitigate public health risks.

This chapter is based on: Scoullos, I.M., Lopez Vazquez, C.M., van de Vossenberg, J., Hammond, M. and Brdjanovic, D.: Effect of artificial solar radiation on the inactivation of pathogen indicator organisms in urban floods. Int. J. Environ. Res. 13, 107-116, doi:10.1007/s41742-018-0160-5, 2019. (IF=1.5)

2.1 INTRODUCTION

In this chapter, the effects on the inactivation of *E. coli* under various light conditions and in the presence of different concentrations of suspended solids were examined in an open reactor using an artificial light source with a spectrum similar to solar light. The experimental setting allowed to simulate different solar light intensities and durations that can occur at different times of the year and at different places around the globe. These conditions can be applied to all kinds of shallow water bodies, including multifunctional urban flood retention and detention basins often used as sport facilities or playgrounds during dry weather. Contamination could originate from wastewater from surcharged combined sewers, pit latrines, cross connections, open defecation, animal litter, or municipal and livestock operations. The selection of the light intensity and duration parameters was based on the conditions at two different latitudes: the equator and 60°. The equator zone was selected because it gets the maximum light intensity at a stable daytime length of 12 h throughout the year. 60° N and S were selected because of the daylight duration of 6 and 18 h respectively in winter and summer solstice in the North and the opposite in the South.

2.2 MATERIALS AND METHODS

2.2.1 Experimental reactor

Batch experiments took place in an open cylindrical stirred reactor of 40 cm internal diameter and 80 cm height. For each batch the reactor was filled with 75 L of demineralised water (up to around 60 cm); suspended solids and *E. coli* were added, according to each experimental phase. Continuous stirring at 95 rpm (IKA® dual-speed overhead stirrer, Staufen, Germany) was used to avoid settling. For maintaining the temperature of the reactor adequately constant at around 25 °C (fluctuating between 20 and 28 °C) and to neutralise the heating effect of the lamp, the reactor was placed in a larger tank containing water as a buffer while a ventilator was also employed. The experimental setup can be seen in Fig. 2.1. Each batch experiment lasted around one week, depending on the inactivation rate and until the microorganisms were not fully detected.

Figure 2.1. The experimental setup used.

2.2.2 Indicator organisms

E. coli ATCC 25922 was chosen as the faecal indicator bacterium of study, because it has been studied thoroughly and can be grown easily under laboratory conditions. Before each batch experiment *E. coli* was incubated in 1.3% *w/v* Oxoid CM0001 Nutrient Broth (Oxoid Ltd., Basingstoke, United Kingdom) solution in Erlenmeyer flasks for 24 h at 37 °C. After incubation the concentration of the inoculum was around $3 \cdot 10^9$ CFU.mL^{-1}. The inoculation took place in the reactor by adding 280 mL of inoculum in demineralised water, to achieve an initial concentration of *E. coli* of around 10^7 CFU.mL^{-1}. The initial concentration of each batch was measured by sampling 15 minutes after the inoculation to allow full mixing. The selection of the initial concentration of 10^7 CFU.mL^{-1} was based on the typical high concentrations of *E. coli* in raw municipal wastewater of around 10^6 CFU.mL^{-1} according to Mark et al., (2015) or $5 \cdot 10^6$ CFU.mL^{-1} when taking into account minor contributions of industrial wastewater (Henze and Comeau, 2008).

The enumeration of *E. coli* in the samples was performed by counting the number of colony forming units (CFU) on Chromocult® Coliform Agar (CCA) (Merck KGaA, Darmstadt, Germany) plates after 24 h of incubation at 37 °C. Appropriate 10-fold dilution steps in 0.1% peptone physiological salt solution were used. The plates were spread in triplicate.

27

2.2.3 Suspended solids

To study the effects of different total suspended solids (TSS) concentrations, soil was used obtained from a previously flooded excavation site in Delft, The Netherlands. The soil was dried, ground and sieved. The particles selected for the experiment after sieving varied between sizes of 38 and 106 μm (silt and fine sand). This fraction corresponds to frequently found particle sizes in urban environments, for example around 28% of the particle size distribution that was measured in a parking lot and at a residential basin (Selbig et al., 2016) and to around 50% of particles measured in asphalt road runoff (Charters et al., 2015). Subsequently, the solids were heated in a muffle furnace at 520 °C for 6h in order to remove all organic compounds. The solids were added in the reactor in each batch experiment prior to the inoculation. The studied TSS concentrations varied from 0 to 200 mg.L^{-1}, because 200 mg.L^{-1} is the reference value of US EPA for event mean concentration of TSS for urban runoff, while the similar ones according to NURP and to USGS and NPDES are lower, at 174 and 78.4 mg.L^{-1}, respectively (Acharya et al., 2010). The measurement of TSS was performed in triplicate with vacuum filtration using Whatman® Grade 1 (GE Healthcare, Little Chalfont, UK) filter papers (11 μm pore size), so that solid particles with attached bacterial cells are retained while non-attached bacteria can pass through. Prior and after the filtration the filter papers were dried at 105 °C for 2 hours and the weight was determined.

2.2.4 Light source and parameters

Simulated sunlight was produced using an OSRAM HQI-BT 400 W/D PRO metal halide lamp with built-in UV filter at around 320 nm. Apart from a few peaks in the area of visible light (with the highest one at 540 nm, see Fig. 2.2) the spectrum of these metal halide lamps is continuous (Calin and Parasca, 2008). The light intensity was regulated by changing the distance of the light source from the water surface. The duration of light/darkness hours was set with a timer. The photon flux, in μmol m^{-2}.s^{-1}, was measured with a LI-250A Light Meter equipped with an underwater quantum sensor (LI-COR® Biosciences, Inc., Lincoln, Nebraska, USA). This sensor has a uniform sensitivity in the range of wavelengths between 400 and 700 nm; this was taken into account when calculating the lamp irradiance spectrum. The measurements were performed in duplicate, in different sides of the reactor.

Figure 2.2 Relative spectral power distribution of the metal halide lamp used (OSRAM GmbH, Munich, Germany).

The total light intensity, I (W.m^{-2}), in the area of wavelengths (320-805 nm) for which the spectral power distribution of the lamp is provided (Fig. 2.2), at any location where the sensor is placed, was calculated with the Planck relation, taking into account the spectral power distribution of the lamp, using Eq. 2.1, adapted from Silverman and Nelson (2016):

$$I = \Sigma_{\lambda=320}^{805} I(\lambda) = \varphi \cdot h \cdot c \cdot N_A \cdot 10^3 \cdot \frac{\Sigma_{\lambda=320}^{805} M(\lambda)}{\Sigma_{\lambda=400}^{700}[M(\lambda)\cdot\lambda]} \qquad (2.1)$$

where φ is the photon flux (μmol m^{-2}.s^{-1}) measured by the sensor in the area of 400-700 nm (presented in Fig. 2.3), h is Planck's constant (6.63·10^{-34} J.s), c is the speed of light (3·10^8 m.s^{-1}), N_A is Avogadro's constant (6.022·10^{23} mol^{-1}), λ is wavelength (nm), $I(\lambda)$ (W.m^{-2}) is light intensity at a specific wavelength and $M(\lambda)$ (%) is the relative power of the lamp at a specific wavelength as provided by the relative spectral power distribution.

2.2.5 Experimental design

Assessment of light attenuation

In order to address the phenomenon of light attenuation and calculate the light attenuation coefficient in the different conditions of the experiments, the photon flux was measured at different depths, with increments of 5 cm, in the reactor filled with demineralised water inoculated with *E. coli* and with the addition of soil particles of different concentrations (0, 25, 50, 100, 150 and 200 mg TSS.L^{-1}).

Effect of light intensity, duration of exposure to light and different concentrations of TSS on the viability of E. coli

The effects of different light intensities (190 and 320 W.m^{-2}, for 12 hours per day, without any solids added), periods of exposure to light (0, 6, 12, 18 and 24 hours per day, under 320 W.m^{-2}, without addition of any solids) and different concentrations of TSS (0, 25, 50, 100, 150 and 200 mg.L^{-1}, for continuous exposure to light of 320 W.m^{-2} for 24 hours per day) on the inactivation of *E. coli* in demineralised water were studied. The viability of the organisms was also studied in dark conditions as a control.

Assessment E. coli attachment on the suspended particles

The attachment of *E. coli* on the suspended particles and particle-related shielding was tested in a batch experiment in Erlenmeyer flasks in an orbital shaker (150 rpm) with 200 mg TSS.L^{-1} of solids in 250 mL of demineralised water by filtering samples of 25 mL and enumerating the bacteria in the filtered and unfiltered samples, in triplicate, at time zero and throughout the duration of the experiment. The soil particles used in the experiment were all selected, using sieving, to be larger than 38 μm and smaller than 106 μm. The bacteria that could have been adhered or attached to particles were retained together with the particles on the filter paper of 11 μm. A *t*-test for paired samples was performed to determine whether the concentrations of *E. coli* before and after filtration are likely to have come from distributions with equal population means. In addition to that, samples were taken from three different depths in the reactor; close to the surface (2 cm), middle (30 cm) and close to the bottom (58 cm).

Comparison of results by testing the inactivation of E. coli in artificial floodwater

The inactivation of *E. coli* was studied in artificial floodwater. Domestic wastewater was used in order to mimic flood water. It was diluted 10 times with demineralised water to simulate dilution by storm water runoff. The wastewater was collected from the influent stream of Harnaschpolder wastewater treatment plant, Den Hoorn, The Netherlands, after the first screening filters. Before dilution, the wastewater had a concentration of 29.6 mg TSS.L^{-1} and 8.5·10^4 *E. coli* CFU.mL^{-1}. In order to study separately the effect of the wastewater's suspended solids, the experiment was performed by removing the wastewater's TSS with filtration using Whatman® GF/C (GE Healthcare, Little Chalfont,

UK) filters (1.2 µm pore size) and replacing the TSS with 150 mg.L^{-1} of the treated soil particles, as described before. *E. coli* was added in the reactor prior to the experiment to achieve an initial concentration of 10^7 CFU.mL^{-1}. Wastewater quality has a high variability and the dilution factor from storm water increases the uncertainty even more. Therefore, the scope of this paper is limited in comparing demineralised water with diluted wastewater, both with the addition of soil particles. A comparison is made between reported characteristics of flooding water and the synthetic flood water used in this research (see Table 2.1.).

Table 2.1 Comparison of TSS and E. coli reported concentrations in different waters with the synthetic flood water used in this research.

Water quality	TSS (mg.L^{-1})	*E. coli* (CFU.ml^{-1})	References
Raw municipal wastewater with minor contributions of industrial wastewater	250 (low) 300 (medium) 600 (high)	10^4 (low) 5·10^6 (high)	Henze and Comeau, 2008
Raw municipal wastewater		10^6 (high)	Mark et al., 2015
Raw influent Harnaschpolder	29.6	8.5·10^4	This research
Artificial flood water	0-200	10^7	

2.2.6 Sampling and physicochemical parameters of study

Samples of 20 mL were taken from the reactor at periodic intervals; (i) approximately every 12 hours when the light was on, (ii) at the beginning and end of each dark and light phase during the experiment of light duration, (iii) every 24 hours for the dark control. The samples were taken from the top of the reactor with a 30 mL syringe. For the experiments involving suspended solids where it was important to assess stirring efficiency and sedimentation, samples were also taken from the middle and close to the bottom of the reactor by connecting a metallic tube to the syringe. The measurement of pH took place at the same intervals, with a handheld pH-meter (WTW pH 323, WTW GmbH, Weilheim, Germany). Temperature and DO were measured every hour (fluctuating between 20 and 28 °C and between 6.5 and 8.5 mg O$_2$.L^{-1}, respectively, data not shown) with a handheld DO-meter (WTW Oxi 3310, WTW GmbH, Weilheim, Germany).

2.2.7 Data analysis

Light attenuation due to different concentrations of suspended solids was addressed by measuring the vertical distribution of light in air and water with different concentrations of solids. This was described using an exponential function equivalent to Beer's law (Eq. 2.2), which states that the light intensity decreases exponentially as a function of depth:

$$I_{(z)} = I_{(surface)} e^{-\mu z} \quad (2.2)$$

where, z is the depth (m), μ is the light attenuation coefficient (m^{-1}), $I_{(z)}$ is light intensity (W.m^{-2}) at depth z and $I_{(surface)}$ is the light intensity on the surface ($z=0$ m) (Kirk, 1994; Morris et al., 1995).

In all the experiments where the inactivation of *E. coli* was addressed, the inactivation rates were calculated using the first order exponential inactivation "Chick-Watson" model (Eq. 1.1). In all experiments the inactivation was considered to start at time 0 with no lag period. Time 0 was considered as 15 minutes after inoculation to allow full mixing in the reactor before measuring C_0. For each batch experiment the first-order inactivation rate was assumed to be constant.

Maraccini et al. (2016) and Davies-Colley et al. (2000) used the following linear model (Eq. 2.3) to express the total inactivation under light conditions as the sum of the photo-inactivation plus the "dark" inactivation, due to stresses other than light, such as osmotic stress, predation and starvation:

$$k = k_D + k_L \cdot I \quad (2.3)$$

where k_D is the first order dark inactivation rate constant (d^{-1}), k_L is a pseudo second order photo-inactivation rate constant (m^2 W^{-1}.d^{-1}) and I is light intensity (W.m^{-2}). In the present study, a similar expression was also presented for the relation of k to the duration of exposure to light (T). Subsequently, k_D was calculated experimentally as the mean of the values corresponding to the intercept of linear regressions, for zero intensity and exposure to light.

2.3 RESULTS AND DISCUSSION

2.3.1 Light attenuation

The vertical distribution of light intensity followed an exponential pattern, which corresponds to Beer's law, with values of $R^2 > 0.98$ for all the different TSS concentrations studied in the reactor (see Fig. 2.3).

Figure 2.3 The values of photon flux measured by the sensor at different depths in the reactor filled with demineralised water (DW) or artificial flood water (FW), with E. coli and different concentrations of TSS. The light attenuation coefficients, μ, obtained from these curves are also presented.

The light attenuation coefficient was proportional to the concentration of suspended solids and Eq. 2.4 was determined from a linear regression with $R^2=0.94$:

$$\mu=2.59+0.02\cdot[TSS] \quad (2.4)$$

where μ is the light attenuation coefficient (m^{-1}) and $[TSS]$ is the concentration of total suspended solids ($mg.L^{-1}$). The minimum theoretical value of μ, corresponding to pure water is 2.59 m^{-1}.

2.3.2 Inactivation of *E. coli* under different light intensities

The inactivation of *E. coli* was studied in the reactor under exposure to light for 12 h per day followed by 12 h in dark, without any solids added. The light intensities studied were 190 and 320 $W.m^{-2}$, with a control of 24 h in dark. The inactivation was described with exponential inactivation kinetics and the resulting inactivation rates are presented in

Table 2.2. The inactivation rate increases with an increase in light intensity and the data fit well (R^2=0.99) to the linear Equation 2.3 *with* k_D=0.19 d^{-1} and k_L=0.03 m^2.W^{-1}.d^{-1}.

2.3.3 Inactivation of *E. coli* under different periods of duration of exposure to light

The inactivation of *E. coli* was studied in the reactor, without any solids added, under exposure to light of 320 W.m^{-2} for 0, 6, 12, 18 and 24 h per day followed by 24, 18, 12, 6 and 0 h in dark, respectively per day. The inactivation was described with exponential inactivation kinetics for the overall curves and the resulting inactivation rates are presented in Table 2.2. The inactivation rate increases with an increase in exposure time and the data fit well (R^2=0.98) to the linear Equation 2.5:

$$k=k_D+k_L{'}{\cdot}T \quad (2.5)$$

where k_D=0.54 d^{-1} is the first order dark inactivation rate constant corresponding to 0 h of exposure to light, $k_L{'}$=0.65 h^{-1} is a pseudo second order photo-inactivation rate constant, and T is the duration of exposure to light per day (h.d^{-1}).

Table 2.2 Inactivation rates of E. coli in the reactor without addition of any solids, for different light intensities (with T=12 h of light per day) and for different periods of exposure to light (with I= 320 W.m^{-2}). Temperature fluctuated between 20 °C in dark conditions and 28 °C under light of 320 W.m^{-2}.

Different light intensities (*I*)		Different periods of exposure to light (*T*)	
I (W.m^{-2})	*k* (d^{-1}) (*R^2*)	*T* (h)	*k* (d^{-1}) (*R^2*)
0 (dark control)	0.08 (0.76)	0 (dark control)	0.04 (0.47)
190	5.66 (0.86)	6	5.07 (0.92)
320	8.91 (0.92)	12	8.91 (0.92)
		18	10.96 (0.95)
		24 (light control)	16.59 (0.99)

2.3.4 Determination of k_D

The inactivation rate coefficient in dark conditions, k_D, represents the effect of inactivation due to stresses other than light, such as osmotic stress, predation and starvation. Its theoretical value was calculated as the average of the intercepts resulting from the previous two linear regressions, $k_D=(0.19+0.54)/2=0.37$ d^{-1}. Although experimentally it was measured at a lower value in the dark control experiments conducted in this research (0.08 d^{-1} and 0.04 d^{-1}, respectively), the calculated value is broadly consistent with the average values of dark inactivation rates for *E. coli*, of 0.55 d^{-1} (ranging from 0.34 to 0.79 d^{-1}) in cold season and 0.79 d^{-1} (from 0.60 to 1.10 d^{-1}) in warm season reported by Maïga et al. (2009), with those of 0.48 and 0.55 d^{-1} at average temperature of 20 °C reported by Craggs et al. (2004), 0.41 and 0.55 d^{-1} calculated for waste stabilisation pond and raw sewage in fresh river water, respectively, by Sinton et al. (2002) and with 0.50 d^{-1} at around 24 °C, pH 8.0 and 7.5 $mg.L^{-1}$ DO in environmental water reported by Gutiérrez-Cacciabue et al. (2016). Although the calculated values of k_D are around 6 times higher than the measured values, the discrepancy is small in comparison to the high values of the inactivation rates under exposure to light, which affect the coefficients of the linear Equations 2.3 and 2.5.

2.3.5 Inactivation of *E. coli* under different concentrations of TSS

In order to assess the role of suspended solids in the inactivation of *E. coli*, a series of batch experiments took place using different concentrations of TSS. The results of this experimental phase can be seen in Fig. 2.4.

Compared to the experiment in absence of suspended solids ($k=16.59$ d^{-1}) it is clear from Fig. 2.4 that the inactivation rate decreased with increasing TSS concentration. With absence of suspended solids the inactivation curve was log_{10}-linear, which was not the case in the presence of particles. In fact, in presence of solids, especially over 25 $mg.L^{-1}$, there was a one-day slow phase, followed by more rapid inactivation in the next couple of days. For concentrations of 50 and 100 mg $TSS.L^{-1}$ the inactivation curves were almost identical, with inactivation rates of 5.21 and 5.78 d^{-1}, respectively, and with the levels of *E. coli* falling under 10^1 $CFU.mL^{-1}$ before the 4th day of the experiment. The steepest parts of these curves had inactivation rates of 9.24 d^{-1} for 50 mg $TSS.L^{-1}$ and 8.10 d^{-1} for 100 mg $TSS.L^{-1}$, which were comparable to the one without solids (16.50 d^{-1}) and to that of 25 mg $TSS.L^{-1}$ (11.03 d^{-1}). For solids' concentrations of 150 and 200 $mg.L^{-1}$ the rapid inactivation rates were lower (4.46 and 3.05 d^{-1}, respectively), followed by a plateau between day 3 and day 5, before the concentration of indicator organisms started falling more rapidly again, reaching levels below 10^1 $CFU.mL^{-1}$. The *E. coli* bacteria in the experiment came from a pure culture and the experiment was performed in separate batches for 150 and 200 mg $TSS.L^{-1}$, with sampling in triplicate; therefore, the observed plateau was obtained in a reproducible manner. Brouwer et al. (2017) also observed such a deviation from the simple exponential inactivation, and modelled that with biphasic inactivation dynamics (fast inactivation followed by a phase of slow inactivation) attributed to population heterogeneity, hardening off (a conversion of the pathogens to a hardier phenotype through altered gene expression or other mechanisms, when exposed to the environment), viable-but-not-culturable states and/or density effects. This raised the question of whether the bacteria were eventually attached to the TSS particles. This was examined in the experiments that follow.

Figure 2.4 Concentration, $\log_{10}$ inactivation and inactivation rate (k) of E. coli under exposure to artificial sunlight (320 W.m^{-2}, 24h per day) with TSS 0, 25, 50, 100, 150 and 200 mg.L^{-1}, compared to dark control (no light and no solids). These curves correspond to the samples taken at the top of the reactor prior to filtration. Temperature fluctuated between 20 °C at the beginning and 28 °C.

The inactivation rates were calculated assuming the traditional model of monophasic inactivation for the overall period of the experiment, including the lag phase and the plateau, where it exists. The correlation between TSS concentration and overall inactivation rate (k) can be seen in Fig. 2.5.

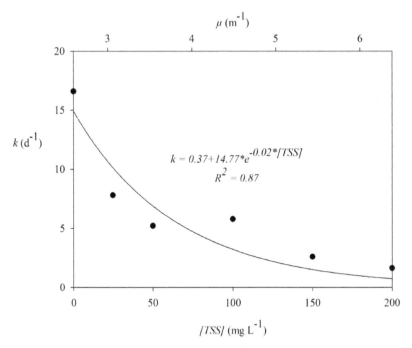

Figure 2.5 The effect of TSS concentration (mg.L⁻¹, lower horizontal axis) and of light attenuation coefficient μ (m⁻¹, upper horizontal axis) on inactivation rate k (d⁻¹) of E. coli measured in the reactor under continuous exposure to artificial sunlight (320 W.m⁻²). The coefficient μ was calculated based on Equation 2.5.

The inactivation rate k (d⁻¹) decreases with an increase in the concentration of TSS (mg.L⁻¹) in an exponential way ($R^2=0.87$) as described by the Equation 2.6:

$$k=k_D+14.77e^{-0.02\cdot[TSS]} \quad (2.6)$$

It can be assumed that for very high TSS concentrations light cannot penetrate in the water, so inactivation rate is similar to that of dark conditions, therefore the $k_D=0.37$ d⁻¹ parameter, as calculated previously, has been added to the exponential inactivation model.

The combination of this with the results of the first experiment lead to the formulation of an exponential relation (Eq. 2.7) between the inactivation rate k (d⁻¹) and the light attenuation coefficient μ (m⁻¹):

$$k=k_D+65.48e^{-0.65\mu} \quad (2.7)$$

for $\mu \geq 2.59$ m⁻¹, which is the minimum theoretical value corresponding to the light attenuation coefficient of water without any solids, as calculated previously. Again, the k_D parameter has been added because it is assumed that for very high values of μ the

37

inactivation rate is similar to that of dark conditions. This relation can be very useful as an indirect method for predicting the inactivation rate of *E. coli*-like pathogens in waters by measuring the vertical distribution of light, without the need of measuring the concentration of suspended solids or turbidity. More research on relevant methodologies has been conducted by Nguyen et al. (2015), Silverman and Nelson (2016) and presented in a critical review paper by Nelson et al. (2018).

At this point, it is interesting to note that although the irradiance of the UV-A band of wavelengths of this lamp accounts for 5.1% of the total irradiance, as measured from the relative spectral power distribution, biological weighting functions show that the inactivation effect of wavelengths higher than 400 nm on *E. coli* is negligible (Nelson et al., 2018). Therefore, as the lamp has a UV filter at 320 nm, we can assume that the total inactivation caused by this lamp corresponds to the UV-A part of the spectrum. The values of the average UV-A irradiance transmitted through the water column for the different TSS concentrations studied are presented in Table 2.3.

Table 2.3 Inactivation rates of E. coli in the reactor with demineralised water (DW) and artificial flood water (FW), at different concentrations of TSS. The average total irradiance, I, transmitted through the water column (calculated from Eq. 2.9, see Annex), as well as the average UV-A irradiance (320-400 nm) transmitted through the water column, $I_{UVA}=5.1\%\cdot I$, and the light attenuation coefficient, μ (measured values), are also presented.

[TSS] (mg.L^{-1})	I (W.m^{-2})	I_{UVA} (W.m^{-2})	μ (m^{-1})	k (d^{-1})	k (h^{-1})
0 (DW)	156.8	8.0	2.2	16.59	0.69
25 (DW)	125.3	6.4	3.3	7.81	0.33
50 (DW)	85.4	4.4	3.6	5.21	0.22
100 (DW)	65.7	3.4	4.9	5.78	0.24
150 (DW)	57.4	2.9	4.9	2.60	0.11
200 (DW)	37.3	1.9	6.4	1.63	0.07
150 (FW)	53.0	2.7	6.5	4.40	0.18

2.3.6 Attachment and particle-related shielding

The results can be seen in Fig. 2.6.

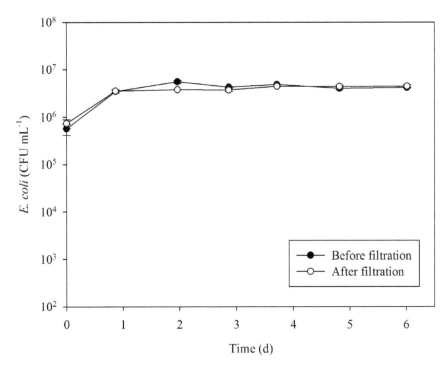

Figure 2.6 The concentration of E. coli in demineralised water with 200 mg TSS.L⁻¹,
before and after filtration (11 μm) in a batch experiment in Erlenmeyer flasks.

The value of t-statistic (0.93) was lower than the critical value (2.45), so the null hypothesis (that the mean difference in concentration between filtered and unfiltered samples was zero) was accepted. Therefore, it is concluded with 95% confidence that the amount of *E. coli* associated with the suspended solids was not significant.

The concentration of suspended solids was found to be higher near the bottom of the reactor than at the middle and at the top of it because of sedimentation, with some fluctuations over time. The stirring speed was not increased, to avoid the creation of a vortex that would affect the water surface and the uniform distribution of light in the reactor. This inevitably caused the formation of a vertical TSS concentration gradient. Even so, *E. coli* was found to be uniformly distributed at different depths in the reactor (data not shown). This was observed both before and after filtration of the samples, supporting the finding that significant attachment did not occur for any of the TSS concentrations studied.

The absence of irreversible adsorption may be attributed to short incubation time with clean particles, stirring velocity, grain characteristics (size, size uniformity and surface roughness), solution chemistry (zeta potential, ionic strength), geochemical heterogeneity and lipopolysaccharide composition in the bacterial outer membrane (Foppen and Schijven, 2006). According to Cantwell and Hofmann (2008) particle association affects the kinetics of inactivation by shielding the bacteria from exposure to light only when the number of particle-associated bacteria is significantly higher than the number of free-floating bacteria. The results indicate that also free floating bacteria could be protected for some time due to shading under high concentrations of TSS and this is a conclusion of practical importance when strategies are conducted and measures are taken for appropriate protection of population in cases of floods.

2.3.7 Comparison of results by testing the inactivation of *E. coli* in simulated floodwater

The inactivation rate of *E. coli* bacteria in the reactor filled with artificial floodwater containing 150 mg TSS.L^{-1} of treated solids and an initial *E. coli* concentration of $1.4 \cdot 10^7$ CFU.mL^{-1}, under continuous exposure to light of 320 W.m^{-2}, was k_{FW}=4.40 d^{-1}. The predicted value of the inactivation rate at 150 mg TSS.L^{-1} for the same light conditions, according to Eq. 2.7 would have been k_{DW}=1.52 d^{-1}. Part of this difference between the predicted and the measured value may be explained by the effect of factors other than light, related to the components of wastewater. Therefore, the experiment was repeated in dark conditions to identify the effect of dark inactivation. In dark conditions, the inactivation rate in artificial floodwater was $k_{D,FW}$=1.07 d^{-1}, which was higher than the one calculated in demineralised water ($k_{D,DW}$=0.37 d^{-1}). This confirms that the wastewater, although diluted, had a negative effect on the survival of *E. coli* in dark conditions. The inactivation rate solely due to light, by subtracting the effect of dark inactivation from the overall inactivation rate, was equal to $k_{L,DW}$=k_{DW}-$k_{D,DW}$=1.15 d^{-1} for demineralised water and $k_{L,FW}$=k_{FW}-$k_{D,FW}$=3.33 d^{-1} for floodwater. The results can be seen in Fig. 2.7. This means that the negative effect of light was stronger in the floodwater than in demineralised water.

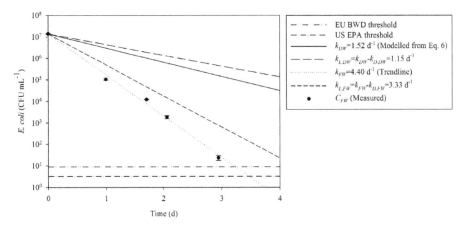

Figure 2.7 The concentration of E. coli in artificial floodwater (C_{FW}, with inactivation rate kFW) measured in the reactor, compared to the theoretical curve for demineralised water with kDW=1.52 d^{-1} calculated from Eq. 2.7 for 150 mg TSS.L^{-1}. The inactivation rates without the effect of dark inactivation for the two water qualities ($k_{L,FW}$ and $k_{L,DW}$, respectively) were calculated by subtracting the dark inactivation rate from the overall inactivation rate. The curves were plotted using Eq. 2.3. The threshold concentrations of E. coli according to US EPA Recreational Water Quality Criteria and EU Bathing Water Directive are also presented. Temperature was around 20 °C in dark conditions and increased up to 28 °C under exposure to light.

The threshold concentration of *E. coli* according to the EU Bathing Water Directive for sufficient water quality of inland waters is 9 CFU mL^{-1}, based upon a 90-percentile evaluation of the $\log_{10}$ normal probability density function of microbiological data acquired from the particular bathing water (Directive 2006/7/EC). The 2012 Recreational Water Quality Criteria by US EPA recommend a 90-percentile statistical threshold value of 3.2 CFU.mL^{-1} for *E. coli* in fresh water for primary contact recreation, including swimming, bathing, water play by children and similar water activities where a high degree of bodily contact with the water, immersion and ingestion are likely (US EPA, 2012). According to the previous results, for an initial concentration of 10^7 CFU *E. coli*.mL^{-1}, in floodwater the EU threshold would be met in 4.2 d and the US threshold would be met in 4.5 d if dark inactivation is not taken into account. In demineralised water the two thresholds will be met only after 12.1 and 13.0 d, respectively.

2.3.8 Strengths and limitations

Although this work provides a better understanding of the inactivation of faecal indicator *E. coli* in shallow water bodies, like urban retention and detention basins, further research has to be undertaken employing a wider range of water-borne pathogens or surrogate organisms and environmental conditions to be able to predict inactivation of pathogens under natural conditions. Also, the lack of replicate experiments and the fact that temperature was not constant are limitations that need to be mentioned.

2.4 CONCLUSIONS

Based on the results of the experiments, the following conclusions can be made:

- The light attenuation coefficient obtained in this study, which is proportional to the concentration of suspended solids, is: $\mu=2.59+0.02*[TSS]$, where μ is the light attenuation coefficient (m^{-1}) and *[TSS]* is the concentration of total suspended solids (mg.L^{-1}).

- The inactivation rate decreases exponentially with an increase in TSS concentration, which can be described in this case as $k=k_D+14.77e^{-0.02 \cdot [TSS]}$, where k is the inactivation rate in d^{-1} and *[TSS]* is the concentration of total suspended solids in mg.L^{-1}.

- In general, it was demonstrated under lab and controlled conditions that inactivation of faecal indicators *E. coli* is higher under higher solar irradiance, longer duration of daylight and low TSS concentrations.

- The results indicate that under high TSS concentrations the bacteria, even if not attached on particles, are protected from photo-inactivation for a period of a few days.

- It is also noteworthy that the negative effect of light was stronger in the flood water experiments than in those with demineralised water.

- The results can be useful for numerical flood modelling and quantitative microbial risk assessment. Further research is needed, combining real environmental and operating conditions, like temperature and various water quality parameters, as well as with more resilient pathogens such as *Cryptosporidium*.

2.5 ANNEX - CALCULATION OF AVERAGE IRRADIANCE SPECTRA TRANSMITTED THROUGH THE WATER COLUMN

The absorbance spectra of all the different water quality solutions used $\alpha_s(\lambda)$ (cm^{-1}) were measured with LAMBDA 365 UV/Vis Spectrophotometer (PerkinElmer, Waltham, MA, USA) and are presented in Fig. 2.8.

Figure 2.8 Absorbance spectra of demineralised water (DW) and artificial flood water (FW) with E. coli (7.8·10^6 CFU.mL^{-1}) and different concentrations of TSS.

This data was used to calculate the average irradiance spectra transmitted through the water column, $I_0(z,\lambda)$ (W.m^{-2}), using the following Equation 2.8 (Silverman and Nelson, 2016):

$$I_0(z, \lambda) = I_d(0, \lambda) \cdot \left(\frac{1-10^{-\alpha_s(\lambda)\cdot z}}{2.303\cdot\alpha_s(\lambda)\cdot z}\right) \qquad (2.8)$$

where $I_d(0,\lambda)$ (W.m^{-2}) is the solar simulator irradiance incident spectrum on the water surface and z is the depth of the water column (cm). The results are presented in Fig. 2.9.

Figure 2.9 Lamp irradiance spectrum and average irradiance transmitted through the water column of demineralised water (DW) and artificial flood water (FW) with E. coli (7.8·10⁶ CFU.mL⁻¹) and different concentrations of TSS.

3

ASSESSMENT OF MICROBIAL SAMPLING METHODS FOR FLOOD-PRONE URBAN SURFACES

The main aim of this chapter was to identify the most reliable method for the recovery of water-borne indicator organisms from flood-prone urban surfaces. Pathogens can occur after the surcharge of sewer networks on surfaces commonly found in the urban environment such as concrete, asphalt, pavement blocks and grass. Four different sampling methods were compared: swabbing, direct agar contact, stamping and adhesive tape-lifting. The surfaces were inoculated with known amounts of *E. coli*. A glass surface was used as control. Contact plating had the highest log recovery ratio, 96.1% on glass, for up to 10^3 CFU.100 cm^{-2} of *E. coli*, but this method has a limited range of bacterial numbers because it is not possible to dilute or concentrate the samples. Swabbing was the most reliable technique because it could be used for a wide range of concentrations with high recovery ratios of up to 96.2% for 10^5 CFU.100 cm^{-2} of *E. coli*. Comparatively, the indirect methods of stamping and tape had no additional advantages. Further experiments using the swabbing technique revealed that the water accumulated on rougher surfaces affected the swabbing recovery ratios when the samples got diluted. Swabbing any amount of sample higher than what the swab heads could absorb (0.15 mL) reduced the recovery ratio. Furthermore, swabbing was more efficient without the use of a nonionic surfactant (Tween 80) in wetting solution and eluent. After the recession of an artificial flood, swabbing and contact plating were confirmed as reliable methods to sample and enumerate the presence of *E. coli* on different urban surfaces, with the log recoveries ranging from 21.0% to 59.0%, depending not only on the sampling method, but also on the actual amount of bacteria on the surface. Standardisation of sampling methods in the future will be helpful in assessing public health risks after floods.

This chapter is based on: Scoullos, I.M., van de Vossenberg, J., Lopez Vazquez, C.M. and Brdjanovic, D.: Assessment of microbial sampling methods for flood-prone urban surfaces. In preparation.

3.1 INTRODUCTION

Several studies have shown that, after the end of floods, water-borne pathogens from human and animal sources are present on the flooded surfaces (ten Veldhuis et al., 2010), a phenomenon directly linked with the amount of urbanization and density of impervious surfaces in watersheds (Olds et al., 2018). Chapter 2 showed that *E. coli* can survive in flood water for at least three days under continuous exposure to artificial solar light. Although bacterial, protozoan and viral pathogens, such as *Campylobacter, Salmonella, Cryptosporidium, Giardia* and adenovirus have been detected in urban storm water runoff, the quality of storm water in terms of microbial contaminants is poorly understood and it is imperative to determine the sources of contamination (Ahmed et al., 2019). In order to reliably assess the concentration of water-borne pathogens on urban surfaces after the recession of flood water and to obtain a more realistic assessment of potential health risks of floods, it is necessary to have reliable sampling methods for recovering indicator organisms from these surfaces in a quantitative way. Most methods for recovery of microorganisms from surfaces found in the literature are designed for smooth and non-porous surfaces, such as stainless steel surfaces used in the food industry (Garayoa et al., 2017; Maunula et al., 2017; Moore and Griffith, 2007), and in healthcare environments (Rawlinson et al., 2019; Claro et al., 2015). At the same time, most methods referring to porous and rough surfaces include either the collection of bulk samples of the material with methods such as scrapping and grinding or collection of water samples after immersion or rinsing. There is a need to adapt and standardise sampling methods for rough, porous and dusty materials (Verdier et al., 2014).

This study was carried out with the aim of defining a reliable sampling method for the microbiological assessment of surfaces after floods. The recovery ratios of swabbing and direct agar contact, widely used for smooth and non-porous surfaces, are compared with the indirect sampling methods of stamping and adhesive tape. Stamping was studied as a method to enumerate high numbers of bacteria by using consecutive imprints. Adhesive tape, a method previously used to sample bacteria from monument surfaces (Urzì and De Leo, 2001), meat surfaces (Fung et al., 2000) and also for collection of touch DNA from fabrics in forensic biology (Verdon et al., 2014) was adapted and tested as an alternative way to recover bacteria attached on surfaces.

3.2 MATERIALS AND METHODS

3.2.1 Indicator organisms

E. coli ATCC 25922 was chosen as the organism of study. Before each batch experiment *E. coli* was incubated in 1.3% w/v sterile Oxoid CM0001 Nutrient Broth (Oxoid Ltd., Basingstoke, UK) solution in Erlenmeyer flasks for 24 h at 37 °C. The concentration of

E. coli bacteria after incubation was around 10^9 CFU.mL^{-1} (stock solution). This was diluted with 0.1% peptone physiological salt solution to different working concentrations (inoculum). The concentration of the inoculum was measured by plating serial dilutions. Drops of the inoculum were transferred on to the surfaces with a micropipette and spread with a sterile glass spreader to achieve a uniform distribution. The enumeration of *E. coli* in the samples was performed by counting the number of CFU on Chromocult Coliform Agar (CCA) (Merck KGaA, Darmstadt, Germany) plates after 24 h of incubation at 37 °C. Appropriate 10-fold dilution steps in 0.1% peptone physiological salt solution were used. All the experiments were performed in triplicate.

3.2.2 Surfaces tested

Permeable pavement blocks (10 x 20 x 8 cm3), pieces of asphalt (3 cm high) and concrete (6 cm high), and grass together with its soil (4.5 cm high) were collected from construction sites in Delft, The Netherlands. A glass panel, (0.4 cm thick), being non-porous and pH neutral, was used as control. All surface areas sampled were 100 cm^2.

3.2.3 Sampling methods

Swabbing

Sterile cotton tipped wooden stick swabs COPAN CLASSIQSwabs 150C (Copan Italia S.p.A., Brescia, Italy) were used to collect samples. Each swab was applied 12 times in one direction (each time from left to right and then from right to left) and 12 times in a 90° perpendicular direction with regard to the first swabbing direction (each time from the front to the back and vice versa). The swabs were held so that the handle created about a 30° angle with respect to the surface, they were rotated slowly each time and, as much as possible, a similar pressure was continuously applied (slightly bending the swab stick without breaking it). The swab tubes were filled with 2 mL sterile 0.1% peptone physiological salt (PPS) solution that used both as wetting solution (except when the swabs were applied dry) and as eluent. In some experiments, different concentrations of nonionic surfactant Tween 80 (Sigma-Aldrich Co., St. Louis, MO, USA) were included. After swabbing, the swabs were submerged in the eluent and vortexed for 5-10 s. The suspension was diluted, if needed, in appropriate 10-fold dilution steps in PPS solution, and the samples were plated for colony counting.

Direct agar contact

Contact plates (Simport Scientific, Beloeil, Canada) of 60 mm diameter, were filled with 18 mL of CCA to create a convex surface and let dry. Samples were taken by applying the plates gently on the sample area. The enumeration of *E. coli* was performed by counting the number of CFU after 24h of incubation at 37 °C (Fig. 3.1).

Figure 3.1 Contact plates after sampling and incubation for control (left) and different concentrations of E. coli.

Indirect Contact

Adhesive tape

A solvent-free, non-toxic paper adhesive tape (Eco Premium Masking Tape, Tesa SE, Norderstedt, Germany) was used. The width of the tape was 3.8 cm, so by selecting a length of 4.2 cm, a comparable sampling area of 16 cm^2 was chosen for enumeration, which is similar to the area of the contact plates. The tape was applied on the sampling area and then the adhesive side of the tape was kept in contact with the agar surface of CCA plates for incubation and enumeration (Fig. 3.2). Random pieces of the tape were sampled as negative controls without sterilisation, consequently no colonies were detected on the CCA plates.

Figure 3.2 Petri dishes with adhesive tape on CCA medium after sampling and incubation. The first column is the control and the other columns contain different concentrations of E. coli. Sampling was performed in triplicate (three rows).

Stamp

Rubber caps of 7.7 cm diameter (46.6 cm^2) were used as stamps. Their flat surface was pressed on to the sample surfaces and then softly applied on the agar layer of a CCA plate (first imprint) and subsequently immediately applied on a second plate (second imprint) (Fig. 3.3). The stamps were initially autoclaved, and then sterilised with ethanol and dried before each experiment.

Figure 3.3 Petri dishes after stamping and incubation of E. coli samples.

3.2.4 Experimental design

Comparison of sampling methods

Glass surfaces of 10 x 10 cm, sterilised with ethanol were inoculated with 0.1 mL of three different concentrations of inoculum to achieve initial concentrations of 10^1, 10^3 and 10^5 CFU.100 cm^{-2}. The surfaces were sampled in triplicate with different sampling methods: swabbing, contact plating, adhesive tape and stamping.

Swabbing on different surfaces

Swabbing was tested on different surfaces of glass, pavement, concrete and asphalt. 1 mL of inoculum was spread on each surface of 100 cm^2 to achieve an initial concentration of around 10^5 CFU.100 cm^{-2}. The test was performed on dry surfaces (room temperature and humidity) and was repeated on water saturated surfaces, each time in triplicate. Saturation was achieved by immersing the surfaces in demineralised water and pumping out the water until it decreased to about 1 cm below the sample surface.

Wet and dry swabbing

The maximum volume of wetting solution absorbed by the swab heads (flocks) was measured by dipping dry swabs in demineralised water and weighing the difference. The importance of using the swabs dry or immersed in the PPS wetting solution was tested on glass inoculated with two different volumes of inoculum, always with a concentration of *E. coli* of around 10^5 CFU.100 cm^{-2}. Sample volumes of 0.1 mL and 1 mL were tested because they are lower and higher, respectively, than the amount of water that can be absorbed on the swabs.

Wetting solution for swabbing

The effect of Tween 80 on *E. coli* viability was tested in batch experiments in Erlenmeyer flasks, in duplicate, kept at an orbital shaker in room temperature for two days. The flasks contained 250 mL of different concentrations of Tween 80 (0, 1, 15 and 50 g.L^{-1}) in PPS and were inoculated with 10^7 CFU.mL^{-1} of *E. coli*. The experiments were performed with flasks sterilised prior to the addition of Tween 80 and repeated with flasks sterilised after the addition of Tween 80 (by autoclaving at 121 °C for 15 min) to rule out any effect of the autoclaving process on the surfactant's properties.

Subsequently, the effect of Tween 80 on swabbing efficiency was tested. Glass panels were inoculated with 0.1 mL of around 10^6 CFU.mL^{-1} of *E. coli*, to obtain an initial surface concentration of around 10^5 CFU.100 cm^{-2}. PPS with different concentrations of Tween 80 (0, 1, 15 and 50 g.L^{-1}) was used as wetting solution for the swabs.

To conclude this set of experiments, the effect of Tween 80 on the of *E. coli* recovery efficiency from swabs was tested separately by directly inoculating dry swabs with 0.1 mL of around 10^5 CFU.mL^{-1} with the use of a micropipette. The inoculated swabs were introduced in their tubes filled with different concentrations of Tween 80, as eluent solution. Thereafter, the suspension was plated after vortexing as described previously.

Contact plating on different surfaces

Different surfaces (glass, pavement, concrete, asphalt and grass) were sampled with direct agar contact plating. Dry surfaces were inoculated with around 10^3 CFU.100 cm^{-2}. The surfaces were not sterilised, thereby negative controls were obtained prior to the inoculation to assess the potential presence of *E. coli*.

Swabbing and contact plating on different surfaces after an artificial flood

The efficiency of swabbing and contact plating on surfaces was studied under simulated flood conditions. The surfaces were not sterilised and negative controls were sampled with both methods before the flooding. The surfaces were placed in plastic tanks cleaned with ethanol and filled with artificial floodwater consisting of demineralised water inoculated with *E. coli* at a concentration of 10^4 CFU.mL^{-1} up to 1 cm higher than each surface level. Samples of the floodwater were taken immediately after the flood. The flood was followed by the recession of water using a peristaltic pump at a flowrate of 1 L.h^{-1}, until the water level reached 1 cm below the surface's level. This stage lasted 1 h and the samples were collected after another 1 h. Samples were taken from different areas of the surfaces with contact plates and swabs. Wet swabs were used for the surfaces that dried faster (glass, pavement and concrete) and dry swabs were used for asphalt and grass that were still wet.

Data analysis

Log recovery was calculated as the ratio of the $\log_{10}$ concentration recovered to the $\log_{10}$ concentration of the inoculum, both concentrations being in CFU.100 cm^{-2}. To investigate whether the results were statistically significant, two-sample t-tests assuming unequal variances were performed, to compare the surface concentrations (CFU.100 cm^{-2}) of $E.$ $coli$ recovered in dry and saturated surfaces (Section 3.3.2.), or observed before and after the experiment (Section 3.3.4.), or recovered with swabbing and contact plating (Section 3.3.6.). Two-factor (sample volume and wet/dry swabs) ANOVA with replication was used in Section 3.3.3., to test if sample volumes affect the recovery ratios of wet and dry swabs. Also, a two-factor (Tween 80 concentration and time) ANOVA with replication was used in Section 3.3.4., to compare the effect of different Tween 80 concentrations on $E.$ $coli$ concentrations at different stages of the experiment. In Section 3.3.4, single factor ANOVA tests were performed, to compare the recovered $E.$ $coli$ concentrations with the use of different Tween concentrations. In all cases, the significance level, α, selected was 0.05. It was assumed that microbial concentrations were normally distributed.

3.3 RESULTS AND DISCUSSION

3.3.1 Comparison of sampling methods

In all cases the control samples on glass had no colonies, which means that sterilisation of the glass with ethanol reached $E.$ $coli$ numbers below the detection limits. The results with the use of all the different methods tested are presented in Fig. 3.4.

The contact plates' method had the highest log recovery, of 77.6% and 96.1%, at the low surface concentrations of $E.$ $coli$ of 10^1 and 10^3 CFU.100 cm^{-2}, respectively. However, the colonies were too many to be enumerated on the contact plates at the concentration of 10^5 CFU.100 cm^{-2} (Fig. 3.1). Another disadvantage was that the area that can be sampled is small (16 cm^2). Since there were limited colony counts at 10^1 CFU.100 cm^{-2}, and dilutions or concentrations (e.g. with filtration) can hardly be made with this method, it can be suggested that the range of concentrations to apply the contact plates' method appears to be limited to between 10^1-10^3 CFU.100 cm^{-2}.

Similarly, with the use of adhesive tape the highest surface concentration of $E.$ $coli$ was not possible to be enumerated (Fig. 3.2). Even at lower concentrations (log recoveries of 67.1% and 79.2% at 10^1 and 10^3 CFU.100 cm^{-2}, respectively) it was more difficult to count colonies, compared to the contact plates, because the colonies, which had to grow between the tape and the agar surface, spread and covered each other. Furthermore, the handling of the tape in a sterile way was difficult. Therefore, the use of this method appeared impractical and did not offer any significant advantages, but there is still room for optimisation. For example, the tape could be placed on the agar plate for a few seconds and then lifted and discarded instead of remaining on the surface during incubation. Such a method was used by Fung et al. (2000) to sample meat surfaces with an easy to handle "pop-up" Scotch tape that can be operated with one hand. Although the recovery was 0.91 times higher compared to swabs, the problem of colonies growing too close to each other was also present because dilution was not possible. The same research observed that sampling of moist or uneven surfaces affected the method. It is interesting to note that adhesive sheets have also been used for direct counting of bacteria from solid surfaces with epifluorescent microscopy (Yamaguchi et al., 2003) and scanning electron microscopy (Urzì and De Leo, 2001).

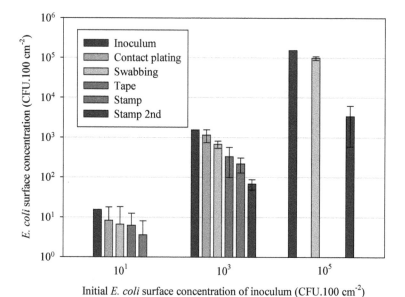

Figure 3.4 E. coli surface concentrations recovered with different methods from a glass surface compared to the surface concentration of the inoculum initially spread over of around 10^1 (left cluster), 10^3 (central cluster) and 10^5 (right cluster) CFU.100 cm^{-2}.

On the other hand, the swabbing method was reliable for the whole range of surface concentrations because the samples could be diluted after being transferred to the eluent. The log recovery percentages for this method were 69.4% for the initial surface concentration of 10^1 CFU.100 cm^{-2}, 88.9% for 10^3 CFU.100 cm^{-2} and 96.2% for 10^5 CFU.100 cm^{-2}. Since there are limited colony counts at 10^1 CFU.100 cm^{-2} (30-300 are needed on a plate) the minimum amount of *E. coli* swabbed should be 10^3 CFU.100 cm^{-2}, unless a larger area is swabbed or the membrane filtration method is used for plating low surface concentrations. Due to the advantages of swabbing, it was studied more in depth in the following sections.

Recovery using rubber stamps was lower compared to the other methods. For an initial level of 10^1 CFU.100 cm^{-2} the first imprint had a log recovery of only 46.7% with regard to the initial bacteria and the second one was below the detection limit (suggesting that no bacteria were recovered). At the surface concentration of 10^3 CFU.100 cm^{-2}, the first imprint had a log recovery of 73.4% and the second one of 57.5%, whereas at 10^5 CFU.100 cm^{-2} the first imprint was too concentrated and the second one had a log recovery of 68.1%. The reason for the low efficiency of this indirect contact method was mainly the poor contact of the rubber on the sampled surfaces. This could be seen from the fact that the distribution of colonies on the plates was not uniform (Fig. 3.3). Previous applications of stamps faced this by using softer stamp materials, such as polyurethane foam cylinders (Tresner and Hayes, 1970) or wooden cylinders with a latex layer covered by velvet (Elek and Hilson, 1954). The advantage of the stamps was that the enumeration of higher concentrations was possible by consecutive imprints, even if the accuracy was lower. This was demonstrated in the case of the highest level of inoculum that was possible to be counted using the second imprint, while the colonies on the first one were too many to count. At 10^3 CFU.100 cm^{-2}, the second imprint had 69% less colonies than the first one. Elek and Hilson (1954) determined an average transfer factor of 1 to 40 on the primary imprint (19.9% log recovery). Furthermore, it is important to note that the stamps' surface was larger (46.5 cm^2) compared to the contact plates, but different sizes can be used. Based on these observations, the detection range of this method is 10^2-10^4 CFU.100 cm^{-2} for the first imprint and 10^3-10^5 CFU.100 cm^{-2} for the second one.

3.3.2 Swabbing on different surfaces

First of all, although the surfaces were not sterilised (except glass that was cleaned with ethanol), the concentration of E. coli recovered by swabbing prior to inoculation was under the detection limit. The results from the swabbing experiments are presented in Fig. 3.5.

In the tests on dry surfaces, the initial level of E. coli was $1.2 \cdot 10^5$ CFU.100 cm^{-2}. The log recovery on glass was 77.0%, on pavement 78.6% and on asphalt 84.8%, while no E. coli colonies were detected by swabbing concrete. The recovery on glass was lower compared to the previous experiment, because the volume of the inoculum was 10 higher (1 mL) and it exceeded the amount of liquid that can be absorbed by the swab. This raised the question of whether swabs are better to be used dry or to be first immersed in a wetting solution. This question is investigated in the next sections.

In the tests on saturated surfaces, the initial concentration of E. coli was $7.7 \cdot 10^4$ CFU.100 cm^{-2}. The log recovery on glass, pavement, concrete, asphalt and grass was 78.2%, 76.5%, 40.9%, 76.1% and 62.8%, respectively. Moore and Griffith (2007) measured an *E. coli* sampling efficiency of 21.8% (log recovery of 66.9%) with cotton swabs on a wet stainless steel surface. Through statistical analysis (using two-sample *t*-tests assuming unequal variances), it was observed that the difference of the recovered concentrations between dry and saturated surfaces was not statistically significant for glass ($p=0.13$) and pavement ($p=0.08$), while for asphalt ($p=0.01$) it was significant. Saturated asphalt has lower recovery ratios than unsaturated, probably because of the excess of water accumulated and "trapped" on the voids of the rough surface; the inoculum is diluted and the volume recovered by the swab is limited. In the case of concrete, as all values on the dry surface are equal to 0, a *t*-test could not be performed, but saturation increased the recovery ratio because the dry surface is permeable and the inoculum percolated through the surface. The effect of high pH of concrete on the survival of *E. coli*, especially when concrete is new and has not undergone carbonation, is examined in Chapter 4. At the same time, the low recovery ratio on concrete could be related to the roughness of the concrete surface, twisting and compressing the cotton tip and reducing its ability to hold sample volume and entrap bacteria between the cotton fibres (Ahnrud et al., 2018).

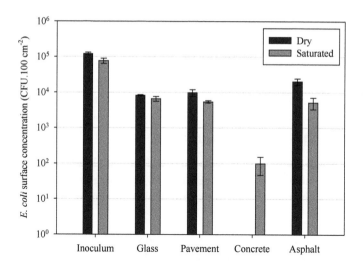

Figure 3.5 E. coli surface concentrations of the inoculum and as recovered by swabs from different surfaces, dry and saturated.

3.3.3 Wet and dry swabbing

The maximum amount of water that can be absorbed by the swab heads was estimated to be around 0.15 mL on average. It was assumed that higher sample volumes would lead to lower recovery ratios. This was confirmed with dry swabs that had a log recovery of

90.1% of *E. coli* in 0.1 mL volume samples and 81.4% in 1 mL volume samples on glass. Similarly, swabs that had been immersed in a wetting solution had a log recovery of 88.5% in 0.1 mL volume samples and 74.0% in 1 mL volume samples, both at the initial concentration of $1.3 \cdot 10^5$ CFU.100 cm^{-2}. A statistical analysis of two-factor ANOVA with replication showed that the volume of the sample was a significant factor affecting recovery ($F=100.04$, $p=0.001$), confirming the hypothesis that sample volumes higher than what the swabs can absorb lead to a lower recovery. However, the same analysis showed that immersing the swabs in a wetting solution did not have a statistically significant effect, nor the interaction between sample volumes and dryness of the swabs was significant. Therefore, it is suggested to use dry swabs on wet surfaces in order to maximise the sample volume absorbed and wet swabs to be used on dry surfaces if needed.

3.3.4 Wetting solution for swabbing

In the case of the experiments of viability of *E. coli* at different concentrations of Tween 80, a two-factor ANOVA with replication showed that there were no significant differences between the *E. coli* concentrations at different concentrations of Tween 80. The results were similar when the flasks were autoclaved prior to ($F=0.57$, $p=0.65$) and after ($F=1.91$, $p=0.18$) the addition of Tween 80. Also, two-sample *t*-tests assuming unequal variances showed that the surface concentration of *E. coli* was constant throughout the experiment in both cases ($p=0.09$ and $p=0.03$, respectively). Therefore, neither the different concentrations of Tween 80 tested had a statistically significant effect on *E. coli* survival, nor this was affected by autoclaving.

With an initial concentration of $2.9 \cdot 10^5$ CFU.100 cm^{-2}, the log recovery of *E. coli* without addition of Tween 80 was 85.6% and it was found to be linearly decreasing as the Tween 80 concentration increased (log recovery=$0.85-0.002 \cdot C_T$, $R^2=0.99$, where C_T is the concentration of Tween 80 in g.L^{-1}), reaching the lowest recovery of 75.2% at 50 g.L^{-1} of Tween 80, the highest concentration tested. This decrease was found to be statistically significant by a single factor ANOVA ($F=4.65$, $p=0.04$). This finding is interesting given the fact that Tween has been proposed by many as one of the best wetting solutions for swabbing and wiping surfaces. For example, You et al. (2019) observed that cotton swabs had an overall higher recovery efficiency (up to three times) of dried *E. coli* and *Staphylococcus aureus* samples from a desk surface with the use of 1% Tween 20 and 1% glycerol on phosphate-buffered saline (PBS) than with plain PBS or with a commercial solution. Valentine et al. (2008) studied different wetting solutions for wiping and swabbing *B. subtilis* bacterial endospores from stainless steel and vinyl surfaces. They concluded that phosphate-buffered saline with 0.3% Tween was a better collection solution than sterile water or phosphate-buffered saline without Tween for all the tested swab and wipe types used, except for the Dacron / polyester swab, which had the same results with the use of water. Therefore they used the Tween solution for swabbing sterile porous (polyester upholstery fabric and a commercial carpet) and nonporous (plastic laminate countertop, finished oak wood flooring and an activated computer monitor screen) surfaces, with the highest recovery being around 8% (CFU.mL^{-1}) on laminate.

With an initial amount of $2.3 \cdot 10^4$ CFU inoculated per swab, a single factor ANOVA, showed (F=4.39, p=0.09) that Tween 80 did not have a statistically significant effect on the recovered concentrations from swabs when used as eluent. The average log recovery ratio was 90.6%. This suggests that the negative effect of Tween on the *E. coli* recovery was not related to the detachment of bacteria from the swabs, but rather to the swabbing process. It is possible that the drying of the inoculum may have caused sub-lethal damage to the cellular membranes, which increases the sensitivity of cells to substances and conditions (Moore and Griffith, 2007) such as agitation during swabbing.

3.3.5 Contact plating on different surfaces

The surface concentration of the inoculum was $2.3 \cdot 10^3$ CFU.100 cm^{-2}. The log recovery ratios were 61.3% on glass, 73.2% on pavement, 63.5% on asphalt and 73.1% on glass. Recovery on concrete was below the detection limit. It was noted that the agar surface could not reach some of the voids, especially on the asphalt surface, that swabs could reach.

3.3.6 Swabbing and contact plating on different surfaces after an artificial flood

The concentration of *E. coli* in flooding water was between 1.6 and $2.3 \cdot 10^4$ CFU.mL^{-1}. The negative control was below detection limit for the swabbing method, while a few colonies were recovered by contact plating. The log recoveries on glass, pavement, concrete, asphalt and grass were 43.3%, 58.9%, 21.0%, 59.0% and 44.8% respectively for the swabbing method and 41.0%, 55.6%, 37.3%, 57.6% and 37.0% respectively for contact plating. It must be noted here, that as the *E. coli* were in suspension in the flooding water the recovery ratios in this experiment were calculated by assuming that the bacteria settled vertically in a uniform way on top of the sampled surfaces when the water receded. Therefore, these recovery ratios cannot be compared to the previous experiments in absolute terms because the hydraulic characteristics of the different surfaces could also play a role on the amount of bacteria that were actually present on the sampled surfaces. For example, the relatively low recoveries on glass could be attributed to the fact that the amount of water that remained on the glass was low, although it was horizontal. What can be concluded from this experiment is that similar trends were observed by both methods. In fact, two-sample *t*-tests assuming unequal variances showed that the means were equal for both methods on each of the surfaces (p=0.76 for glass, p=0.32 for pavement, p=0.18 for concrete, p=0.13 for asphalt and p=0.09 for grass). The highest recoveries were achieved on pavement and asphalt. Concrete had the lowest recovery, probably because it dried fast. The recovery on grass was not high because only the top layer was being sampled. In previous research on different surfaces in hospital rooms, such as floor, bed linen, door handles, and toilet seats, Lemmen et al. (2001) detected *E. coli* contamination with a frequency of 30.0% with the use of swabs, while with contact plates only 5.0% contamination was detected. Similar results were found for other gram-negative rods, while for gram-positive cocci, contact plates showed a higher recovery compared to swabbing. Therefore, a combination of methods is suggested for the assessment of overall contamination.

3.3.7 Strengths and limitations

This work provides data on the recovery ratios of different methods on materials like concrete, asphalt and grass, for which the scientific literature is very limited.

A limitation of this research was the comparison of only four methods. For example, alternative types of swabs, some of which are presented in section 1.5., could possibly be sturdier and provide better results that cotton tipped swabs, especially on rough surfaces.

3.4 CONCLUSIONS

Based on the results of the tests, the following conclusions can be made:

- On a flat, smooth and non-porous surface like glass, swabbing was the best of the tested methods for a wide range of concentrations. Contact plates had the highest recovery, but a low range of inoculum levels. Tape and stamps had technical challenges and no advantages over contact plating or swabbing.

- Moisture on glass and pavement did not affect swabbing recovery ratios, but the accumulation and "trapping" of water in the voids of asphalt surface resulted in dilution of the samples, while the fact that concrete surfaces did not trap water led to no recovery of *E. coli*.

- The maximum amount of water absorbed by the swabs was 0.15 mL. Swabbing any amount of sample higher than that reduced the recovery ratio. Dry swabs are proposed for wet surfaces and wet swabs for dry surfaces.

- The addition of Tween 80 in wetting solution and eluent did not result in higher recovery. It did not have any statistically significant effects on the survival of *E. coli* and on its recovery from swabs, while it decreased the overall swabbing efficiency.

- Swabbing and contact plating were confirmed as reliable methods to sample and enumerate the presence of *E. coli* after flooding on different urban surfaces.

4

INACTIVATION OF *E. COLI* AS FAECAL INDICATOR ORGANISM ON DIFFERENT SURFACES AFTER URBAN FLOODS

A better understanding of the effects of different urban and recreational surfaces on the inactivation of water-borne pathogens that can cause infections after urban floods if released from surcharged combined sewers and other sources of faecal contamination is needed. The inactivation of faecal indicator *E. coli* was studied under controlled exposure to simulated sunlight on a range of different surfaces found in urban environments: gravel, sand, asphalt, pavement blocks, concrete, playground rubber tiles and grass, using glass as control. The surfaces were inoculated with artificial flooding water containing 10^5 CFU.mL^{-1} of *E. coli* and sampled periodically using the sterile cotton swab technique, after lowering the water level. The results show that dark inactivation was not statistically significant for any surface, suggesting that chemical composition and pH (varying between 6.5±0.8 and 9.2±0.4) did not affect the inactivation rates. The highest light-induced inactivation rates for *E. coli* after the floodwater recession, observed on rubber (>3.46 h^{-1}) and asphalt (2.7 h^{-1}), were attributed to temperature stress and loss of surface moisture.

This chapter is based on: Scoullos, I.M., Lopez Vazquez, C.M., van de Vossenberg, J. and Brdjanovic, D.: Die-off of E. coli as fecal indicator organism on different surfaces after urban floods. J. Environ. Manage., 250, 109516, doi:10.1016/j.jenvman.2019.109516, 2019. (IF=4.9)

4.1 INTRODUCTION

In the previous chapter, it was demonstrated that the die-off of *E. coli* is governed by sunlight; die-off is higher under higher solar irradiance, longer duration of daylight and low TSS concentrations. The effect of different surfaces on the solar inactivation of faecal indicators has not been studied in depth in the literature. The need for further research on the influence of real-world conditions on the inactivation processes has been identified (Nelson et al., 2018).

The objective of this chapter is to better understand and quantify the effects of different surface types found in urban and sub-urban open areas, including storm water detention and retention basins, on the inactivation of indicator organisms under different light conditions, immediately after floods. The research hypothesis is that light irradiation affects bacterial growth, because of damage of light to the cells, but also because light affects the temperature and moisture of different surfaces. The novelty of this research is the focus on the importance of the surface type on the aforementioned processes in order to predict the concentration of pathogens and ultimately to facilitate the decision-making process to assess public health risk and safety conditions.

4.2 MATERIALS AND METHODS

4.2.1 Experimental setup

Samples of gravel of two particle sizes (sieved between 2.5-4.0 mm as very fine gravel and 1.4-2.24 mm as very coarse sand, respectively), pavement blocks (10 x 20 x 8 cm3), pieces of asphalt (3 cm high) and concrete (6 cm high), and grass together with its soil (4.5 cm high) were collected from construction sites in Delft, The Netherlands. Additionally, new playground rubber tiles (2.5 cm high) were used, consisting of styrene-butadiene rubber granulate, heat-bonded with moisture resistant polyurethane binder (Granuflex, Amsterdam, The Netherlands). A glass panel (0.4 cm thick), being non-porous and pH neutral, was used as control. All the above mentioned surfaces were fitted in plastic containers of 27×37 cm2 and 12 cm height. These containers were placed in larger containers with a temperature control system set at 20 °C. The inner containers were covered with a Perspex lid to eliminate evaporation. The experimental setup is presented in Fig. 4.1 and in Fig. 4.2.

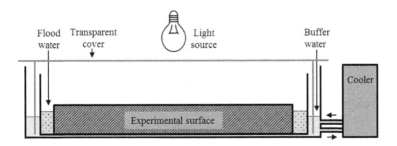

Figure 4.1. Schematic representation of a reactor exposed to artificial light. Two such reactors were used in parallel.

Figure 4.2 Experimental setup used for the study of the inactivation of E. coli under artificial light and dark conditions. Two such reactors were used in parallel.

4.2.2 Indicator organism

E. coli ATCC 25922 was chosen as a proxy of faecal microorganisms. Before each batch experiment *E. coli* was incubated in 1.3% w/v sterile Oxoid CM0001 Nutrient Broth (Oxoid Ltd., Basingstoke, UK) solution in Erlenmeyer flasks for 24 h at 37 ˚C. After incubation, the concentration of the inoculum was estimated at $3 \cdot 10^9$ CFU.mL^{-1}. The initial concentration of *E. coli* for the experiment was around 10^5 CFU.mL^{-1}, based on the typical concentrations of *E. coli* and enterococci observed in raw sewage: between 10^4 CFU.mL^{-1}, the maximum event mean concentration of the first flush of urban storm water runoff after dilution with storm water (Hathaway and Hunt, 2010), and 10^6 CFU.mL^{-1} (Mark et al., 2015).

The enumeration of *E. coli* in the samples was performed by counting the number of CFU on Chromocult® Coliform Agar (CCA) (Merck KGaA, Darmstadt, Germany) plates after 24 h of incubation at 37 °C. 10-fold dilution steps were performed using 0.1% peptone physiological salt solution. The plates were spread in triplicate.

4.2.3 Swabbing

Sterile cotton tipped wooden stick swabs COPAN CLASSIQSwabs 150C (Copan Italia S.p.A., Brescia, Italy) were used to collect samples from designated 10×1 cm^2 areas on all surfaces, in triplicate. Each swab was applied 20 times, while being rotated. For sampling dry surfaces, the swabs were moistened with sterile 0.1% peptone physiological salt solution prior to swabbing. For wet surfaces, swabbing was performed with dry swabs. In both cases, after swabbing, the swabs were submerged in 2 mL of 0.1% peptone physiological salt eluent and vortexed for 5-10 s. The suspension was diluted and/or plated for colony counting. In the case of grass, swabbing was performed, to the extent possible, only on the grass leaves, not touching the soil.

4.2.4 Light source and parameters

Simulated sunlight was produced using an OSRAM HQI-BT 400 W/D PRO (OSRAM GmbH, Munich, Germany) metal halide lamp with built-in UV filter that blocks wavelengths shorter than 320 nm. Apart from a few peaks in the area of visible light (with the highest one at 540 nm) the spectrum of these metal halide lamps is continuous (Calin and Parasca, 2008). The light intensity was fixed at 690 W.m^{-2}. The photon flux, in µmol.m^{-2}.s^{-1}, was measured with a LI-250A Light Meter equipped with an underwater quantum sensor (LI-COR® Biosciences, Inc., Lincoln, Nebraska, USA). This sensor has a uniform sensitivity in the range of wavelengths between 400 and 700 nm; this limitation was taken into account when calculating the lamp irradiance spectrum. The light intensity was calculated as discussed in Section 2.2.4, taking into account the spectral power distribution of the lamp.

4.2.5 Experimental design

Batch experiments in flasks

The effect of different surface materials on *E. coli* inactivation was tested in Erlenmeyer flasks in dark condition. The flasks contained 10 g of the different surface materials and 250 mL of demineralised water and were inoculated with *E. coli* at a concentration of 10^5 CFU.mL^{-1}. In a second set of flasks 25 mM HEPES Buffer Solution (Life Technologies Europe B.V., Bleiswijk, The Netherlands) were added as buffering agent to achieve an initial pH of around 7.0 in order to eliminate the effect of pH of the different materials. The *E. coli* concentration and pH were measured daily.

Flood cycle on different surfaces

Each container was filled on a pulse with the flooding water up to a height of 2 cm above the surface. Water was pumped out with the use of a peristaltic pump at a rate of 1 L.h^{-1} until the water surface reached 1 cm below the surface, for approximately 2 h, depending on the surface subject to study. At this point (time 0) the lamp was turned on for 6 h followed by a dark period of 18 h.

4.2.6 Sampling and physicochemical parameters of study

pH was measured in the water phase with a handheld pH-meter (WTW pH 323, WTW GmbH, Weilheim, Germany) and on the experimental surfaces with pH paper (Fisherbrand pH-Fix Test Strips, Fisher Scientific, Landsmeer, The Netherlands). Temperature was measured every hour (fluctuating between 20 and 28 °C, data not shown) with a glass thermometer.

The total organic carbon (TOC) concentration of samples was determined using a carbon analyser TOC-L$_{CPN}$ (Shimadzu, Tokyo, Japan). Porosity (%) and gravimetric water content (%) of the surfaces were derived by measuring the mass and volume of the surfaces in dry condition (heated at 105 °C for 24 h) and after saturation in water. For measuring the dry volume, the surfaces were covered watertight with paraffin film and immersed in water; the mass and volume of the paraffin used were taken into account.

4.2.7 Data analysis

In all the experiments where the inactivation of *E. coli* was addressed, the inactivation rates were calculated using the first order exponential inactivation Chick-Watson equation (Eq. 1.1). The inactivation was considered to start at time 0 (time of inoculation) with no lag period. The inactivation rate was assumed to be constant in each of the batch experiments because of their short duration, adopting the traditional approach (Wu et al., 2016). In the experiments with light and dark phases, the inactivation rates were calculated separately for each phase.

To investigate whether the difference of concentrations at the beginning and the end of each experimental phase was statistically significant, two-sample *t*-tests assuming unequal variances were performed, with a significance level, α, of 0.05. The null hypothesis was that the concentrations of *E. coli* before and after each phase (light or dark) came from distributions with equal population means. It was assumed that microbial concentrations are normally distributed.

To describe the effect of pH on the maximum specific growth rate $k_{max}(pH)$ (h^{-1}) the Cardinal pH Model equation (CPM, Eq. 1.3, Rosso et al., 1995) was used. The parameters were calculated for the best fit to the experimental data with a minimum sum of squared errors.

4.3 RESULTS AND DISCUSSION

4.3.1 Batch experiments in flasks

The initial pH of water in flasks without buffer, measured 24 h after the immersion of the surface materials in water, ranged from pH 5.4±0.4 for the control to 10.9±0.5 for pavement. In HEPES buffer all the initial pH values were between 7.0±0.1 and 8.2±0.4.

The data from the experiment in the pH unbuffered flasks fitted best to Equation 2 for parameters with the values of pH_{min}=4.6, pH_{max}=6.8, pH_{opt}=5.8 and k_{opt}=0.01 h^{-1}; the square of the Pearson product moment correlation coefficient was equal to r^2=0.96. It is known that *E. coli* species are sensitive to changes in internal pH in an alkaline environment and can maintain pH homeostasis in the external pH range of 4.5 to 7.9 (Booth, 1985). The pH of the surrounding environment affects the net charge of the organism's interface with the environment. This may lead to denaturing proteins and nucleic acids or affect the organisms' net charge and transport. Most organisms exhibit the greatest stability at near-neutral pH (Sinclair et al., 2012). The data from the experiment with HEPES did not fit well to the curve using the same parameters (r^2=0.09). This shows that pH is not the only parameter affecting the concentration of *E. coli* in the flasks and underlines the need to better understand the effect of the materials on the cells. The regression curve is presented together with the two experiments' data in Fig. 4.3.

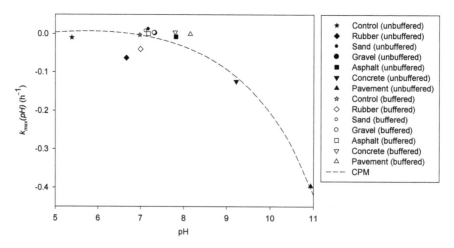

Figure 4.3 The effect of pH of water with different surface materials in batch experiments in flasks with and without the addition of HEPES, on k_{max}(pH), of E. coli, compared to the CPM curve that best fits the data (r^2=0.96). The experimental data for gravel with and without HEPES coincide. The experiments took place in room temperature of 20 °C.

The only material that did not fit well to the CPM curve, either with or without buffer, was rubber, which had a higher inactivation rate with (0.04 h^{-1}) and without (0.06 h^{-1}) the use of HEPES. The level of TOC for rubber, 7.46 mg.L^{-1}, standing out from the rest, which were all between 0.92 and 1.83 mg.L^{-1}, accounts in large part for organic compounds that are inhibitory to *E. coli*, as toxicity of styrene-butadiene rubber leachate has been observed on different organisms (Krüger et al., 2013).

4.3.2 Flood cycle on different surfaces

The results from sampling in the water phase around the surfaces (Fig. 4.4) and from swabbing the surfaces (Fig. 4.5) are presented in Table 4.1.

The value of pH measured around the surfaces (Table 4.1) was around 7.0 for most materials, except pH 7.9 ± 0.1 for pavement and pH 9.2 ± 0.4 in the case of concrete. It has to be noted that in the flask experiment (Fig. 4.3) the pH of pavement was higher, 10.9 ± 0.5, because the pavement blocks were freshly ground and the interior of the blocks had not undergone carbonation, the process of hydroxide anions of calcium hydroxide being replaced over time by carbonate anions resulting in the formation of calcium carbonate (Pade and Guimaraes, 2007). In the case of concrete, as opposed to pavement blocks, no difference between the pH of water around the surface and the pH of water in the flasks with ground sample was observed.

In the case of transparent glass used as a control surface, there were no viable cells recovered with swabbing after three hours of exposure to light and the surface was fully dry. The experiment was performed separately in dark and there was no statistical significant difference between the initial and the final concentrations, therefore the inactivation was solely caused by light. Also, in the floodwater below and around the glass the bacteria died more rapidly ($k_D > 2.27$ h^{-1}) during the light phase compared to all other materials, because the glass is transparent and thin and the whole reactor was shallow and directly exposed to light.

On rubber, there were no viable cells recovered by swabbing after three hours of exposure to light, which could be partly attributed to evaporation as the surface appeared dry. However, in the surrounding floodwater a relatively high inactivation rate of *E. coli* was observed ($k_D = 0.96$ h^{-1}), suggesting an effect of the rubber's constituents on the removal of the bacteria, as also observed in the experiment in flasks. Canepari et al. (2018) confirmed the presence of significant bio-accessible amounts of toxic metals in styrene-butadiene rubber. Furthermore, rubber was the only one among the tested surfaces that had a statistically significant dark inactivation rate in swabbed samples, and the highest

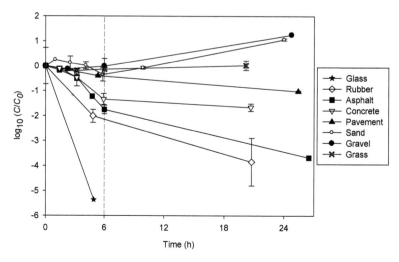

Figure 4.4 Inactivation of E. coli in water samples around different flooded surfaces. The vertical line at 6 h signifies the transition from the light to the dark phase. The detection limit corresponds to $log_{10}(C/C_0)$ = -4.

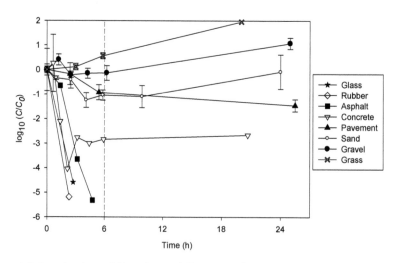

Figure 4.5 Inactivation of E. coli on different surfaces, recovered by swabbing. The vertical line at 6 h signifies the transition from the light to the dark phase. The detection limit corresponds to $log_{10}(C/C_0)$ = -4.

Table 4.1 Experimental parameters on surfaces under artificial sunlight of 690 $W.m^{-2}$ (two-sample t-tests assuming unequal variances, α=0.05, comparing log_{10} values of microbial concentrations at the beginning and end of each phase). Values noted with an * were measured in separate experiments. The number of samples (n) is indicated.

Surface	Porosity	pH	Surface k_D (h^{-1})		Floodwater k_D (h^{-1})	
			Light	Dark	Light	Dark
Glass	0%	7.1±0.1	>2.44 (n=2)	0.17* (p=0.23, n=2)	>2.27 (n=2)	-0.01* (p=0.32, n=2)
Rubber	15%	6.8±0.5	>3.46 (n=2)	0.19* (p=0.04, n=2)	0.96 (p=0.01, n=2)	0.27 (p=0.01, n=2) (0.01*) (p=0.60, n=2)
Asphalt	4%	6.8±0.3	2.70 (n=4)	-0.02* (p=0.30, n=3)	0.68 (p=0.00, n=5)	0.22 (p=0.00, n=2) (0.03*) (p=0.09, n=3)
Concrete	14%	9.2±0.4	0.90 (p=0.04, n=7)	-0.03 (p=0.62, n=2)	0.54 (p=0.01, n=4)	0.05 (p=0.15, n=2)
Pavement	1%	7.9±0.1	0.40 (p=0.03, n=3)	0.06 (p=0.10, n=2)	0.17 (p=0.00, n=2)	0.07 (p=0.00, n=2)
Sand	9%	6.8±0.3	0.46 (p=0.00, n=5)	-0.13 (p=0.23, n=3)	0.16 (p=0.96, n=5)	-0.18 (p=0.01, n=3)
Gravel	15%	6.9±0.5	0.12 (p=0.36, n=5)	-0.15 (p=0.01, n=2)	-0.002 (p=0.70, n=3)	-0.16 (p=0.00, n=2)
Grass	-	6.5±0.8	-0.22 (p=0.22, n=3)	-0.23 (n=2)	0.05 (p=0.04, n=3)	-0.03 (p=0.37, n=2)

dark inactivation rate (k_D=0.27 h^{-1}) in the water samples, although a repetition of the same experiment in dark (without prior exposure to light) showed no significant inactivation. This observation, together with the difference of inactivation rates between dark and light conditions in the water phase, suggest that exposure of rubber tiles to light and consequently to a high temperature, induced by its dark colour, increased the stronger inactivation of *E. coli*.

The effect of asphalt was similar to the effect of rubber, but on a lower magnitude. Under exposure to light, *E. coli* had an inactivation rate of 2.70 h^{-1} (R^2=0.92) on the surface and 0.68 h^{-1} (R^2=0.94) in the surrounding water. In the dark phase, there was no significant inactivation on the surface, but an inactivation rate of 0.22 h^{-1} was observed in the water. In a separate experiment without prior exposure to light, no statistical significant difference was found between the initial and the final concentrations in dark conditions in the surrounding water. Therefore, it is suggested that the chemical composition of asphalt plays a minor role. The effect of its dark colour, which increased the surface temperature up to a maximum of 60.7±1.3 °C after 285 min of exposure to light was stronger (despite that the surrounding floodwater temperature did not exceed 23±1.4 °C). The temperature dependence of *E. coli* has been studied thoroughly and is commonly expressed as an exponential increase of the inactivation rate with increasing temperature, assuming applicability of the first order exponential inactivation model (Blaustein et al., 2013). This assumption is also supported by Mendez et al. (2011), who suggest that low concentrations of FIB in water collected from a metal roof might be related to the low emissivity of the metal, leading to higher surface temperatures on the roof, which increase the inactivation rate of these organisms.

In the case of concrete, although the pH was 9.2±0.4, much higher compared to the other surfaces, we did not observe dark inactivation on the surface or in the surrounding water. Light inactivation is, however, significant on the surface (maximum temperature 44.6±0.9 °C at the end of the light phase) and around it (30.5±0.7 °C), as can be seen in Table 4.1 and in Fig. 4.4 and 4.5. Knowledge on the impact of cement chemistry and surface characteristics such as roughness and texture (controlled by chemical composition of mortar admixtures as well as the type of curing applied) on the adhesion and growth of microorganisms is very limited (Grengg et al., 2018).

While in the dark there was an insignificant or low inactivation, the light inactivation rate on and around the pavement blocks was less than half of the rate for concrete, as shown in Table 4.1. It was concluded that this was related to the observation that the surface of concrete dried fast under exposure to light, while the surface of pavement remained wet. The difference in drying rates can be attributed to faster draining of concrete after the recession of floodwater due to the difference in porosity of the two surfaces. The coefficient of permeability increases with an increase in porosity and the exact relationship is under debate, presented either as a power function (Xu et al., 2018; Kayhanian et al., 2012), a linear relationship (Zhong and Wille, 2015; Bhutta et al., 2012) or an exponential relationship (Sata et al., 2013).

On sand and gravel, there was no significant inactivation, except on the surface of sand under exposure to light (maximum surface temperature 46.6±1.1 °C at the end of the light phase, compared to 28.5±0.7 °C measured in the water). Some growth of *E. coli* in dark condition was observed (Fig. 4.4 and Fig. 4.5), which was however not always statistically significant according to the *t*-test (see Table 4.1). Both surface samples, especially sand, have a high standard error. The cause of the variation was the practical limitations of the swabbing method. Even though swabbing was performed gently in order to sample only the top layer that was relatively dry, it was unavoidable to also sample lower grains that were both protected from the light and wet. The presence of organic compounds was not higher than in the other surfaces (TOC of 1.25 mg.L^{-1} for sand and 1.38 mg.L^{-1} for gravel). The results agree with previous research showing that *E. coli* can survive for many weeks in sand, especially in cooler fine-grain sand (Staley et al., 2016), if the conditions and/or their physiological capabilities allow them to replicate (Whitman et al., 2014). Also, Beversdorf et al. (2007) observed that high doses of UV irradiation had no effect on *E. coli* in 30 cm deep sand plots, as opposed to a 94-99% reduction in cell numbers in control liquid cultures.

Although the concentration of *E. coli* had an apparent growth in dark conditions on the surface of grass leaves, which all kept some droplets of water until the end of the experiment, the increase of concentrations was not statistically significant in the light phase. The measurement of TOC on the grass surface was not possible as it was affected by the soil below. Page et al. (2015) calculated a mean inactivation rate of 0.169 d^{-1} (or 0.007 h^{-1}) by combining the data for *E. coli* and *E. faecalis* in summer and winter on grass surfaces of a sporting oval in Perth, Australia, irrigated with secondary treated effluent seeded with microorganisms. The solar radiation was 1440-1580 W.m^{-2} during winter and 1440-1640 W.m^{-2} during summer, much higher compared to 690 W.m^{-2} used in our research. The high variability (standard deviation of 0.126 d^{-1}) was attributed to the environmental conditions of temperature and shading, including the shading created by the green grass leaves, which can absorb more than 90% of radiation (Sidhu et al. 2008). These observations are in agreement with the lower inactivation rates observed in this study, indicating that the shading effects caused by the grass leaves decreased considerably the inactivation effect of solar radiation due to light absorption.

The experimental results suggested a link between the surface moisture and the concentration of bacteria. Sudden changes in moisture are stressful to microbes because they must expend energy to regulate osmotic pressure to their microenvironment (Csonka, 1989). Although gravimetric water content in dark and light conditions was investigated as a measure of surface moisture, the mass of water in the surfaces above the water level (1 cm) did not exceed 4% of the overall mass in any sample, therefore this method was not reliable. The possibility to measure the electric resistance of the surfaces to quantify the surface moisture level was also explored, but when the surfaces dried out, the resistance was too high to measure, except in the case of concrete, which is also conductive when dry. Therefore, it was not possible to determine a quantitative relation between moisture and inactivation rate. Taking these limitations into account, future research can be conducted by keeping the moisture of the surfaces constant throughout the experiment. For example, with the use of a bioclimatic chamber, Ordaz et al. (2019) observed a faster inactivation of *E. coli* inoculated on cellulose filters and on flat cork coupons for 25 days at 25 °C and 65% relative humidity (5.8 log CFU and 1.8 log CFU, respectively) than at 10 °C and 95% relative humidity (2.9 log CFU and 1.2 log CFU, respectively).

Furthermore, the effect of porosity was investigated and the data, presented in Table 4.1, suggest that, apart from the differences between concrete and pavement blocks due to the indirect effect of porosity on surface moisture, as discussed above, there was no direct relation between porosity and *E. coli* concentration. Therefore, the concentration of the indicator organism depends on the moisture, temperature and pH of different surfaces.

4.3.3 Strengths and limitations

This work provides data and qualitative observations related to the inactivation of *E. coli* on different surfaces, for which the scientific literature is very limited.

Apart from the limitations related to the lack of a reliable means of measuring surface moisture that were discussed above, the temperature control system did not maintain a constant temperature throughout the experiments that took place under exposure to light. As a result, temperature rose up to 60.7±1.3 °C on asphalt, and the maximum temperatures differed depending on the surfaces, making comparison difficult. Also, it was difficult to achieve a uniform temperature distribution due to the presence of the surrounding cooling water and to the small distance of the light source. A proposed solution to that would be the use of bioclimatic chambers with controlled temperature and relative humidity conditions. This would also allow the removal of the Perspex lid that was necessary to eliminate evaporation. Although light irradiance was measured below the lid in order to compensate for the irradiance that was reflected or attenuated, it is probable that the absorbance spectrum of Perspex had an effect on the incident radiation. This could have lead to a reduced penetration of UV wavelengths, which in turn could lead to an underestimation of the inactivation rates under the studied light intensity.

Other limitations were the lack of replicate experiments, the short duration of the experiments for 24 h and the small sample size.

4.4 CONCLUSIONS

Based on the examination of *E. coli* inactivation on surfaces after floodwater recession, the following conclusions can be made:

- Photo-inactivation was minor compared to the temperature stress and evaporation reducing surface moisture.

- The impact of high pH was significant only in the case of ground pavement blocks (pH 10.9±0.5), corresponding to fresh concrete, while the outer surface had lower pH (7.9±0.1) due to carbonation.

- Porosity was not directly related to *E. coli* inactivation, apart from affecting soil moisture.

5

INACTIVATION OF SURROGATE ORGANISMS ON DIFFERENT URBAN SURFACES AFTER URBAN FLOODS

This chapter contributes to a better understanding of the effects of urban and recreational surfaces on the inactivation of surrogates for enteric pathogens under controlled exposure to simulated sunlight for 6 h followed by 18 h in dark conditions. Concrete, asphalt, pavement blocks and glass as control were inoculated with artificial floodwater containing faecal indicator bacteria *E. coli*, *B. subtilis* spores chosen as surrogates for *C. parvum* oocysts and *Giardia* cysts, and bacteriophages MS2 as surrogates for viral contamination. On practically all the surfaces in this study, *E. coli* had the highest inactivation under light conditions followed by MS2 and *B. subtilis*, except asphalt where MS2 was inactivated faster. The highest inactivation under light conditions was seen with *E. coli* on a concrete surface (pH 9.6) with an inactivation rate of 1.85 h^{-1}. However, the pH of the surfaces (varying between 7.0 and 9.6) did not have any influence on inactivation rates under dark conditions. MS2 bacteriophage had the highest inactivation under light conditions on asphalt with a rate of 1.29 h^{-1}. No inactivation of *B. subtilis* spores was observed on any of the surfaces during the experiment, neither in light nor in dark conditions. This study underpins the need to use different indicator organisms to test the inactivation rate after flooding. It also suggests that given the sunlight conditions, concentration of indicator organisms and type of surface, the fate of water-borne pathogens after a flood could be estimated.

This chapter is based on: Scoullos, I.M., Adhikari, S., Lopez Vazquez, C.M., van de Vossenberg, J. and Brdjanovic, D.: Inactivation of indicator organisms on different surfaces after urban floods. Sci. Total Environ., 704, 135456, doi:10.1016/j.scitotenv.2019.135456, 2020. (IF=5.6)

5.1 INTRODUCTION

Urban and recreational surfaces have a possibility to act as a reservoir of faecal contamination after urban floods. The objective of this chapter was to better understand and quantify the effects of different surface types on the inactivation of surrogate organisms after urban floods. The inactivation of *E. coli*, *B. subtilis* spores, and bacteriophage MS2 was studied under controlled exposure to simulated sunlight on artificially flooded concrete, asphalt, pavement blocks and glass as control. Similar conditions can be met in all kinds of shallow water bodies in urban and sub-urban open areas, including multifunctional storm water retention and detention basins often used as sport facilities or playgrounds during dry weather. The research took into account the pH of the water that is exposed to these surfaces and its role on the survival of the organisms.

5.2 MATERIALS AND METHODS

5.2.1 Experimental setup

The experimental setup consisted of two open batch reactors in which different surfaces were tested (Fig. 5.1 and Fig. 5.2). Pieces of asphalt and concrete and pavement blocks were collected from construction sites in Delft, The Netherlands. A glass panel, being nonporous and pH neutral was used as control. The reactors contained the tested materials submerged in demineralised water, and tubing around the inner side of the walls, connected to a cooler to maintain a stable temperature on the surfaces (at 25 ± 5 °C). Sampling areas were defined on each surface in triplicate by glass walls, creating water wells of 100 cm^2 area each. The glass walls were glued watertight with aquarium silicone sealant that was tested for absence of biocidal activity before use. The sampling wells were filled with 100 mL (equivalent to 1 cm of water height in each well) of demineralised water to mimic rain, spiked with the initial concentration of organisms according to each phase. This was the minimum amount of water necessary to keep the surfaces moist throughout the experiments because it has been shown in Chapter 4 that loss of surface moisture leads to high *E. coli* inactivation rates. Similar conditions can be met in all types of shallow water bodies in urban and sub-urban open areas, including multifunctional storm water retention and detention basins often used as sport facilities or playgrounds during dry weather.

Figure 5.1. Schematic representation of a reactor exposed to artificial light. Two such reactors were used in parallel. In the experiments under real sunlight the cooler was not used.

Figure 5.2. Experimental setup of the reactors used for the study of the inactivation of surrogate organisms under artificial light and dark conditions.

5.2.2 Surrogate organisms

The organisms studied were *E. coli* (ATCC 25922), *B. subtilis* (12.01.31, GAP Lab, Canada) chosen as a non-pathogenic surrogate for *C. parvum* oocysts and *Giardia* cysts, and *E. coli* bacteriophage MS2 (ATCC 15597-B1) as a surrogate for viral contamination.

E. coli

Before each batch experiment *E. coli* was incubated in 1.3% w/v sterile Oxoid CM0001 Nutrient Broth (Oxoid Ltd., Basingstoke, UK) solution in Erlenmeyer flasks for 24 h at

37 °C. After incubation, the concentration of the inoculum was around $3 \cdot 10^9$ CFU.mL^{-1}. The initial concentration of *E. coli* for the experiment was around 10^5 CFU mL^{-1}, based on the typical concentrations of *E. coli* and enterococci observed in raw sewage: between 10^4 CFU.mL^{-1}, the maximum event mean concentration of the first flush of urban storm water runoff after dilution with storm water (Hathaway and Hunt, 2010), and 10^6 CFU.mL^{-1} (Mark et al., 2015). The enumeration of *E. coli* in the samples was performed by counting the number of colony forming units (CFU) on Chromocult® Coliform Agar (CCA) (Merck KGaA, Darmstadt, Germany) plates after 24 h of incubation at 37 °C. Appropriate 10-fold dilution steps in 0.1% peptone physiological salt solution were used. The plates were spread in triplicate.

B. subtilis

The *B. subtilis* spores were enumerated with the spread plate method. The spores' propagation was adapted from US EPA (2006), UV Disinfection Guidance Manual For the Final Long Term 2 Enhanced Surface Water Treatment Rule (LT2ESWTR). The resulting culture was stained and examined under the microscope (Fig. 5.3) to confirm the presence of free spores without the vegetative cells (Chang et al., 1985). The *B. subtilis* spore suspension was assayed in triplicates by spreading it over plate count (PC) agar plates. Appropriate dilutions of the samples were made with 0.001 M phosphate buffer solution (PBS). They were plated and spread uniformly on PC plates and incubated for 24 hours at 37 °C.

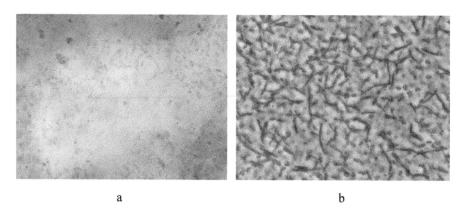

a   b

Figure 5.3. B. subtilis spores under 100x magnification. Before pasteurization (a), the spores can be seen inside vegetative cells. After pasteurization and sonication (b), spores (green in colour) are released from inactivated cells (brown in colour).

MS2 bacteriophages

The enumeration of MS2 phage was done by double layer agar method following ISO 10705-1:1995. *E. coli* ATCC 15597 was used as a host organism. The host culture was grown to exponential phase (concentration 10^8 CFU.mL^{-1}) and mixed with the dilute sample. The solution was mixed and poured uniformly over the surface of Tryptone Yeast Glucose Agar (TYGA) plates and incubated for 18h ± 2h at 37 ˚C. Sterile filter pipette tips were used in all steps to avoid contamination.

5.2.3 Light source and parameters

Simulated sunlight was produced using an OSRAM HQI-BT 400 W/D PRO (OSRAM GmbH, Munich, Germany) metal halide lamp with built-in UV filter that blocks wavelengths shorter than 320 nm. The cut-off filter was used because lower wavelengths are highly attenuated by the terrestrial atmosphere, especially in a cloudy sky (Calbó et al., 2005). The same lamps were used in Chapters 2 and 4 and a similar cut-off filter was used in a solar disinfection study on poliovirus and *Acanthamoeba polyphaga* cysts by Heaselgrave et al. (2006). These metal halide lamps were used because their spectrum is continuous, like sunlight, with a few peaks in the area of visible light, the highest one being at 540 nm (Calin and Parasca, 2008). The light intensity was fixed at 320 W.m^{-2}, approximating the extreme conditions of minimal daily direct solar irradiance at a latitude of 60° N, observed at winter solstice (Arabali et al., 2017), as very few cities are located at higher latitudes. The photon flux, in μmol.m^{-2}.s^{-1}, was measured with a LI-250A Light Meter equipped with an underwater quantum sensor (LI-COR® Biosciences, Inc., Lincoln, Nebraska, USA). This sensor has a uniform sensitivity in the range of wavelengths between 400 and 700 nm; this limitation was taken into account when calculating the lamp irradiance spectrum. Light intensity was calculated considering the spectral power distribution of the lamp, as done in Chapter 2 using the same lamp. In the tests performed with natural sunlight, the light intensity was calculated in a similar manner, taking into account the spectral power distribution of solar light from reference spectra (Air Mass 1.5) (ASTM, 2012), which has 5.9% UV-A and 0.2% UV-B. The average intensity was calculated as a weighted average of the measured values.

5.2.4 Experimental design

Inactivation of E. coli, B. subtilis spores and MS2 bacteriophages on urban surfaces under artificial light and dark conditions

This phase was carried out in two batches. In the first batch, the inactivation of *E. coli* and *B. subtilis* was measured in the water wells set up on the pavement blocks, concrete, asphalt, and glass. The inactivation was studied on a light phase of 6 h followed by 18 h of dark phase, representing the minimal daylight duration at a latitude of 60° N and 60° S, observed at winter and summer solstice, respectively. In addition, the inactivation of *E. coli* was studied separately on the concrete surface for 24 h in dark conditions to test whether the high pH of concrete affects inactivation in these conditions. The experiments with MS2 were carried out separately (second batch) under the same conditions because high concentrations of bacteria other than the host for MS2 could hinder plaque enumeration. The effect of temperature on the inactivation of all three organisms was studied separately, by keeping all the organisms on a glass surface at a constant temperature of 33 °C for 24 h in dark conditions, in the incubator.

Inactivation of E. coli, B. subtilis spores and MS2 bacteriophages on pavement under natural sunlight and dark conditions

The inactivation of *E. coli*, *B. subtilis*, and MS2 was studied outdoors on pavement, under natural sunlight, in two open batch reactors, in duplicate. Pavement was tested in real conditions to evaluate whether the real sunlight would accelerate inactivation, because the lowest, most critical inactivation under artificial light was observed on pavement. Each reactor contained two pavement blocks of the same type as in the previous experiment and water around the tiles was kept as a buffer to minimize the heating effect from sunlight, but without the use of a cooler. The pavement blocks in the first reactor were inoculated with *E. coli* and *B. subtilis* together, while in the second reactor they were inoculated with MS2. During the night, the reactors were covered using a transparent glass cover to be protected from wind and other interferences. The experiment was carried out on February 26, 2019, a dry day with continuous sunlight (apparent sunrise at 7:36 and sunset at 18:16 (NOAA, 2019)). As "light phase" was considered a period of 8 h, from 9:00 to 17:00, when the setup was exposed to direct sunlight. The following 16 h were considered as part of the "dark phase".

5.2.5 Sampling and physicochemical parameters

Samples of 2.5 mL were taken from each water well on the surfaces. During the light phase, hourly samples were taken in the first part, while in the second part samples were taken every 2h. Samples were taken once in two hours in both parts under dark conditions. Samples were taken from the centre of each water well after mixing the water with the help of a pipette by drawing up and releasing the sample from the pipette tip for around 3-5 times. The temperature of water on the surfaces was measured by using an infrared thermometer (FERM, ITM1001, The Netherlands). pH was measured with a handheld pH meter (WTW pH 323, WTW GmbH, Weilheim, Germany) and with pH paper (Fisherbrand™ pH indicator sticks). The data for each test sample was taken in triplicate for the first part while for the rest it was taken in duplicate.

5.2.6 Data analysis

The inactivation of organisms was calculated based on the "Chick-Watson" first-order exponential inactivation equation (Eq. 1.1). The sample for initial concentration of the organisms was measured immediately after the sample was poured into the water pool on the surfaces. The log survival rate of organisms $\ln(C_t/C_0)$ was plotted against time (t) and the inactivation rate was calculated for each experiment as the value corresponding to the slope of linear regression.

The statistical significance of the data in all experiments was analysed using t-test for paired two samples for means with a significance level of $\alpha=0.05$, comparing the $\log_{10}$ values of microbial concentrations at the beginning and the end of each experimental phase. In the case of *B. subtilis* spores, the test was carried out for the overall experiment, between the beginning of the light phase (0 h) and the end of the dark phase (24 h). To compare inactivation rates between *E. coli* and MS2 under exposure to light, two-factor ANOVA tests with replication were used ($\alpha=0.05$). In these tests, the $\log_{10}$ values of the concentrations of *E. coli* at the beginning and the end of the experimental phase were compared to the respective values of MS2. It was assumed that microbial concentrations were normally distributed.

5.3 RESULTS

5.3.1 Inactivation of *E. coli* under artificial light and dark conditions

The inactivation profiles of *E. coli* on all the tested surfaces is shown in Fig. 5.4.a. The graph shows that there was no significant inactivation under dark conditions on any of the surfaces over the measurement period. The t-test for the dark phase confirmed that there was no significant difference in the concentration of *E. coli* before and after the dark phase. In the case of concrete the last point (24 h) was under the detection limit, but a repetition of the experiment in dark conditions confirmed that the inactivation was not statistically significant. Under exposure to light, the inactivation was significant on all surfaces: the highest one took place on concrete with an inactivation rate of 1.85 h^{-1} followed by asphalt (1.01 h^{-1}), glass (0.78 h^{-1}) and the lowest was on pavement (0.36 h^{-1}). The summary can be seen in Table 5.1.

The difference between the concentrations before and the ones after 24 h in dark conditions at 33 °C on glass was not statistically significant for any of the three organisms ($p=0.19$ and sample size $n=8$ for *E. coli*, $p=0.43$ and $n=8$ for *B. subtilis*, $p=0.78$ and $n=5$ for MS2, data not shown). Therefore, since in dark conditions there was no inactivation at any of the tested temperatures, the increase of temperature alone did not affect inactivation.

5.3.2 Inactivation of *B. subtilis* under artificial light and dark conditions

No significant inactivation under light or dark conditions was observed for *B. subtilis* spores on any of the surfaces in this study (Fig. 5.4.b). The concentration of *B. subtilis* spores was constant, regardless of the different surfaces used. Statistical *t*-tests conducted for *B. subtilis* spores confirmed for all surfaces except pavement that any differences between the concentrations at 0 h and 24 h were not statistically significant, implying that no inactivation was induced on the spores. In the case of pavement, a minor overall inactivation rate of 0.03 h^{-1} was observed ($p=0.03$). However, *t*-tests performed separately on pavement for the data of the light phase (0 h and 6 h) and for the data from the beginning of the dark phase (6 h) until 12 h, showed no statistical significance (with $p=0.052$ and $p=0.77$, respectively). Therefore, the only statistical difference is caused by the last experimental point at 24 h, and can be considered to be an outlier.

Table 5.1 Summary of all inactivation rates under artificial light, sunlight and dark conditions for all organisms (t-test for paired two samples for means, α=0.05, comparing log₁₀ values of microbial concentrations at the beginning and end of each phase, or of the overall experiment). The number of samples (n) is indicated.

Surfaces	Average pH	E. coli			B. subtilis spores	MS2 bacteriophages		
		Light		Dark	Overall	Light		Dark
		Inactivation rate (h⁻¹)	R^2	Inactivation rate (h⁻¹)	Inactivation rate (h⁻¹)	Inactivation rate (h⁻¹)	R^2	Inactivation rate (h⁻¹)
Artificial light (320 W.m⁻² for 6 h)								
Glass	6.9	0.78 ($p=0.01$, $n=7$)	0.90	0.02 ($p=0.59$, $n=3$)	0.02 ($p=0.47$, $n=9$)	0.64 ($p=0.08$, $n=4$)	0.99	0.04 ($p=0.12$, $n=3$)
Pavement	8.1	0.36 ($p=0.04$, $n=7$)	0.78	0.07 ($p=0.27$, $n=5$)	0.03 ($p=0.03$, $n=11$)	0.18 ($p=0.77$, $n=4$)	0.91	0.05 ($p=0.51$, $n=3$)
Concrete	9.6	1.85 ($p=0.04$, $n=8$)	0.96	0.02 ($p=0.19$, $n=4$)	-0.01 ($p=0.17$, $n=12$)	0.39 ($p=0.06$, $n=4$)	0.99	0.05 ($p=0.07$, $n=4$)
Asphalt	7.6	1.01 ($p=0.047$, $n=7$)	0.93	0.01 ($p=0.65$, $n=4$)	-0.05 ($p=0.19$, $n=11$)	1.29 ($p=0.02$, $n=4$)	0.96	0.03 ($p=0.01$, $n=4$)
Natural sunlight (512 W.m⁻² for 8 h)								
Pavement	8.0	1.03 ($p=0.01$, $n=5$)	0.84	0.08 ($p=0.97$, $n=2$)	-0.01 ($p=0.93$, $n=6$)	0.11 ($p=0.15$, $n=4$)	0.91	0.02 ($p=0.50$, $n=2$)

Figure 5.4 Inactivation of E. coli (a), B. subtilis (b) and MS2 bacteriophages (c) on different surfaces. The white area represents the light phase and the grey area represents the dark phase.

5.3.3 Inactivation of MS2 bacteriophages under artificial light and dark conditions

The inactivation of MS2 bacteriophages on all the surfaces is depicted in Fig. 5.4.c. The *t*-tests performed for MS2 bacteriophages in dark and in light conditions, for all surfaces except asphalt, indicated that the difference between the initial and the final MS2 concentrations at each phase was not statistically significant. In the case of asphalt, the inactivation in dark was statistically significant (p=0.01), but the rate (0.03 h^{-1}) is an order of magnitude lower than the inactivation under light conditions, making it negligible in practice. On the other hand, under light conditions the highest inactivation was observed on asphalt, followed by glass, concrete and the lowest one was on pavement blocks. The summary of all the inactivation rates can be seen in Table 5.1.

The trends followed by the inactivation rate of all three organisms under light conditions were the same on all surfaces (glass, pavement, concrete), except on asphalt, *E. coli* being the most sensitive organism, followed by MS2 and *B. subtilis* spores. In the case of asphalt, the inactivation under light conditions of *E. coli* and MS2 were similar followed by the inactivation of *B. subtilis* spores.

5.3.4 Inactivation of *E. coli*, *B. subtilis* spores and MS2 bacteriophages on pavement under natural sunlight and dark conditions

The results of the study performed under natural sunlight on pavement can be seen in Fig. 5.5. The average pH measured on the water well on the pavement was 8.2. The weighted average light intensity during the light phase was calculated to be 515 W.m^{-2}. Also, the average temperature during the light and dark phase was 15.5 °C and 4.5 °C, respectively.

The results show that the difference between the initial and the final concentrations was not statistically significant for any of the three organisms under dark or light conditions, except *E. coli* having an inactivation rate of 1.03 h^{-1} under exposure to light. There was no significant change in the concentration of *B. subtilis* spores even after a second day of exposure (data shown only for the first 24 h). Table 5.1 summarises all the inactivation rates of the three organisms for the experiment conducted under natural sunlight and dark conditions.

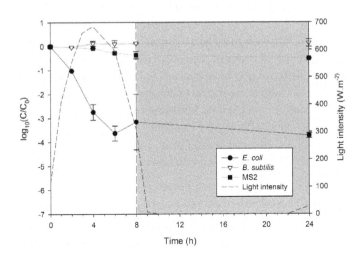

Figure 5.5 Inactivation of E. coli, B. subtilis and MS2 on pavement under natural sunlight (white area) and dark (grey area).

5.4 DISCUSSION

5.4.1 *E. coli*

The highest inactivation under light conditions was observed on *E. coli*. Biological weighting functions of *E. coli* show that the inactivation effect of wavelengths higher than 400 nm is negligible (Nelson et al., 2018). Therefore, as the lamp has a cut-off filter at 320 nm, the total inactivation under artificial light was caused by UV-A radiation (Chapter 4) which can cause damage to *E. coli* cells when the photons are absorbed by sensitizer molecules and induce the formation of photo-reactive intermediates, within or outside the cells (Nelson et al., 2018). Nevertheless, photosynthetically active radiation (400-700 nm) can still be important for *E. coli* inactivation (Dias and von Sperling, 2018), especially in deeper waters or higher turbidity levels where UV is attenuated (Dias and Sperling, 2017).

On concrete the highest inactivation under light conditions of *E. coli* observed may be explained due to the combined effect of UV from the light source and pH of concrete (average pH 9.6). This was interpreted based on the result that all the surfaces, except concrete, had neutral pH ranging between 6.5 and 8.0 and the same light intensity was applied to all surfaces. On the other hand, *E. coli* was not inactivated when kept for 24 h in dark, therefore the pH of concrete alone did not lead to the inactivation of *E. coli*.

83

Further research is needed on the combined effect of concrete samples with different properties (with varying composition and wearing) and light intensity.

The inactivation of *E. coli* seen on asphalt could be explained due to the combined effect of light and the presence of complex aromatic hydrocarbons, extracts of crude oils, in asphalt (Xia et al., 2019) that may be detrimental to *E. coli*. However, the composition of asphalt does not have a significant impact on inactivation on its own, as the inactivation under dark conditions was negligible. Furthermore, there is rather limited literature regarding the inactivation effects of *E. coli* on asphalt. Hence, the effects of the composition of asphalt on the inactivation of *E. coli* (and other organisms) under light conditions can be the starting point for future studies.

The inactivation of *E. coli* was lower on pavement (0.36 h^{-1}) than on the glass control (0.78 h 1). This may be attributed to the fact that the glass surface was shiny and acted as a reflector, increasing the exposure to light, in a similar way to solar mirrors and reflectors used in solar photo-reactors (Nalwanga et al., 2014). The neutral role of the surface of the pavement blocks on *E. coli* inactivation is also supported by the fact that the inactivation rate on pavement is relatively similar to the rates measured in Chapter 2 in clear artificial flood water under exposure to the same lamp and light intensity for a duration of 6 hours per day (0.21 h^{-1}) or 12 hours per day (0.37 h^{-1}).

5.4.2 *B. subtilis* spores

B. subtilis spores had a constant concentration on all surfaces (except a small inactivation on pavement) regardless the light or dark conditions, including the effects of natural sunlight on pavement. These results reflect that *B. subtilis* spores are much more resistant to environmental stresses such as UV light and pH as compared to *E. coli* and MS2 viruses. Several studies confirm the endurance of *B. subtilis* spores against various environmental stresses. For instance, Chang et al. (1985) found that *B. subtilis* spores were more resistant to UV doses than vegetative bacteria like *E. coli*, *Staphylococcus aureus* and *Shigella sonnei*. Nicholson et al. (2000) also concluded that the decrease in concentration of *B. subtilis* spores was low or non-detectable even after exposure to solar heating (but protected from UV primary effects) at a temperature higher than 70 °C. The high resistance of *B. subtilis* spores has been attributed by Nicholson et al. (2000) to different factors like presence of spore coat and low relative permeability of spore core, among other characteristics and features. The layers present within the spore coat were found to be responsible for the resistance of spores in exposure to UV for both sunlight and artificial radiation (Mamane et al., 2007; Riesenman and Nicholson, 2000). This makes *B. subtilis* spores suitable surrogates for persistent organisms like *Cryptosporidium* (Headd et al., 2016).

5.4.3 MS2 bacteriophages

During the dark phase, the concentration of MS2 remained constant on all surfaces, with a negligible inactivation on asphalt (0.03 h^{-1}), therefore the composition of the materials alone did not affect MS2.

The inactivation rate of MS2 under light conditions increased from pavement to glass and concrete, reaching the highest with asphalt. Furthermore, the ANOVA tests with replication showed that on concrete and glass under light conditions MS2 was more resistant than *E. coli* ($F=38.4$, $p=0.004$ for concrete and $F=110.2$, $p=0.001$ for glass), and there was no statistical significance when comparing the decrease of concentration of the two organisms on pavement ($F=0.68$, $p=0.46$). However, the inactivation rate of MS2 was slightly higher than the one of *E. coli* on asphalt, 1.29 h^{-1} and 1.01 h^{-1}, respectively ($F=128.8$, $p=0.0003$). The possibility of a substantial attachment of MS2 viruses on the hydrophobic surface of asphalt (Farkas et al., 2015; Dika et al., 2013; Hefer et al., 2006) was ruled out because of the absence of significant decrease of MS2 concentration in the water phase in dark conditions. The same applies to the other organisms as well because in all cases the samples were taken from the bulk water phase and any organisms attached would not be included in the sample. Moreover, the phenomenon of aggregation of bacteriophages, which would lead to fewer plaque counts than the actual viruses, was unlikely because the pH of the samples on all the surfaces was higher than the isoelectric point of MS2 bacteriophages (pH 3.9) at which this phenomenon occurs (Furiga et al., 2011; Gassilloud and Gantzer, 2005; Langlet et al., 2007).

Bacteriophage inactivation has also been attributed to the exposure of bacteriophages to interfaces like air-water interface (AWI) or gas, liquid and solid interface known as triple-phase boundary (TPB), due to the lethal unfolding of the hydrophobic part of the viral capsid on the interface, an effect which is stronger on hydrophobic surfaces (Furiga et al., 2011; Thompson et al., 1998; Thompson and Yates, 1999; Trouwborst et al., 1974). However, in the experimental setup no inactivation caused by exposure to an AWI or TPB interface was observed in dark conditions, therefore the above is likely not a significant factor.

From all these observations, it can be concluded that a detailed quantification of the contribution of each process (inactivation, adhesion, exposure to TPB, etc.) on the concentration of MS2 bacteriophage is complex (Dika et al., 2013), as is for bacteria. Hence, subsequent studies can include further analysis at a microscopic level, including the nature of the viruses (such as surface charge and hydrophobicity, among others), properties of the surfaces where viruses were exposed to (like roughness, hydrophobic/hydrophilic nature), properties of the medium and interactions of viruses with particles and with other organisms (Verbyla and Mihelcic, 2015).

5.4.4 Inactivation of *E. coli*, *B. subtilis* spores and MS2 bacteriophages on pavement under natural sunlight and dark conditions

The inactivation of *E. coli* on pavement was almost triple under natural sunlight than in the tests exposed to the artificial light source. This can be explained owing to the fact that the weighted average sunlight intensity measured was 512 W.m^{-2}, which was around 1.5 times higher than the light intensity used in the setup with artificial light source (i.e. 320 W.m^{-2}). Also, the duration of the light phase was 8 h under sunlight while it was only 6 h for artificial light. Thus, both the intensity and exposure time were higher in the experiment carried out under natural sunlight. In addition, even though sunlight comprises only small amounts (0.2% of reference spectrum) of UV-B radiation that was cut-off in the lamp, UV-B can have detectable effects on survival of microorganisms (Nelson et al., 2018; Dias et al., 2017; US EPA, 2010). The inactivation of MS2 under exposure to natural sunlight was similar to the inactivation under artificial light. The effect of UV-B on MS2 was little compared to *E. coli*, as was also observed by Lian et al. (2018) when both were exposed to 4.5 W.m^{-2} of UV-B radiation, probably because of the larger physical size and amount of genetic material in *E. coli* compared to MS2.

5.4.5 General observations

E. coli and MS2 followed similar trends of inactivation under light conditions on the different surfaces tested (with the exception of asphalt), *E. coli* having the highest inactivation, followed by MS2, while *B. subtilis* practically did not show any inactivation. In the case of asphalt, the inactivation rates of *E. coli* and MS2 under light conditions were similar and that of *B. subtilis* spores was the lowest. A similar trend was observed for UV disinfection by Chang et al. (1985) where viruses (polio and rota virus), spores (*B. subtilis*) and cysts (*Acanthamoeba castellanii*) were found to be 3-4, 9 and 15 times more resistant than vegetative bacterial cells (*E. coli*, *S. aureus* and *S. sonnei*). This again brings up an important conclusion that in order to assess the public health risk after urban floods, it is necessary to monitor viral and spore forming organisms that can act as indicators for environmentally stable pathogens capable of surviving as spores or oocysts (Dias et al., 2018; Headd et al., 2016). For this reason, it is also necessary to develop low cost, fast and reliable detection techniques for these indicators.

Further research can be undertaken with different kinds of organisms, in particular helminth eggs because they are very resistant (Nelson, 2003), cysts and oocysts, or with a combination of organisms, in real environmental conditions, with different levels of turbidity and/or in a wider variety of surfaces found in the urban environment with different levels of hydrophobicity and pH, as well as different weathering conditions. This research, in combination with QMRA and mapping of urban surfaces, and taking into consideration the influence of the geographical location and time of the year on surface solar irradiance, can be used to develop policy-making tools for the implementation of measures to mitigate public health risks after flooding. Based on the behaviour and survival of the indicator organisms on those surfaces, and with the help of GIS maps, precautionary and public awareness measures can be taken. This can be a valuable input to plan immediate actions regarding the microbial risk and safety, to concerned authorities, after urban flooding.

5.4.6 Strengths and limitations

Although this study provides a better understanding of the inactivation of surrogates for bacterial and viral contamination and parasite oocysts on different surfaces after urban floods, for which the scientific literature is very limited, the persistence of the surrogate organisms has not been examined side-by-side with the persistence of the respective pathogens. Therefore, future experiments must study how well these organisms represent the persistence of pathogens under identical experimental conditions.

Similarly to the previous chapter, the temperature control system did not maintain a constant temperature throughout the experiments that took place under exposure to light. On the positive side, as expected, the maximum temperatures were lower in this setup thanks to the direct contact of the surface to the buffer water and the constant presence of water in the sampling wells on top of the surfaces. The presence of this water provided also a solution to the problem of evaporation that was observed in the previous chapter, which allowed the removal of the Perspex cover in this setup. Of course, it was still difficult to achieve a uniform temperature distribution due to the presence of the surrounding cooling water and to the small distance of the light source and an ideal solution would be the use of bioclimatic chambers with controlled temperature and relative humidity conditions.

Also, the lack of replicate experiments and the short duration of the experiments for 24 h were limitations that need to be mentioned.

5.5 CONCLUSIONS

After testing different surfaces and sunlight exposures to *E. coli*, *B. subtilis* and MS2, the following conclusions can be made:

- *E. coli* had the highest inactivation under artificial light exposure followed by MS2 bacteriophages, while *B. subtilis* spores were stable, practically not inactivated on any of the tested surfaces, with the exception of asphalt. On asphalt the inactivation of *E. coli* and MS2 under light conditions was similar, due to much faster inactivation of MS2.

- The highest inactivation under light conditions was that of *E. coli* on concrete, with an inactivation rate of 1.85 h⁻¹, attributed to the synergetic effect of light and high pH (pH 9.6) of concrete.

- MS2 bacteriophages had the highest inactivation under light conditions on asphalt with a rate of 1.29 h⁻¹.

- No inactivation of *B. subtilis* spores was observed in any of the experiments conducted, indicating the resistance of *B. subtilis* spores to the conditions applied.

- No inactivation under dark conditions of *E. coli*, *B. subtilis*, and MS2 bacteriophages was observed, concluding that the sole effect of pH and other properties of the surfaces (pavement, concrete and asphalt) was not significant.

- *E. coli* cannot be used as a suitable indicator for human viral, spore forming and protozoan pathogens. Hence, it is always necessary to monitor a combination of indicator organisms that resemble more to environmentally stable pathogens, like those surviving as spores or oocysts, in order to assess the public health risk after urban floods.

6

OUTLOOK

Knowledge and understanding of the parameters and processes that affect the inactivation of water-borne pathogens after floods, can lead to better estimation of public health risks.

6.1 REFLECTIONS

In the context of adaptation to climate change and the related to it extreme events it is necessary to better understand the phenomena linked to contamination of waters with water-borne pathogenic microorganisms after urban floods. Three main research questions were posed at the beginning of this thesis and were addressed in the previous chapters.

6.1.1 Sampling methods

The question of selecting a reliable and simple method for collecting samples from different urban surfaces after the recession of flood water was studied in Chapter 3. Although several methods are described for smooth surfaces like stainless steel benches in the food industry and in hospitals, literature about rough and porous surfaces was limited. The experiments showed that swabbing and contact plating are reliable methods to sample and enumerate the presence of *E. coli* on different urban surfaces after flooding. Swabbing was the best of the tested methods for a wide range of concentrations. Contact plates had the highest recovery, but only for a limited range of *E. coli* concentrations. Tape and stamps had technical challenges and no advantages over contact plating or swabbing. These conclusions lead to the use of the swabbing method in Chapter 4. Moreover, these methods are simple, of low cost and already commercially available, and can therefore be easily used *in situ* for the collection of samples in tests after floods. Although swabbing and contact plating are methods widely in use already, the novelty of this study was to assess them on different surfaces of the urban environment for which literature was extremely limited.

This study can be expanded to include more surface types. A table of recovery ratios for swabbing on different surfaces can allow direct conversion of plate counts into actual surface concentrations of pathogens. Also, as attachment, detachment and inactivation depend on the organisms, the study of further indicator organisms, such as protozoan cysts or oocysts, bacterial spores, bacteriophages or other viruses and helminth eggs will be useful. Different methods, such as sponges, wipes, vacuum methods and different kinds of swab tips can also be tested. Standardisation of sampling methods in the future, considering the relevant species, the aim of the study, the limits of detection and other parameters will be helpful in assessing microbial proliferation on different surfaces (Verdier et al., 2014) and in addressing public health risks after floods.

Of course, these results are not relevant only in the case of urban floods, but also in any other case where the detection of pathogens is required on urban surfaces such as areas around pit latrines or septic tanks, especially after emptying operations, with an impact on the society and the environment.

6.1.2 Inactivation of indicator organisms in water

The second research question was related to the effect of solar radiation and total suspended solids on the inactivation of indicator organisms that originate from the surcharging of combined sewer systems and was studied in Chapter 2. Under laboratory controlled conditions it became evident that inactivation of faecal indicators *E. coli* in waters was linearly proportional to light intensity and duration of exposure to light. As light attenuation coefficient was proportional to the concentration of suspended solids, the inactivation rate decreased exponentially with an increase in TSS concentration. It is noteworthy that even if bacteria were not found to be attached on particles, they were protected from photo-inactivation thanks to the presence of particles for a period of a few days. The outcome of these findings is that by measuring the initial concentration of *E. coli* and the pH of flooding water, and sunlight conditions, it is possible to predict the fate of faecal coliforms after a flooding event. Of course, water quality was also confirmed as an important parameter that needs to be taken into account, underlining the increased complexity of natural systems. The experimental results provided a good understanding of the inactivation of *E. coli* in the water phase that was a necessary step before studying the effect of different surface types.

6.1.3 Inactivation of surrogate organisms on surfaces

The third research question was about the effects of solar radiation on the inactivation of different organisms on urban surfaces. As no inactivation of *E. coli*, *B. subtilis*, and MS2 bacteriophages was observed under dark conditions, it was concluded that the sole effect of pH and other physical or chemical properties of the urban surfaces (pavement, concrete and asphalt) was not significant (Chapter 5). Under dark conditions only the composition of playground rubber tiles had a minor negative effect on *E. coli*. Also, the impact of high pH was significant only in the case of ground pavement blocks (pH 10.9 ± 0.5) corresponding to fresh concrete, while in older concrete the outer surface had lower pH (Chapter 4). This underlines the high risk of water-borne diseases in places that are not exposed directly to sunlight, such as in flooded buildings, between tall buildings, in the shadow and in cloudy weather. Extreme local values of pH can increase inactivation, but these are not so common and the use of materials with a permanent high pH or other antimicrobial properties is something that needs to be further investigated.

Examination of the inactivation of *E. coli* after the recession of flood water in Chapter 4 indicated that photo-inactivation was minor compared to temperature stress and evaporation that reduced the surface moisture of the tested surfaces. For this reason, in Chapter 5 the surfaces were kept flooded with 1 cm of flood water, under controlled temperature. *E. coli* had the highest inactivation under light exposure followed by MS2 bacteriophages, while *B. subtilis* spores were stable, practically not inactivated on any of the tested surfaces, with the exception of asphalt. In fact on asphalt the inactivation of *E.*

coli and MS2 under light conditions was similar, due to much faster inactivation of MS2. This can be attributed to the combined impact of light and the leachate of asphalt, which is much more evident on MS2.

The high resistance of *B. subtilis* spores to the conditions applied leads to the conclusion that *E. coli* is not the most suitable indicator for all human pathogens. Hence, it is always necessary to monitor a combination of indicator organisms that resemble more to environmentally stable pathogens, like those surviving as spores, oocysts or helminth eggs, in order to assess the public health risk and safety conditions after urban floods.

This thesis examined the contribution of urban surfaces to pathogen inactivation under different conditions of exposure to light, for which literature is extremely limited. The sets of inactivation rates provided in the thesis can be used for the assessment and reduction of health risks through better planning. Of course this is an extremely complex issue with a variable set of situations, depending on the location, the climate characteristics, the institutional frameworks present, etc. Although many aspects still remain to be studied, this thesis provided a new approach by correlating the issue of public health risks due to water-borne diseases with the presence of different urban surfaces.

6.1.4 Strengths and limitations of this thesis

The novelty of this thesis was the study of the inactivation of surrogate organisms for bacterial and viral contamination and parasite oocysts after urban floods, for which the scientific literature is very limited. A better understanding of the complex processes taking place was achieved, but some research gaps still remain. The following section attempts to describe only a few of the potential practical applications and further research directions that can follow up this thesis.

Of course, it is also important to indicate the limitations of this work. In terms of the setups used, the main limitation was the difficulty to maintain constant temperature under exposure to light. Although in Chapter 5 this was improved compared to Chapter 4, an ideal solution would be the use of bioclimatic chambers with controlled temperature and relative humidity conditions. Although some tests at a constant temperature of 33°C for 24 h in dark conditions in the incubator showed that all three organisms maintained constant concentrations, the effect of temperature needs to be taken into account and studied separately. Another limitation was the absence of replicate experiments in most cases, the small duration of experiments (in Chapters 4 and 5), and the small sample sizes used.

6.2 RECOMMENDATIONS FOR PRACTICAL APPLICATIONS AND SUGGESTIONS FOR FURTHER RESEARCH

Further research is necessary on the chemical effects of different materials on faecal pathogens. The effect of asphalt constituents on the inactivation of pathogens, for instance, needs to be studied further, as asphalt is a very common surface in urban areas and it is known that leaching is a common process related to it.

The effect on pathogen removal of carbonation and other processes related to the weathering of materials and in particular of concrete should be further investigated. Water in the pores of fresh concrete is saturated with calcium hydroxide, and contains sodium hydroxide and potassium hydroxide, having a typical pH of around 13-14. On the contrary, a fully carbonated part has usually a pH as low as 8.4. An easy method to assess the pH of the concrete other than pH paper is by spraying a phenolphthalein indicator solution which turns purple at pH higher than 9. The denser is the colour, the higher is the pH. Using this method, automatic detection of carbonated regions is possible with the use of image-processing algorithms (Choi et al., 2017). A non-toxic curcumin-based solution that changes from red to yellow in carbonated zones has also been proposed (Chinchón-Payá et al., 2016). These tests can be applied as part of on-site collection of data, especially in urban areas where concrete of different ages is an abundant surface. There are still many research gaps concerning the combined effect of the above and other factors, such as pH, and sunlight irradiation.

Future research is also necessary on the processes taking place within the sewer system, including decentralised, on-site sanitation systems such as pit latrines or septic tanks, as well as on the spread of indicator organisms from the sewerage network to urban surfaces and to the water sources and supply infrastructures under various conditions in both developed and developing countries. All that can provide valuable information on the initial presence of different concentrations of pathogens, nutrients available, pH and other water quality characteristics immediately after extreme rainfall events. In combination with the results of the present work, this can lead to a more holistic approach in addressing the critical issue of contamination of the environment with pathogens during and after urban/peri-urban floods.

Furthermore, many elements of the present thesis could be combined with research on irrigation of agricultural or recreational/forested lands and gardens by using non-conventional water resources such as storm waters stored in various types of reservoirs or treated wastewaters. In conclusion, the present work offers many opportunities for further research projects of high application potential.

6.2.1 Contribution to public health measures, policy frameworks and the SDGs

This thesis has showed that the use of FIBs alone for the assessment of public health risks is not sufficient. For instance, the viruses MS2 and spores of *B. subtilis* were not always inactivated, while *E. coli* concentration was within the required limits. Therefore, it is necessary to take additional measures, by introducing more resilient indicator organisms for the safety standards. For this reason, more categories of organisms need to be studied, such as helminth eggs and parasite cysts/oocysts that were not included in the present research. The detection of any organisms used as indicators, however, needs to be low cost, reliable and fast. The development of commercial kits for their detection in the field, similar to the kits used for *E. coli* detection, is an alternative that needs to be considered.

Moreover, additional practices can be suggested in order to allow public access to certain areas. Such examples may be chemical or physicochemical disinfection of areas, including through natural inactivation mechanisms allowing for certain number of days after the surfaces dry out, or even more advanced methods such as bacteriophage therapy, which has the advantage of reducing only specific targeted host bacteria (Ye et al. 2019). For this, the design/development and use of models that could predict possible risks would be extremely useful. Such an approach is explained in the next section. The above measures have to be complimented by policy frameworks, education and training in order to achieve behavioural change.

The results of the present thesis and the models to be developed can be very useful if properly taken into account for drafting new or amending existing policies connected to floods, covering a wide spectrum of plans related directly or indirectly to protection and amelioration of public health. Such plans and policy frameworks are the following: planning for adaptation to climate change; urban planning; physical planning; integrated urban water management; integrated water resources management; integrated coastal zone management; ecosystem based approach; environmental planning; sustainable development planning including contribution to achieve the SDGs; "nexus planning" for security of the water-energy-food-ecosystems nexus; and combinations of the above as suggested in the integrative methodological framework .

More specifically related to the SDGs, this work can be actually applied to various of the target communities of SDG 13 on climate action, as well as on a number of targets of SDG 3 on good health and well-being, and more specifically on SDG 3.3 on water-borne diseases, on SDG 6 on clean water and sanitation, on SDG 4 on education, and more specifically on target 4.7 on education for sustainable development and many more.

The results allow for a series of proposals for further work and useful applications addressing different target groups in both developing and developed countries. The following three can be regarded as the most important target groups. First, decision makers in cooperation with those directly responsible for addressing risks from floods and extreme weather events. They could use the results in predicting the needed time for safe use of water in infrastructures and the environment. Second, local authorities, designers of new cities or new neighbourhoods who may incorporate, if not already in practice, or prioritise the use of appropriate drainage schemes and construction materials that allow for a rapid drying of surfaces and inactivation of microbes. Third, formal, non-formal and informal educators, such as schools, local authorities and NGOs, who could inform the public on how to enhance safety and avoid disasters related to the impact of floods. This may be part of an education for sustainable development (ESD) intervention. The role of ESD is fundamental in reducing health risks related to floods. Basic elements should be integrated in the relevant education for ESD curricula. A variety of projects could be proposed for formulating appropriate formal and non-formal education messages or courses, identifying simultaneously the suitable subjects in the curricula to make the interventions. Training for trainers on the interventions may be also considered in cooperation with academic or in-service training institutions.

However the difficulty of obtaining data in the field is an important limiting factor. In view of the above the lack of data, particularly for specific districts, towns, or even neighbourhoods could be complemented by "citizen science" on the basis of some protocols and guidelines to be issued in synergy among local authorities, academic institutions and other stakeholders such local NGOs. These data could include qualitative and quantitative elements e.g. on surfaces, plant cover, paving materials, open spaces, vulnerable surface and/or shallow underground detention and retention systems, wetlands etc.

6.2.2 Urban planning

Contemporary flood management strategies try to go beyond floodwater control by integrating multi-faceted methods in multi- and intra-disciplinary approaches (O'Neill, 2018). In this framework it is important to use sustainable blue-green storm water and nature-based solutions (Bush and Doyon, 2019) and to design appropriate urban surfaces using sustainable pavement materials with low environmental impact and with overall societal benefit (Plati, 2019), such as permeable concrete (Chandrappa and Biligiri, 2016). Also, it is necessary to consider the selection process as a high impact city-level driver (Berndtsson et al., 2019) when planning measures to address urban flood risk. Based on the present work, the chosen materials in flood prone areas or areas with higher public attendance or interaction with vulnerable groups such as children and old people, or places such as schools and hospitals, should not only be based on their hydraulic characteristics concerning the management of flood water quantity, but also based on their role in removing pathogens. This could be, for instance, the use of darker materials that induce higher surface temperatures, materials with higher pH and materials that dry faster, keeping in mind of course all other safety and environmental regulations connected with materials used. This perspective should also be incorporated in sustainable approaches to integrated urban flood management, as one of the criteria for the selection of appropriate sustainable drainage systems, based on local factors (Lashford et al., 2019).

6.2.3 Flood-related public health risk mapping

One of the goals of this work was to provide a better understanding of the concentration of water-borne pathogens after floods, eventually allowing the development of mathematical expressions that can be incorporated into existing flooding models. An effective overall model or combination of models would describe and ultimately predict the fate of the indicator organisms and their potential dispersion in urban environments leading to public health risks taking into account the inactivation of pathogens on specific tested surfaces (Yakirevich et al., 2013; Jonsson and Agerberg, 2014). The design of a model and a resulting warning system should take into account not only the local climate, surface types, plant cover, hydraulic characteristics and latitude of a region but also periodically altering factors like cloud cover.

An integrated approach would consist of the following components:

- **Surface flood model**. There is a wide variety of one- (1D), two- (2D) and three-dimensional (3D) hydraulic flood models for river channels and floodplains. It is worthy to note that most of these models take into account not only the topography, but also surface roughness, because both topography and roughness coefficients affect the flow area and velocity (Liu et al., 2019). The new generation of urban flood simulations aim to not only accurately replicate the water flows, but also to include associated events, such as flows on street networks, water entering buildings, transport of sediment and pollutants in urban environments, for which dedicated experimental data are virtually unavailable (Mignot et al., 2019). Some of these models take into account the vertical flow interactions between the sewer system and the urban floodplain (Seyoum et al., 2012).

- **Pathogen transport model**. Transport of sediment and contaminants in porous media and in rivers has been studied (Walters et al., 2014), but applications on non-permeable urban flood plains are very limited.

- **Inactivation model**. Several existing models can be adapted to take into account the different inactivation rates of pathogens on different surfaces. For example, it is important to know with relative accuracy the presence of concrete surfaces, both relatively recent (high pH) and older ones. Surface maps may be taken from land use maps and other GIS applications. In this approach it is also important to include soil moisture as an important parameter reducing inactivation. Advancements in infrared thermography, allowing high resolution spatio-temporal mapping of surface temperature and moisture by remote sensing

(Schwarz et al., 2018) give in the long term further potential to a real-time public health warning and safety system.

- **Quantitative Microbial Risk Assessment (QMRA).** The health risks of extreme flood events can be assessed by QMRA, a tool to estimate the risk of disease. It relies on several assumptions, especially regarding water quality and hydrological conditions due to insufficient quantitative data. The results of the present work may reduce the uncertainty of some of the model's predictions. However, often, the lack of epidemiological data also prevents the validation of QMRA (Curriero et al., 2001; ten Veldhuis et al., 2010; Andersen et al., 2013).

The outcomes of the above models, combining the public health components (contamination, spread and inactivation of water-borne and vector-borne pathogens) and monitoring modules, after being calibrated and validated *in situ*, could be extremely useful in risk mapping and classification of surfaces and areas into zones of different risk levels in an integrated model. This system could easily be applied in cities and peri-urban or rural areas for the prediction of the minimum time needed for safe inactivation of pathogens. This can be a very useful component of adaptation to climate change practices, from forecasting of risks connected to extreme events under different climate change scenaria, to early warning for implementation of upfront decisions to minimize the potential public health issues, or providing guidance for the management of the situation during or after floods and properly educating and informing key stakeholders and the public at large on health risks associated with flooding events and ways to reduce them.

In conclusion, the predicted by all climatic models increase of the frequency and intensity of floods connected to climate change, on the one hand, and the rapid increase of population, particularly in developing countries and in poor urban and peri-urban areas, on the other, are expected to lead to higher exposure to water-borne pathogens and to enhanced public health risks. The aforementioned conditions and factors make the issue of assessing the fate of pathogens after floods increasingly important, requiring an in-depth understanding. The obtained through the present thesis knowledge of the relevant processes constitutes a concrete contribution towards this direction. Based on this knowledge, no regret measures (physical/technical, institutional and educational/cultural) could be proposed. These may include, *inter alia*, both nature-based solutions and appropriate infrastructures for retention/detention of flood water, improved capacity of drainage systems, guidelines, regulations and incentives for use of suitable surfaces and employment of efficient procedures allowing more rapid inactivation of pathogens, followed and combined with controlled access to potentially contaminated areas. At the same time, capacity building of different stakeholders involved through training and education together with advocacy are necessary in order to achieve behavioural changes and meaningful public participation in proactive and contingency management schemes. Due to the complexity of the phenomena and management activities involved it is necessary to adopt an integrated, multi- and trans-disciplinary approach. A model-based methodology capitalising on all the aspects investigated through the present thesis can offer a useful next step.

REFERENCES

Abraham, W. -R. and Wenderoth, D. F.: Fate of facultative pathogenic microorganisms during and after the flood of the Elbe and Mulde rivers in August 2002. Acta Hydrochim. Hydrobiol., 33, 449–454, doi:10.1002/aheh.200400587, 2005.

Acharya A., Piechota T. C. and Acharya K.: Characterization of first flush phenomenon in an urban stormwater runoff: a case study of Flamingo Tropicana watershed in Las Vegas valley. American Society of Civil Engineers (ASCE), pp. 3366–3375, doi:10.1061/41114(371)347, 2010.

Ahern, M., Kovats, R. S., Wilkinson, P., Few, R. and Matthies, F.: Global Health Impacts of Floods: Epidemiologic Evidence. Epidemiol. Rev., 27, 36–46, doi:10.1093/epirev/mxi004, 2005.

Ahmed, W., Hamilton, K., Toze, S., Cook, S. and Page, D.: A review on microbial contaminants in stormwater runoff and outfalls: Potential health risks and mitigation strategies. Sci. Total Environ., 692, 1304–1321, doi:10.1016/j.scitotenv.2019.07.055, 2019.

Ahnrud, G. P., Mendoza, A. J., Hurley, M. J. and Marek, P. J.: Efficacy of a sonicating swab for removal and capture of microorganisms from experimental and natural contaminated surfaces. Appl. Environ. Microbiol., 84, e00208-18, doi:10.1128/AEM.00208-18, 2018.

Alderman, K., Turner, L. R. and Tong, S.: Floods and human health: A systematic review. Environ. Int., 47, 37–47, doi:10.1016/j.envint.2012.06.003, 2012.

Andersen, S. T., Erichsen, A. C., Mark, O. and Albrechtsen, H. -J.: Effects of a 20 year rain event: a quantitative microbial risk assessment of a case of contaminated bathing water in Copenhagen, Denmark. J. Water Health, 11, 636, doi:10.2166/wh.2013.210, 2013.

Andrade, L., O'Dwyer, J., O'Neill, E. and Hynds, P.: Surface water flooding, groundwater contamination, and enteric disease in developed countries: A scoping review of connections and consequences. Environ. Pollut., 236, 540–549, doi:10.1016/j.envpol.2018.01.104, 2018.

Arabali, A., Ghofrani, M., Bassett, J. B., Pham, M. and Moeini-Aghtaei, M.: Optimum sizing and siting of renewable-energy-based DG units in distribution systems, in: Erdinç, O. (Ed.), Optimization in renewable energy systems, Butterworth-Heinemann, Boston, 233–277, doi:10.1016/B978-0-08-101041-9.00007-7, 2017.

Arnell, N. W. and Gosling, S. N.: The impacts of climate change on river flood risk at the global scale. Clim. Change, 134, 387–401, doi:10.1007/s10584-014-1084-5, 2016.

Ashbolt, N. J., Grabow, W. O. K. and Snozzi, M.: Indicators of microbial water quality, in: Fewtrell, L. and Bartram, J. (Eds.): Water quality: Guidelines, standards and health: Assessment of risk and risk management for water-related infectious disease, World Health Organization, 289–316, available at: https://apps.who.int/iris/handle/10665/42442, 2001.

ASTM: ASTM G173-03(2012), Standard tables for reference solar spectral irradiances: Direct normal and hemispherical on 37° tilted surface, ASTM International, West Conshohocken, PA, doi:10.1520/G0173-03R12, 2012.

Bae, S., Maestre, J. P., Kinney, K. A. and Kirisits, M. J.: An examination of the microbial community and occurrence of potential human pathogens in rainwater harvested from different roofing materials. Water Res., 159, 406–413, doi:10.1016/j.watres.2019.05.029, 2019.

Baffico, G. D.: Optical properties and light penetration in a deep, naturally acidic, iron rich lake: Lago Caviahue (Patagonia, Argentina). Limnologica - Ecology and Management of Inland Waters, 43, 475–481, doi:10.1016/j.limno.2013.03.003, 2013.

Berndtsson, R., Becker, P., Persson, A., Aspegren, H., Haghighatafshar, S., Jönsson, K., Larsson, R., Mobini, S., Mottaghi, M., Nilsson, J., Nordström, J., Pilesjö, P., Scholz, M., Sternudd, C., Sörensen, J. and Tussupova, K.: Drivers of changing urban flood risk: A framework for action. J. Environ. Manage., 240, 47–56, doi:10.1016/j.jenvman.2019.03.094, 2019.

Bertrand, I., Schijven, J. F., Sánchez, G., Wyn-Jones, P., Ottoson, J., Morin, T., Muscillo, M., Verani, M., Nasser, A., de Roda Husman, A. M., Myrmel, M., Sellwood, J., Cook, N. and Gantzer, C.: The impact of temperature on the inactivation of enteric viruses in food and water: A review. J. Appl. Microbiol., 112, 1059–1074, doi:10.1111/j.1365-2672.2012.05267.x, 2012.

Beversdorf, L. J., Bornstein-Forst, S. M. and McLellan, S. L.: The potential for beach sand to serve as a reservoir for *Escherichia coli* and the physical influences on cell die-off. J. Appl. Microbiol., 102, 1372–1381, doi:10.1111/j.1365-2672.2006.03177.x, 2007.

Bhutta, M. A. R., Tsuruta, K. and Mirza, J.: Evaluation of high-performance porous concrete properties. Constr. Build. Mater., 31, 67–73, doi:10.1016/j.conbuildmat.2011.12.024, 2012.

Bitton, G.: Wastewater microbiology. Wiley series in ecological and applied microbiology, Wiley-Liss, New York, 1999.

Blaustein, R. A., Pachepsky, Y., Hill, R. L., Shelton, D. R. and Whelan, G.: *Escherichia coli* survival in waters: Temperature dependence. Water Res., 47, 569–578, doi:10.1016/j.watres.2012.10.027, 2013.

Boehm, A. B., Griffith, J., McGee, C., Edge, T. A., Solo-Gabriele, H. M., Whitman, R., Cao, Y., Getrich, M., Jay, J. A., Ferguson, D., Goodwin, K. D., Lee, C. M., Madison, M. and Weisberg, S. B.: Faecal indicator bacteria enumeration in beach sand: A comparison study of extraction methods in medium to coarse sands. J. Appl. Microbiol., 107, 1740–1750, doi:10.1111/j.1365-2672.2009.04440.x, 2009.

Bonadonna, L., Briancesco, R., Ottaviani, M. and Veschetti, E.: Occurrence of *Cryptosporidium* oocysts in sewage effluents and correlation with microbial, chemical and physical water variables. Environ. Monit. Assess., 75, 241–252, doi:10.1023/A:1014852201424, 2002.

Booth, I. R.: Regulation of cytoplasmic pH in bacteria. Microbiol. Rev. 49, 359–378, 1985.

Bradford, S. A., Kim, H., Headd, B. and Torkzaban, S.: Evaluating the transport of *Bacillus subtilis* spores as a potential surrogate for *Cryptosporidium parvum* oocysts. Environ. Sci. Technol., 50, 1295–1303, doi:10.1021/acs.est.5b05296, 2016.

Brookes, J. D., Antenucci, J., Hipsey, M., Burch, M. D., Ashbolt, N. J. and Ferguson, C.: Fate and transport of pathogens in lakes and reservoirs. Environ. Int., 30, 741–759, doi:10.1016/j.envint.2003.11.006, 2004.

Brouwer, A. F., Eisenberg, M. C., Remais, J. V., Collender, P. A., Meza, R. and Eisenberg, J. N. S.: Modeling biphasic environmental decay of pathogens and implications for risk analysis. Environ. Sci. Technol., 51:2186–2196, doi:10.1021/acs.est.6b04030, 2017.

Brown, L. and Murray, V.: Examining the relationship between infectious diseases and flooding in Europe. Disaster Health, 1, 117–127, doi:10.4161/dish.25216, 2013.

Burkhardt III, W., Calci, K. R., Watkins, W. D., Rippey, S. R. and Chirtel, S. J.: Inactivation of indicator microorganisms in estuarine waters. Water Res., 34, 2207–2214, doi:10.1016/S0043-1354(99)00399-1, 2000.

Bush, J. and Doyon, A.: Building urban resilience with nature-based solutions: How can urban planning contribute? Cities, 95, 102483, doi:10.1016/j.cities.2019.102483, 2019.

Busta, F. F., Suslow, T. V., Parish, M. E., Beuchat, L. R., Farber, J. N., Garrett, E. H. and Harris, L. J.: The use of indicators and surrogate microorganisms for the evaluation of pathogens in fresh and fresh-cut produce. Compr. Rev. Food Sci. F., 2, 179–185, doi:10.1111/j.1541-4337.2003.tb00035.x, 2003.

Calbó, J., Pagès, D. and González, J.-A.: Empirical studies of cloud effects on UV radiation: A review. Rev. Geophys., 43, RG2002, doi:10.1029/2004RG000155, 2005.

Calin, M. A. and Parasca, S. V.: Light sources for photodynamic inactivation of bacteria. Lasers Med. Sci., 24, 453–460, doi:10.1007/s10103-008-0588-5, 2008.

Caminade C., Medlock Jolyon M., Ducheyne E., McIntyre K. M., Leach S., Baylis M. and Morse A. P.: Suitability of European climate for the Asian tiger mosquito *Aedes albopictus*: recent trends and future scenarios. J. Roy. Soc. Interface, 9, 2708–2717, doi:10.1098/rsif.2012.0138, 2012.

Canepari, S., Castellano, P., Astolfi, M. L., Materazzi, S., Ferrante, R., Fiorini, D. and Curini, R.: Release of particles, organic compounds, and metals from crumb rubber used in synthetic turf under chemical and physical stress. Environ. Sci. Pollut. Res., 25, 1448–1459, doi:10.1007/s11356-017-0377-4, 2018.

Cantwell, R. E. and Hofmann, R.: Inactivation of indigenous coliform bacteria in unfiltered surface water by ultraviolet light. Water Res., 42:2729–2735, doi:10.1016/j.watres.2008.02.002, 2008.

Cao, X., Zhang, M., Mujumdar, A. S., Zhong, Q. and Wang, Z.: Effects of ultrasonic pretreatments on quality, energy consumption and sterilization of barley grass in freeze drying. Ultrason. Sonochem., 40, 333–340, doi:10.1016/j.ultsonch.2017.06.014, 2018.

Carroll, B., Balogh, R., Morbey, H. and Araoz, G.: Health and social impacts of a flood disaster: responding to needs and implications for practice. Disasters, 34, 1045–1063, doi:10.1111/j.1467-7717.2010.01182.x, 2010.

Chandrappa, A. K. and Biligiri, K. P.: Pervious concrete as a sustainable pavement material – Research findings and future prospects: A state-of-the-art review. Constr. Build. Mater., 111, 262–274, doi:10.1016/j.conbuildmat.2016.02.054, 2016.

Chang, J. C. H., Ossoff, S. F., Lobe, D. C., Dorfman, M., Dumais, C. M., Qualls, R. and Johnson, J. D.: UV inactivation of pathogenic and indicator organisms. Appl. Environ. Microbiol., 49, 1361–5, 1985.

Characklis, G. W., Dilts, M. J., Simmons III, O. D., Likirdopulos, C. A., Krometis, L. -A. H. and Sobsey, M. D.: Microbial partitioning to settleable particles in stormwater. Water Res., 39, 1773–1782, doi:10.1016/j.watres.2005.03.004, 2005.

Charters, F. J., Cochrane, T. A. and O'Sullivan, A. D.: Particle size distribution variance in untreated urban runoff and its implication on treatment selection. Water Res., 85:337–345, doi:10.1016/j.watres.2015.08.029, 2015.

Chinchón-Payá, S., Andrade, C. and Chinchón, S.: Indicator of carbonation front in concrete as substitute to phenolphthalein. Cement Concrete Res., 82, 87–91, doi:10.1016/j.cemconres.2015.12.010, 2016.

Choi, J. -I., Lee, Y., Kim, Y. Y. and Lee, B. Y.: Image-processing technique to detect carbonation regions of concrete sprayed with a phenolphthalein solution. Constr. Build. Mater., 154, 451–461, doi:10.1016/j.conbuildmat.2017.07.205, 2017.

Cissé, G.: Food-borne and water-borne diseases under climate change in low- and middle-income countries: Further efforts needed for reducing environmental health exposure risks. Acta Trop., 194, 181–188, doi:10.1016/j.actatropica.2019.03.012, 2019.

Claro, T., O'Reilly, M., Daniels, S. and Humphreys, H.: Surface microbial contamination in hospitals: A pilot study on methods of sampling and the use of proposed microbiologic standards. Am. J. Infect. Control, 43, 1000–1002, doi:10.1016/j.ajic.2015.05.009, 2015.

Cooley, M., Carychao, D., Crawford-Miksza, L., Jay, M. T., Myers, C., Rose, C., Keys, C., Farrar, J. and Mandrell, R. E.: Incidence and tracking of *Escherichia coli* O157:H7 in a major produce production region in California. PLoS ONE, 2, e1159, doi:10.1371/journal.pone.0001159, 2007.

Craggs, R. J., Zwart, A., Nagels, J. W. and Davies-Colley, R. J.: Modelling sunlight disinfection in a high rate pond. Ecol. Eng., 22:113–122, doi:10.1016/j.ecoleng.2004.03.001, 2004.

Craik, S. A., Weldon, D., Finch, G. R., Bolton, J. R. and Belosevic, M.: Inactivation of *Cryptosporidium parvum* oocysts using medium- and low-pressure ultraviolet radiation. Water Res., 35, 1387–1398, doi:10.1016/S0043-1354(00)00399-7, 2001.

Csonka, L. N.: Physiological and genetic responses of bacteria to osmotic stress. Microbiol. Mol. Biol. Rev., 53, 121–147, 1989.

Curriero, F. C., Patz, J. A., Rose, J. B. and Lele, S.: The association between extreme precipitation and waterborne disease outbreaks in the United States, 1948–1994. Am. J. Public Health, 91, 1194–1199, doi:10.2105/AJPH.91.8.1194, 2001.

Curtis, T. P., Mara, D. D. and Silva, S. A.: The effect of sunlight on faecal coliforms in ponds: Implications for research and design. Water Sci. Technol., 26, 1729–1738, doi:10.2166/wst.1992.0616, 1992.

Davies, C. M., Long, J. A., Donald, M. and Ashbolt, N. J.: Survival of fecal microorganisms in marine and freshwater sediments. Appl. Environ. Microbiol., 61, 1888–1896, 1995.

Davies-Colley, R. J., Donnison, A. M., Speed, D. J., Ross, C. M. and Nagels, J. W.: Inactivation of faecal indicator micro-organisms in waste stabilisation ponds: Interactions of environmental factors with sunlight. Water Res., 33:1220–1230, doi:10.1016/S0043-1354(98)00321-2, 1999.

Davies-Colley, R. J., Donnison, A. M. and Speed, D. J.: Towards a mechanistic understanding of pond disinfection. Water Sci. Technol., 42:10-11, 149–158, 2000.

Dawson, D. J., Paish, A., Staffell, L. M., Seymour, I. J. and Appleton, H.: Survival of viruses on fresh produce, using MS2 as a surrogate for norovirus. J. Appl. Microbiol., 98, 203–209, doi:10.1111/j.1365-2672.2004.02439.x, 2005.

de Man, H., van den Berg, H. H. J. L., Leenen, E. J. T. M., Schijven, J. F., Schets, F. M., van der Vliet, J. C., van Knapen, F. and de Roda Husman, A. M.: Quantitative assessment of infection risk from exposure to waterborne pathogens in urban floodwater. Water Res., 48, 90–99, doi:10.1016/j.watres.2013.09.022, 2014.

de Wit, M. A., Koopmans, M. P., Kortbeek, L. M., van Leeuwen, N. J., Vinjé, J. and van Duynhoven, Y. T.: Etiology of gastroenteritis in sentinel general practices in The Netherlands. Clin. Infect. Dis. Off. Publ. Infect. Dis. Soc. Am., 33, 280–288, doi:10.1086/321875, 2001.

Dias, D. F. C., Passos, R. G. and von Sperling, M.: A review of bacterial indicator disinfection mechanisms in waste stabilisation ponds. Rev. Environ. Sci. Biotechnol., 16, 517–539, doi:10.1007/s11157-017-9433-2, 2017.

Dias, D.F.C. and von Sperling, M.: Solar radiation (PAR, UV-A, UV-B) penetration in a shallow maturation pond operating in a tropical climate. Water Sci. Technol., 76, 182–191, doi:10.2166/wst.2017.203, 2017.

Dias, D.F.C. and von Sperling, M.: Vertical profiling and modelling of *Escherichia coli* decay in a shallow maturation pond operating in a tropical climate. Environ. Technol., 39, 759–769, doi:10.1080/09593330.2017.1310936, 2018.

Dias, E., Ebdon, J. and Taylor, H.: The application of bacteriophages as novel indicators of viral pathogens in wastewater treatment systems. Water Res., 129, 172–179, doi:10.1016/j.watres.2017.11.022, 2018.

Dika, C., Ly-Chatain, M. H., Francius, G., Duval, J. F. L. and Gantzer, C.: Non-DLVO adhesion of F-specific RNA bacteriophages to abiotic surfaces: Importance of surface roughness, hydrophobic and electrostatic interactions. Colloids Surf. Physicochem. Eng. Asp., 435, 178–187, doi:10.1016/j.colsurfa.2013.02.045, 2013.

Digel, I., Akimbekov, N. S., Kistaubayeva, A. and Zhubanova, A. A.: Microbial sampling from dry surfaces: Current challenges and solutions, in: Artmann, G. M., Artmann, A., Zhubanova, A. A. and Digel, I. (Eds.): Biological, physical and technical basics of cell engineering, Springer Singapore, Singapore, 421–456, doi:10.1007/978-981-10-7904-7_19, 2018.

Directive 2006/7/EC of the European Parliament and of the Council of 15 February 2006 concerning the management of bathing water quality and repealing Directive 76/160/EEC. Official Journal of the European Union, L64/37, available at: http://data.europa.eu/eli/dir/2006/7/oj, 2006.

Donovan, E., Unice, K., Roberts, J. D., Harris, M. and Finley, B.: Risk of gastrointestinal disease associated with exposure to pathogens in the water of the lower Passaic River. Appl. Environ. Microbiol., 74, 994–1003, doi:10.1128/AEM.00601-07, 2008.

Du, W., FitzGerald, G. J., Clark, M. and Hou, X. -Y.: Health impacts of floods. Prehosp. Disaster Med., 25:265–272, 2010.

Eccles, K. M., Checkley, S., Sjogren, D., Barkema, H. W. and Bertazzon, S.: Lessons learned from the 2013 Calgary flood: Assessing risk of drinking water well contamination. Appl. Geogr., 80, 78–85, doi:10.1016/j.apgeog.2017.02.005, 2017.

Elek, S. D. and Hilson, G. R. F.: Combined agar diffusion and replica plating techniques in the study of antibacterial substances. J. Clin. Pathol., 7, 37–44, 1954.

EM-DAT (The Emergency Events Database), Centre for Research on the Epidemiology of Disasters (CRED), Université catholique de Louvain (UCL), Brussels, Belgium: https://www.emdat.be, last access: 8 January 2019.

Euripidou, E. and Murray, V.: Public health impacts of floods and chemical contamination. J. Public Health, 26, 376–383, doi:10.1093/pubmed/fdh163, 2004.

Farkas, K., Varsani, A. and Pang, L.: Adsorption of rotavirus, MS2 bacteriophage and surface-modified silica nanoparticles to hydrophobic matter. Food Environ. Virol., 7, 261–268, doi:10.1007/s12560-014-9171-3, 2015.

Fewtrell, L. and Bartram, J. (Eds.): Water quality: Guidelines, standards and health: Assessment of risk and risk management for water-related infectious disease, World Health Organization, available at: https://apps.who.int/iris/handle/10665/42442, 2001.

Finazzi, G., Losio, M. N. and Varisco, G.: FLOQSwabTM: Optimisation of procedures for the recovery of microbiological samples from surfaces. Ital. J. Food Saf., 5, doi:10.4081/ijfs.2016.5756, 2016.

Fischer, D., Thomas, S. M., Niemitz, F., Reineking, B. and Beierkuhnlein, C.: Projection of climatic suitability for *Aedes albopictus* Skuse (Culicidae) in Europe under climate change conditions. Global Planet. Change, 78, 54–64, doi:10.1016/j.gloplacha.2011.05.008, 2011.

Foppen, J. W. A. and Schijven, J. F.: Evaluation of data from the literature on the transport and survival of *Escherichia coli* and thermotolerant coliforms in aquifers under saturated conditions. Water Res., 40:401–426, doi:10.1016/j.watres.2005.11.018, 2006.

Fries, J. S., Characklis, G. W. and Noble, R. T.: Sediment–water exchange of *Vibrio* sp. and fecal indicator bacteria: Implications for persistence and transport in the Neuse River Estuary, North Carolina, USA. Water Res., 42, 941–950, doi:10.1016/j.watres.2007.09.006, 2008.

Fujioka, R. S. and Yoneyama, B. S.: Sunlight inactivation of human enteric viruses and fecal bacteria. Water Sci. Technol. 46, 291–295, 2002.

Fung, D. Y. C., Thompson, L. K., Crozier-Dodson, B. A. and Kastner, C. L.: Hands-free, "pop-up," adhesive tape method for microbial sampling of meat surfaces. J. Rapid Meth. Aut. Mic., 8, 209–217, doi:10.1111/j.1745-4581.2000.tb00218.x, 2000.

Furiga, A., Pierre, G., Glories, M., Aimar, P., Roques, C., Causserand, C. and Berge, M.: Effects of ionic strength on bacteriophage MS2 behavior and their implications for the assessment of virus retention by ultrafiltration membranes. Appl. Environ. Microbiol., 77, 229–236, doi:10.1128/AEM.01075-10, 2011.

Garayoa, R., Abundancia, C., Díez-Leturia, M. and Vitas, A. I.: Essential tools for food safety surveillance in catering services: On-site inspections and control of high risk cross-contamination surfaces. Food Control, 75, 48–54, doi:10.1016/j.foodcont.2016.12.032, 2017.

Garvey, E., Tobiason, J. E., Hayes, M., Wolfram, E., Reckhow, D. A. and Male, J. W.: Coliform transport in a pristine reservoir: Modeling and field studies. Water Sci. Technol., 37, 137–144, doi:10.1016/S0273-1223(98)00048-1, 1998.

Gassilloud, B. and Gantzer, C.: Adhesion-aggregation and inactivation of Poliovirus 1 in groundwater stored in a hydrophobic container. Appl. Environ. Microbiol., 71, 912–920, doi:10.1128/AEM.71.2.912-920.2005, 2005.

Gaynor, K., Katz, A. R., Park, S. Y., Nakata, M., Clark, T. A. and Effler, P. V.: Leptospirosis on Oahu: an outbreak associated with flooding of a university campus. Am. J. Trop. Med. Hyg., 76, 882–885, 2007.

Giannakis, S., Darakas, E., Escalas-Cañellas, A. and Pulgarin, C.: The antagonistic and synergistic effects of temperature during solar disinfection of synthetic secondary effluent. J. Photoch. Photobio. A., 280, 14–26, doi:10.1016/j.jphotochem.2014.02.003, 2014.

Gloyna, E. F.: Waste stabilization ponds, World Health Organization, Geneva, Switzerland, available at: https://apps.who.int/iris/handle/10665/41786, 1971.

Gray, N. F.: Biology of wastewater treatment, Oxford University Press, Oxford, UK, 1989.

Grengg, C., Mittermayr, F., Ukrainczyk, N., Koraimann, G., Kienesberger, S. and Dietzel, M.: Advances in concrete materials for sewer systems affected by microbial induced concrete corrosion: A review. Water Res., 134, 341–352, doi:10.1016/j.watres.2018.01.043, 2018.

Griffith, C.: Surface sampling and the detection of contamination, in: Lelieveld, H., Holah, J. and Gabrić, D. (Eds.): Handbook of hygiene control in the food industry (2nd Edition), Woodhead Publishing Series in Food Science, Technology and Nutrition,

Woodhead Publishing, San Diego, 673–696, doi:10.1016/B978-0-08-100155-4.00044-3, 2016.

Gutiérrez-Cacciabue, D., Cid, A. G. and Rajal, V. B.: How long can culturable bacteria and total DNA persist in environmental waters? The role of sunlight and solid particles. Sci. Total Environ., 539:494–502, doi:10.1016/j.scitotenv.2015.07.138, 2016.

Harwood, V.J., Levine, A.D., Scott, T.M., Chivukula, V., Lukasik, J., Farrah, S.R. and Rose, J.B.: Validity of the indicator organism paradigm for pathogen reduction in reclaimed water and public health protection. Appl. Environ. Microbiol., 71, 3163–3170, doi:10.1128/AEM.71.6.3163-3170.2005, 2005.

Hathaway, J. M. and Hunt, W. F.: Evaluation of first flush for indicator bacteria and total suspended solids in urban stormwater runoff. Water Air Soil Pollut., 217, 135–147, doi:10.1007/s11270-010-0574-y, 2010.

Headd, B. and Bradford, S. A.: Use of aerobic spores as a surrogate for *Cryptosporidium* oocysts in drinking water supplies. Water Res., 90, 185–202, doi:10.1016/j.watres.2015.12.024, 2016.

Heaselgrave, W., Patel, N., Kilvington, S., Kehoe, S. C. and McGuigan, K. G.: Solar disinfection of poliovirus and *Acanthamoeba polyphaga* cysts in water: A laboratory study using simulated sunlight. Lett. Appl. Microbiol., 43, 125–130, doi:10.1111/j.1472 765X.2006.01940.x, 2006.

Hefer A. W., Bhasin A. and Little D. N.: Bitumen surface energy characterization using a contact angle approach. J. Mater. Civ. Eng., 18, 759–767, doi:10.1061/(ASCE)0899 1561(2006)18:6(759), 2006.

Hellberg, R. S. and Chu, E.: Effects of climate change on the persistence and dispersal of foodborne bacterial pathogens in the outdoor environment: A review. Crit. Rev. Microbiol., 42, 548–572, doi.org/10.3109/1040841X.2014.972335, 2016.

Henze, M. and Comeau, Y.: Wastewater characterization, in: Henze, M., van Loosdrecht, M., Ekama, G. and Brdjanovic, D. (Eds.): Biological wastewater treatment: Principles, modelling and design, IWA Publishing, London, UK, 35, 2008.

Hirsch, P., Eckhardt, F. E. W. and Palmer, R. J.: Methods for the study of rock-inhabiting microorganisms: A mini review. J. Microbiol. Methods, 23, 143–167, doi:10.1016/0167-7012(95)00017-F, 1995.

Hsu, N. -Y., Chen, P. -Y., Chang, H. -W. and Su, H. -J.: Changes in profiles of airborne fungi in flooded homes in southern Taiwan after Typhoon Morakot. Sci. Total Environ., 409, 1677–1682, doi:10.1016/j.scitotenv.2011.01.042, 2011.

Hussey, M. A. and Zayaitz, A.: Endospore stain protocol, American Society for Microbiology, available at: https://www.asmscience.org/content/education/protocol/protocol.3112, 2007.

Ismaïl, R., Aviat, F., Michel, V., Le Bayon, I., Gay-Perret, P., Kutnik, M. and Fédérighi, M.: Methods for recovering microorganisms from solid surfaces used in the food industry: A Review of the literature. Int. J. Environ. Res. Public. Health, 10, 6169–6183, doi:10.3390/ijerph10116169, 2013.

Ismaïl, R., Le Bayon, I., Michel, V., Jequel, M., Kutnik, M., Aviat, F. and Fédérighi, M.: Comparative study of three methods for recovering microorganisms from wooden surfaces in the food industry. Food Anal. Methods, 8, 1238–1247, doi:10.1007/s12161-014-0008-3, 2015.

Jha, A. K., Bloch, R. and Lamond, J.: Cities and Flooding. The World Bank, doi:10.1596/978-0-8213-8866-2, 2012.

Jonkman, S. N. and Kelman, I.: An analysis of the causes and circumstances of flood disaster deaths. Disasters 29, 75–97, doi:10.1111/j.0361-3666.2005.00275.x, 2005.

Jonsson, A. and Agerberg, S.: Modelling of *E. coli* transport in an oligotrophic river in northern Scandinavia. Ecol. Model, doi:10.1016/j.ecolmodel.2014.10.021, 2014.

Kay, D., Crowther, J., Stapleton, C. M., Wyer, M. D., Fewtrell, L., Edwards, A., Francis, C. A., McDonald, A. T., Watkins, J. and Wilkinson, J.: Faecal indicator organism concentrations in sewage and treated effluents. Water Res., 42, 442–454, doi:10.1016/j.watres.2007.07.036, 2008.

Kayhanian, M., Anderson, D., Harvey, J. T., Jones, D. and Muhunthan, B.: Permeability measurement and scan imaging to assess clogging of pervious concrete pavements in parking lots. J. Environ. Manage., 95, 114–123, doi:10.1016/j.jenvman.2011.09.021, 2012.

King, B. J., Hoefel, D., Daminato, D. P., Fanok, S. and Monis, P. T.: Solar UV reduces *Cryptosporidium parvum* oocyst infectivity in environmental waters. J. Appl. Microbiol., 104, 1311–1323, doi:10.1111/j.1365-2672.2007.03658.x, 2008.

Kirk, J. T. O.: Light and photosynthesis in aquatic ecosystems, Cambridge University Press, Cambridge, UK, 1994.

Kovats, R. S., Valentini, R., Bouwer, L. M., Georgopoulou, E., Jacob, D., Martin, E., Rounsevell, M. and Soussana, J. -F.: Europe, in: Barros, V. R., Field, C. B., Dokken, D. J., Mastrandrea, M. D., Mach, K. J., Bilir, T. E., Chatterjee, M., Ebi, K. L., Estrada, Y. O., Genova, R. C., Girma, B., Kissel, E. S., Levy, A. N., MacCracken, S., Mastrandrea, P. R. and White, L. L. (Eds.): Climate change 2014: Impacts, adaptation, and vulnerability: Part B: Regional Aspects: Contribution of Working Group II to the Fifth

Assessment Report of the Intergovernmental Panel on Climate Change Cambridge University Press, Cambridge, UK and New York, NY, USA, 1267-1326, 2014.

Krometis, L. -A. H., Characklis, G. W., Simmons III, O. D., Dilts, M. J., Likirdopulos, C. A. and Sobsey, M. D.: Intra-storm variability in microbial partitioning and microbial loading rates. Water Res., 41, 506–516, doi:10.1016/j.watres.2006.09.029, 2007.

Krüger, O., Kalbe, U., Richter, E., Egeler, P., Römbke, J. and Berger, W.: New approach to the ecotoxicological risk assessment of artificial outdoor sporting grounds. Environ. Pollut., 175, 69–74, doi:10.1016/j.envpol.2012.12.024, 2013.

Lahou, E. and Uyttendaele, M.: Evaluation of three swabbing devices for detection of *Listeria monocytogenes* on different types of food contact surfaces. Int. J. Environ. Res. Public Health, 11, 804–814, doi:10.3390/ijerph110100804, 2014.

Laine, J., Huovinen, E., Virtanen, M. J., Snellman, M., Lumio, J., Ruutu, P., Kujansuu, E., Vuento, R., Pitkänen, T., Miettinen, I., Herrala, J., Lepistö, O., Antonen, J., Helenius, J., Hänninen, M. -L., Maunula, L., Mustonen, J., Kuusi, M. and the Pirkanmaa Waterborne Outbreak Study Group: An extensive gastroenteritis outbreak after drinking-water contamination by sewage effluent, Finland. Epidemiol. Infect., 139, 1105–1113, doi:10.1017/S0950268810002141, 2011.

Langlet, J., Gaboriaud, F. and Gantzer, C.: Effects of pH on plaque forming unit counts and aggregation of MS2 bacteriophage. J. Appl. Microbiol., 103, 1632–1638, doi:10.1111/j.1365-2672.2007.03396.x, 2007.

Lashford, C., Rubinato, M., Cai, Y., Hou, J., Abolfathi, S., Coupe, S., Charlesworth, S. and Tait, S.: SuDS & Sponge Cities: A Comparative Analysis of the Implementation of Pluvial Flood Management in the UK and China. Sustainability, 11, 213, doi:10.3390/su11010213, 2019.

Lemmen, S. W., Häfner, H., Zolldann, D., Amedick, G. and Lutticken, R.: Comparison of two sampling methods for the detection of Gram-positive and Gram-negative bacteria in the environment: Moistened swabs versus Rodac plates. Int. J. Hyg. Envir. Heal. 203, 245–248, doi:10.1078/S1438-4639(04)70035-8, 2001.

Lian, Y., Mai, L., Cromar, N., Buchanan, N., Fallowfield, H. and Li, X.: MS2 coliphage and *E. coli* UVB inactivation rates in optically clear water: dose, dose rate and temperature dependence. Water Sci. Technol., 78(10), 2228–2238, doi:10.2166/wst.2018.509, 2018.

Liu, Z., Merwade, V. and Jafarzadegan, K.: Investigating the role of model structure and surface roughness in generating flood inundation extents using one- and two-dimensional hydraulic models. J. Flood Risk Manag., 12, e12347, doi:10.1111/jfr3.12347, 2019.

Liu-Helmersson, J., Brännström, Å., Sewe, M. O., Semenza, J. C. and Rocklöv, J.: Estimating past, present, and future trends in the global distribution and abundance of the arbovirus vector *Aedes aegypti* under climate change scenarios. Front. Public Health, 7, doi:10.3389/fpubh.2019.00148, 2019.

Loge, F. J., Emerick, R. W., Thompson, D. E., Nelson, D. C. and Darby, J. L.: Factors influencing ultraviolet disinfection performance: Part I: Light penetration to wastewater particles. Water Environ. Res., 71, 377–381, 1999.

Lutz, J. K., Crawford, J., Hoet, A. E., Wilkins, J. R. and Lee, J.: Comparative performance of contact plates, electrostatic wipes, swabs and a novel sampling device for the detection of *Staphylococcus aureus* on environmental surfaces. J. Appl. Microbiol., 115, 171–178, doi:10.1111/jam.12230, 2013.

Maïga, Y., Denyigba, K., Wethe, J. and Ouattara, A. S.: Sunlight inactivation of *Escherichia coli* in waste stabilization microcosms in a sahelian region (Ouagadougou, Burkina Faso). J. Photochem. Photobiol., B., 94, 113–119, doi:10.1016/j.jphotobiol.2008.10.008, 2009.

Mamane, H., Shemer, H. and Linden, K. G.: Inactivation of *E. coli*, *B. subtilis* spores, and MS2, T4, and T7 phage using UV/H_2O_2 advanced oxidation. J. Hazard. Mater., 146, 479–486, doi:10.1016/j.jhazmat.2007.04.050, 2007.

Maraccini, P. A., Mattioli, M. C. M., Sassoubre, L. M., Cao, Y., Griffith, J. F., Ervin, J. S., Van De Werfhorst, L. C. and Boehm, A. B.: Solar inactivation of enterococci and *Escherichia coli* in natural waters: Effects of water absorbance and depth. Environ. Sci. Technol., 50:5068–5076, doi:10.1021/acs.est.6b00505, 2016.

Mark, O., Jørgensen, C., Hammond, M., Khan, D., Tjener, R., Erichsen, A. and Helwigh, B.: A new methodology for modelling of health risk from urban flooding exemplified by cholera – case Dhaka, Bangladesh. J. Flood Risk Manag., 11, S28-S42, doi:10.1111/jfr3.12182, 2015.

Martínez-Bastidas, T., Castro-del Campo, N., Mena, K. D., Castro-del Campo, N., León-Félix, J., Gerba, C. P. and Chaidez, C.: Detection of pathogenic micro-organisms on children's hands and toys during play. J. Appl. Microbiol., 116, 1668–1675, doi:10.1111/jam.12473, 2014.

Maunula, L., Rönnqvist, M., Åberg, R., Lunden, J. and Nevas, M.: The presence of norovirus and adenovirus on environmental surfaces in relation to the hygienic level in food service operations associated with a suspected gastroenteritis outbreak. Food Environ. Virol., 9, 334–341, doi:10.1007/s12560-017-9291-7, 2017.

Mazoua, S. and Chauveheid, E.: Aerobic spore-forming bacteria for assessing quality of drinking water produced from surface water. Water Res., 39, 5186–5198, doi:10.1016/j.watres.2005.09.027, 2005.

McMinn, B. R., Ashbolt, N. J. and Korajkic, A.: Bacteriophages as indicators of faecal pollution and enteric virus removal. Lett. Appl. Microbiol., 65, 11–26, doi:10.1111/lam.12736, 2017.

Medema, G.J., Bahar, M. and Schets, F.M.: Survival of *Cryptosporidium parvum*, *Escherichia coli*, faecal enterococci and *Clostridium perfringens* in river water: Influence of temperature and autochthonous microorganisms. Water Sci. Technol., 35, 249–252, doi:10.2166/wst.1997.0742, 1997.

Melnick, J. L., Gerba, C. P. and Berg, G.: The ecology of enteroviruses in natural waters. Crit. Rev. Environ. Control, 10, 65–93, doi:10.1080/10643388009381677, 1980.

Mendez, C. B., Klenzendorf, J. B., Afshar, B. R., Simmons, M. T., Barrett, M. E., Kinney, K. A. and Kirisits, M. J.: The effect of roofing material on the quality of harvested rainwater. Water Res., 45, 2049–2059, doi:10.1016/j.watres.2010.12.015, 2011.

Menne, B. and Murray, V. (Eds.): Floods in the WHO European Region: Health effects and their prevention, World Health Organization, available at: http://www.euro.who.int/en/publications/abstracts/floods-in-the-who-european-region-health-effects-and-their-prevention, 2013.

Mignot, E., Li, X. and Dewals, B.: Experimental modelling of urban flooding: A review. J. Hydrol., 568, 334–342, doi:10.1016/j.jhydrol.2018.11.001, 2019.

Mihajlovski, A., Seyer, D., Benamara, H., Bousta, F. and Martino, P. D.: An overview of techniques for the characterization and quantification of microbial colonization on stone monuments. Ann. Microbiol., 65, 1243–1255, doi:10.1007/s13213-014-0956-2, 2015.

Monaghan, A. J., Sampson, K. M., Steinhoff, D. F., Ernst, K. C., Ebi, K. L., Jones, B. and Hayden, M. H.: The potential impacts of 21st century climatic and population changes on human exposure to the virus vector mosquito *Aedes aegypti*. Clim. Change, 146, 487–500, 2018.

Moore, G. and Griffith, C.: Problems associated with traditional hygiene swabbing: The need for in-house standardization. J. Appl. Microbiol., 103, 1090–1103, doi:10.1111/j.1365-2672.2007.03330.x, 2007.

Morris, D. P., Zagarese, H., Williamson, C. E., Balseiro, E. G., Hargreaves, B. R., Modenutti, B., Moeller, R. and Queimalinos, C.: The attenuation of solar UV radiation in lakes and the role of dissolved organic carbon. Limnol. Oceanogr., 40(8):1381–91, 1995.

Nalwanga, R., Quilty, B., Muyanja, C., Fernandez-Ibañez, P. and McGuigan, K. G.: Evaluation of solar disinfection of *E. coli* under Sub-Saharan field conditions using a 25L borosilicate glass batch reactor fitted with a compound parabolic collector. Sol. Energy, 100, 195–202, doi:10.1016/j.solener.2013.12.011, 2014.

Nelson, K. L., Boehm, A. B., Davies-Colley, R. J., Dodd, M. C., Kohn, T., Linden, K. G., Liu, Y., Maraccini, P. A., McNeill, K., Mitch, W. A., Nguyen, T. H., Parker, K. M., Rodriguez, R. A., Sassoubre, L. M., Silverman, A. I., Wigginton, K. R. and Zepp, R. G.: Sunlight-mediated inactivation of health-relevant microorganisms in water: a review of mechanisms and modeling approaches. Environ. Sci. Process Impacts, 20:1089–1122, doi:10.1039/C8EM00047F, 2018.

Nelson, K. L.: Concentrations and inactivation of *Ascaris* eggs and pathogen indicator organisms in wastewater stabilization pond sludge. Water Sci. Technol., 48, 89–95, doi:10.2166/wst.2003.0093, 2003.

Nguyen, M. T., Jasper, J. T., Boehm, A. B. and Nelson, K. L.: Sunlight inactivation of fecal indicator bacteria in open-water unit process treatment wetlands: Modeling endogenous and exogenous inactivation rates. Water Res., 83:282–292, doi:10.1016/j.watres.2015.06.043, 2015.

Nicholson, W. L., Munakata, N., Horneck, G., Melosh, H. J. and Setlow, P.: Resistance of *Bacillus* endospores to extreme terrestrial and extraterrestrial environments. Microbiol. Mol. Biol. Rev., MMBR 64, 548–572, doi:10.1128/mmbr.64.3.548-572.2000, 2000.

Nieminski, E., Durrant, G.C., Hoyt, M.B., Owens, M.E., Peterson, L., Peterson, S., Tanner, W.D., Rosen, J. and Clancy, J.L.: Is *E. coli* an appropriate surrogate for *Cryptosporidium* occurrence in water? J. Am. Water Works Assn., 102, 65–78, doi:10.1002/j.1551-8833.2010.tb10073.x, 2010.

Noe, R., Cohen, A. L., Lederman, E., Gould, L. H., Alsdurf, H., Vranken, P., Ratard, R., Morgan, J., Norton, S. A. and Mott, J.: Skin disorders among construction workers following Hurricane Katrina and Hurricane Rita: an outbreak investigation in New Orleans, Louisiana. Arch. Dermatol., 143, 1393–1398, doi:10.1001/archderm.143.11.1393, 2007.

NOAA (National Oceanic and Atmospheric Administration): NOAA Solar Calculator. Earth System Research Laboratory, Global Monitoring Division, Global Radiation Group: https://www.esrl.noaa.gov/gmd/grad/solcalc/, last access: 29 July 2019.

Okaka, F. O. and Odhiambo, B. D. O.: Relationship between flooding and outbreak of infectious diseases in Kenya: A review of the literature. J. Environ. Public Health, 2018, doi:10.1155/2018/5452938, 2018.

Olds, H. T., Corsi, S. R., Dila, D. K., Halmo, K. M., Bootsma, M. J. and McLellan, S. L.: High levels of sewage contamination released from urban areas after storm events: A quantitative survey with sewage specific bacterial indicators. PLOS Medicine 15, e1002614, doi:10.1371/journal.pmed.1002614, 2018.

O'Neill, E.: Expanding the horizons of integrated flood risk management: A critical analysis from an Irish perspective. Int. J. River Basin Manag., 16, 71-77, doi:10.1080/15715124.2017.1351979, 2018.

Oppezzo, O. J.: Contribution of UVB radiation to bacterial inactivation by natural sunlight. J. Photochem. Photobiol. B., 115:58–62, doi:10.1016/j.jphotobiol.2012.06.011, 2012.

Ordaz, G., Merino-Mascorro, J. Á., García, S. and Heredia, N.: Persistence of *Bacteroidales* and other fecal indicator bacteria on inanimated materials, melon and tomato at various storage conditions. Int. J. Food Microbiol., 299, 33–38, doi:10.1016/j.ijfoodmicro.2019.03.015, 2019.

Otter, J. A., Mutters, N. T., Tacconelli, E., Gikas, A. and Holmes, A. H.: Controversies in guidelines for the control of multidrug-resistant Gram-negative bacteria in EU countries. Clin. Microbiol. Infec., 21, 1057–1066, doi:10.1016/j.cmi.2015.09.021, 2015.

Pade, C. and Guimaraes, M.: The CO_2 uptake of concrete in a 100 year perspective. Cem. Concr. Res., 37, 1348–1356, doi:10.1016/j.cemconres.2007.06.009, 2007.

Page, D., Sidhu, J. P. S. and Toze, S.: Microbial risk reduction of withholding periods during public open space irrigation with recycled water. Urban Water J., 12, 581–587, doi:10.1080/1573062X.2014.923474, 2015.

Parhad, N. M. and Rao, N. U.: Effect of pH on survival of *Escherichia coli*. J. Water Pollut. Control Fed., 46, 980–986, 1974.

Peng, X., Murphy, T. and Holden, N. M.: Evaluation of the effect of temperature on the die-off rate for *Cryptosporidium parvum* oocysts in water, soils, and feces. Appl. Environ. Microbiol., 74, 7101–7107, doi:10.1128/AEM.01442-08, 2008.

Plati, C.: Sustainability factors in pavement materials, design, and preservation strategies: A literature review. Constr. Build. Mater., 211, 539–555, doi:10.1016/j.conbuildmat.2019.03.242, 2019.

Pond, K.: Water recreation and disease: Plausibility of associated infections: Acute effects, sequelae and mortality. IWA Publishing, London, UK, doi:10.2166/9781780405827, 2013.

Randolph, S. E. and Rogers, D. J.: The arrival, establishment and spread of exotic diseases: patterns and predictions. Nat. Rev. Microbiol., 8, 361–371, doi:10.1038/nrmicro2336, 2010.

Rauch, W., Bertrand-Krajewski, J. L., Krebs, P., Mark, O., Schilling, W., Schütze, M. and Vanrolleghem, P. A.: Deterministic modelling of integrated urban drainage systems. Water Sci. Technol., 45, 81–94, 2002.

Rawlinson, S., Ciric, L. and Cloutman-Green, E.: How to carry out microbiological sampling of healthcare environment surfaces? A review of current evidence. J. Hosp. Infect., doi:10.1016/j.jhin.2019.07.015, 2019.

Reed, R. H.: The inactivation of microbes by sunlight: Solar disinfection as a water treatment process. Adv. Appl. Microbiol., 54, 333–365, doi:10.1016/S0065-2164(04)54012-1, 2004.

Reiner, R. C., King, A. A., Emch, M., Yunus, M., Faruque, A. S. G. and Pascual, M.: Highly localized sensitivity to climate forcing drives endemic cholera in a megacity. Proc. Natl. Acad. Sci. U. S. A., 109, 2033–2036, doi:10.1073/pnas.1108438109, 2012.

Rice, E. W., Fox, K. R., Miltner, R. J., Lytle, D. A. and Johnson, C. H.: Evaluating plant performance with endospores. J. Am. Water Works Assoc., 88, 122–30, 1996.

Riesenman, P. J. and Nicholson, W. L.: Role of the spore coat layers in *Bacillus subtilis* spore resistance to hydrogen peroxide, artificial UV-C, UV-B, and solar UV radiation. Appl. Environ. Microbiol., 66, 620–626, doi:10.1128/aem.66.2.620-626.2000, 2000.

Robertson, L. J., Campbell, A. T. and Smith, H. V.: Survival of *Cryptosporidium parvum* oocysts under various environmental pressures. Appl. Environ. Microbiol., 58, 3494–3500, 1992.

Roiz, D., Neteler, M., Castellani, C., Arnoldi, D. and Rizzoli, A.: Climatic factors driving invasion of the tiger mosquito (*Aedes albopictus*) into new areas of Trentino, Northern Italy. PLoS ONE, 6, doi:10.1371/journal.pone.0014800, 2011.

Rosso, L., Lobry, J. R., Bajard, S. and Flandrois, J. P.: Convenient model to describe the combined effects of temperature and pH on microbial growth. Appl. Environ. Microbiol., 61, 610–616, 1995.

Rui, Y., Fu, D., Do Minh, H., Radhakrishnan, M., Zevenbergen, C. and Pathirana, A.: Urban surface water quality, flood water quality and human health impacts in Chinese cities. What do we know? Water, 10:240, doi:10.3390/w10030240, 2018.

Sales-Ortells, H. and Medema, G.: Microbial health risks associated with exposure to stormwater in a water plaza. Water Res., 74, 34–46, doi:10.1016/j.watres.2015.01.044, 2015.

Sata, V., Wongsa, A. and Chindaprasirt, P.: Properties of pervious geopolymer concrete using recycled aggregates. Constr. Build. Mater., 42, 33–39, doi:10.1016/j.conbuildmat.2012.12.046, 2013.

Schets, F. M., Wijnen, J. H. van, Schijven, J. F., Schoon, H. and de Roda Husman, A. M.: Monitoring of waterborne pathogens in surface waters in Amsterdam, The Netherlands, and the potential health risk associated with exposure to *Cryptosporidium* and *Giardia* in these waters. Appl. Environ. Microbiol., 74, 2069–2078, doi:10.1128/AEM.01609-07, 2008.

Schijven, J. F. and de Roda Husman, A. M.: Effect of climate changes on waterborne disease in The Netherlands. Water Sci. Technol. 51, 79–87, 2005.

Schultz-Fademrecht, C., Wichern, M. and Horn, H.: The impact of sunlight on inactivation of indicator microorganisms both in river water and benthic biofilms. Water Res., 42, 4771–4779, doi:10.1016/j.watres.2008.08.022, 2008.

Schwarz, K., Heitkötter, J., Heil, J., Marschner, B. and Stumpe, B.: The potential of active and passive infrared thermography for identifying dynamics of soil moisture and microbial activity at high spatial and temporal resolution. Geoderma, 327, 119–129, doi:10.1016/j.geoderma.2018.04.028, 2018.

Sdiri-Loulizi, K., Hassine, M., Aouni, Z., Gharbi-Khelifi, H., Chouchane, S., Sakly, N., Neji-Guédiche, M., Pothier, P., Aouni, M. and Ambert-Balay, K.: Detection and molecular characterization of enteric viruses in environmental samples in Monastir, Tunisia between January 2003 and April 2007. J. Appl. Microbiol., 109, 1093–1104, doi:10.1111/j.1365-2672.2010.04772.x, 2010.

Selbig, W. R., Fienen, M. N., Horwatich, J. A. and Bannerman, R. T.: The effect of particle size distribution on the design of urban stormwater control measures. Water, 8:17, doi:10.3390/w8010017, 2016.

Semenza, J. C. and Menne, B.: Climate change and infectious diseases in Europe. Lancet Infect. Dis., 9, 365–375, doi:10.1016/S1473-3099(09)70104-5, 2009.

Semenza, J. C., Suk, J. E., Estevez, V., Ebi, K. L. and Lindgren, E.: Mapping climate change vulnerabilities to infectious diseases in Europe. Environ. Health Perspect., 120, 385–392, doi:10.1289/ehp.1103805, 2012.

Seyoum, S. D., Vojinovic, Z., Price, R. K. and Weesakul, S.: Coupled 1D and noninertia 2D flood inundation model for simulation of urban flooding. J. Hydraul. Eng., 138, 23–34, doi:10.1061/(ASCE)HY.1943-7900.0000485, 2012.

Shah, A. H., Abdelzaher, A. M., Phillips, M., Hernandez, R., Solo-Gabriele, H. M., Kish, J., Scorzetti, G., Fell, J. W., Diaz, M. R., Scott, T. M., Lukasik, J., Harwood, V. J., McQuaig, S., Sinigalliano, C. D., Gidley, M. L., Wanless, D., Ager, A., Lui, J., Stewart, J. R., Plano, L. R. W. and Fleming, L. E.: Indicator microbes correlate with pathogenic bacteria, yeasts and helminthes in sand at a subtropical recreational beach site. J. Appl. Microbiol., 110, 1571–1583, doi:10.1111/j.1365-2672.2011.05013.x, 2011.

Sidhu, J. P. S., Hanna, J. and Toze, S. G.: Survival of enteric microorganisms on grass surfaces irrigated with treated effluent. J. Water Health, 6, 255–262, doi:10.2166/wh.2008.029, 2008.

Silverman, A. I. and Nelson, K. L.: Modeling the endogenous sunlight inactivation rates of laboratory strain and wastewater *E. coli* and enterococci using biological weighting functions. Environ Sci. Technol., 50:12292–12301, doi:10.1021/acs.est.6b03721, 2016.

Sinclair, R. G., Rose, J. B., Hashsham, S. A., Gerba, C. P. and Haas, C. N.: Criteria for selection of surrogates used to study the fate and control of pathogens in the environment. Appl. Environ. Microbiol. 78, 1969–1977, doi:10.1128/AEM.06582-11, 2012.

Sinton, L. W., Hall, C. H., Lynch, P. A. and Davies-Colley, R. J.: Sunlight inactivation of fecal indicator bacteria and bacteriophages from waste stabilization pond effluent in fresh and saline waters. Appl. Environ. Microbiol., 68, 1122–1131, doi:10.1128/AEM.68.3.1122-1131.2002, 2002.

Staley, Z. R., Robinson, C. and Edge, T. A.: Comparison of the occurrence and survival of fecal indicator bacteria in recreational sand between urban beach, playground and sandbox settings in Toronto, Ontario. Sci. Total Environ., 541, 520–527, doi:10.1016/j.scitotenv.2015.09.088, 2016.

Stelma, G.N.: Use of bacterial spores in monitoring water quality and treatment. J. Water Health, 16, 491–500, doi:10.2166/wh.2018.013, 2018.

Strande, L., Ronteltap, M. and Brdjanovic, D. (Eds.): Faecal sludge management: Systems approach for implementation and operation, IWA Publishing, London, UK, 2014.

Tamburini, E., Donegà, V., Marchetti, M. G., Pedrini, P., Monticelli, C. and Balbo, A.: Study on microbial deposition and contamination onto six surfaces commonly used in chemical and microbiological laboratories. Int. J. Environ. Res. Public Health, 12, 8295–8311, doi:10.3390/ijerph120708295, 2015.

Taylor, J., Lai, K. man, Davies, M., Clifton, D., Ridley, I. and Biddulph, P.: Flood management: Prediction of microbial contamination in large-scale floods in urban environments. Environ. Int., 37, 1019–1029, doi:10.1016/j.envint.2011.03.015, 2011.

Taylor, J., Davies, M., Canales, M. and Lai, K. man: The persistence of flood-borne pathogens on building surfaces under drying conditions. Int. J. Hyg. Environ. Health, 216, 91–99, doi:10.1016/j.ijheh.2012.03.010, 2013.

ten Veldhuis, J. A. E., Clemens, F. H. L. R., Sterk, G. and Berends, B. R.: Microbial risks associated with exposure to pathogens in contaminated urban flood water. Water Res., 44, 2910–2918, doi:10.1016/j.watres.2010.02.009, 2010.

Thompson, S. S., Flury, M., Yates, M. V. and Jury, W. A.: Role of the air-water-solid interface in bacteriophage sorption experiments. Appl. Environ. Microbiol., 64, 304–309, 1998.

Thompson, S. S. and Yates, M. V.: Bacteriophage inactivation at the air-water-solid interface in dynamic batch systems. Appl. Environ. Microbiol., 65, 1186–1190, 1999.

Tresner, H. D. and Hayes, J. A.: Improved methodology for isolating soil microorganisms. Appl. Microbiol., 19, 186–187, 1970.

Trouwborst, T., Kuyper, S., de Jong, J. C. and Plantinga, A. D.: Inactivation of some bacterial and animal viruses by exposure to liquid-air interfaces. J. Gen. Virol., 24, 155–165, doi:10.1099/0022-1317-24-1-155, 1974.

UN (United Nations, General Assembly): Transforming our world: The 2030 agenda for sustainable development, 21 October 2015, A/RES/70/1, 2015.

UNICEF and WHO (World Health Organization): Progress on sanitation and drinking water: 2015 update and MDG assessment, 2015.

Urzì, C. and De Leo, F.: Sampling with adhesive tape strips: An easy and rapid method to monitor microbial colonization on monument surfaces. J. Microbiol. Methods, 44, 1–11, doi:10.1016/S0167-7012(00)00227-X, 2001.

US EPA (US Environmental Protection Agency): Ultraviolet disinfection guidance manual for the final long term 2 enhanced surface water treatment rule. EPA 815-R-06-007, US EPA Office of Water (4601), Washington, DC, USA, 2006.

US EPA (US Environmental Protection Agency): Long term 2 enhanced surface water treatment rule toolbox guidance manual. EPA 815-R-0e16, US EPA, Washington, DC, USA, 2010.

US EPA (US Environmental Protection Agency): Recreational Water Quality Criteria. EPA 820-F-12-058, US EPA Office of Water. Washington, D.C., USA, available at: https://www.epa.gov/wqc/2012-recreational-water-quality-criteria-documents, 2012.

Yamaguchi, N., Ishidoshiro, A., Yoshida, Y., Saika, T., Senda, S. and Nasu, M.: Development of an adhesive sheet for direct counting of bacteria on solid surfaces. J. Microbiol. Methods, 53, 405–410, doi:10.1016/S0167-7012(02)00246-4, 2003.

Valentine, N. B., Butcher, M. G., Su, Y. -F., Jarman, K. H., Matzke, M., Webb-Robertson, B. -J., Panisko, E. A., Seiders, B. A. B. and Wahl, K. L.: Evaluation of sampling tools for environmental sampling of bacterial endospores from porous and nonporous surfaces. J. Appl. Microbiol., 105, 1107–1113, doi:10.1111/j.1365-2672.2008.03840.x, 2008.

Verbyla, M.E. and Mihelcic, J. R.: A review of virus removal in wastewater treatment pond systems. Water Res., 71, 107–124, doi:10.1016/j.watres.2014.12.031, 2015.

Verdier, T., Coutand, M., Bertron, A. and Roques, C.: A review of indoor microbial growth across building materials and sampling and analysis methods. Build. Environ., 80, 136–149, doi:10.1016/j.buildenv.2014.05.030, 2014.

Verdon, T. J., Mitchell, R. J. and van Oorschot, R. A. H.: Evaluation of tapelifting as a collection method for touch DNA. Forensic Sci. Int. Genet., 8, 179–186, doi:10.1016/j.fsigen.2013.09.005, 2014.

Wakuma Abaya, S., Mandere, N. and Ewald, G.: Floods and health in Gambella region, Ethiopia: A qualitative assessment of the strengths and weaknesses of coping mechanisms. Glob. Health Action, 2, doi:10.3402/gha.v2i0.2019, 2009.

Walters, E., Graml, M., Behle, C., Müller, E. and Horn, H.: Influence of particle association and suspended solids on UV inactivation of fecal indicator bacteria in an urban river. Water Air Soil Pollut., 225, 1–9, doi:10.1007/s11270-013-1822-8, 2013.

Walters, E., Schwarzwälder, K., Rutschmann, P., Müller, E. and Horn, H.: Influence of resuspension on the fate of fecal indicator bacteria in large-scale flumes mimicking an oligotrophic river. Water Res., 48, 466–477, doi:10.1016/j.watres.2013.10.002, 2014.

Watson, J. T., Gayer, M. and Connolly, M. A.: Epidemics after natural disasters. Emerg. Infect. Dis., 13, 1–5, doi:10.3201/eid1301.060779, 2007.

Weber, D. J. and Rutala, W. A.: The emerging nosocomial pathogens *Cryptosporidium*, *Escherichia coli* O157:H7, *Helicobacter pylori*, and hepatitis C: Epidemiology, environmental survival, efficacy of disinfection, and control measures. Infect. Control Hosp. Epidemiol., 22, 306–315, doi:10.1086/501907, 2001.

Wegelin, M., Canonica, S., Mechsner, K., Fleischmann, T., Pesaro, F. and Metzler, A.: Solar water disinfection: Scope of the process and analysis of radiation experiments. J. Water Supply Res. T., 43, 154-169, 1994.

Whitman, R. L., Harwood, V. J., Edge, T. A., Nevers, M. B., Byappanahalli, M., Vijayavel, K., Brandão, J., Sadowsky, M. J., Alm, E. W., Crowe, A., Ferguson, D., Ge, Z., Halliday, E., Kinzelman, J., Kleinheinz, G., Przybyla-Kelly, K., Staley, C., Staley, Z. and Solo-Gabriele, H. M.: Microbes in beach sands: Integrating environment, ecology and public health. Rev. Environ. Sci. Biotechnol., 13, 329–368, doi:10.1007/s11157-014-9340-8, 2014.

WHO (World Health Organization): Guidelines on viral inactivation and removal procedures intended to assure the viral safety of human blood plasma products. WHO technical report, Series No. 924, Annex 4, 2004.

WHO (World Health Organization): Flooding and communicable diseases fact sheet: http://www.who.int/hac/techguidance/ems/flood_cds/en/#, last access: 15 November 2019.

Wu, J., Cao, Y., Young, B., Yuen, Y., Jiang, S., Melendez, D., Griffith, J. F. and Stewart, J. R.: Decay of coliphages in sewage-contaminated freshwater: Uncertainty and seasonal effects. Environ. Sci. Technol., 50, 11593–11601, doi:10.1021/acs.est.6b03916, 2016.

Xia, W., Xu, T. and Wang, H.: Thermal behaviors and harmful volatile constituents released from asphalt components at high temperature. J. Hazard. Mater., 373, 741–752, doi:10.1016/j.jhazmat.2019.04.004, 2019.

Xu, G., Shen, W., Huo, X., Yang, Z., Wang, J., Zhang, W. and Ji, X.: Investigation on the properties of porous concrete as road base material. Constr. Build. Mater., 158, 141–148, doi:10.1016/j.conbuildmat.2017.09.151, 2018.

Yakirevich, A., Pachepsky, Y. A., Guber, A. K., Gish, T. J., Shelton, D. R. and Cho, K. H.: Modeling transport of *Escherichia coli* in a creek during and after artificial high-flow events: Three-year study and analysis. Water Res., 47, 2676–2688, doi:10.1016/j.watres.2013.02.011, 2013.

Yamaguchi, N., Ishidoshiro, A., Yoshida, Y., Saika, T., Senda, S. and Nasu, M.: Development of an adhesive sheet for direct counting of bacteria on solid surfaces. J. Microbiol. Methods, 53, 405–410, doi:10.1016/S0167-7012(02)00246-4, 2003.

Ye, M., Sun, M., Huang, D., Zhang, Z., Zhang, H., Zhang, S., Hu, F., Jiang, X. and Jiao, W.: A review of bacteriophage therapy for pathogenic bacteria inactivation in the soil environment. Environ. Int., 129, 488–496, doi:10.1016/j.envint.2019.05.062, 2019.

You, H. S., Lee, S. H., Ok, Y. J., Kang, H. -G., Sung, H. J., Lee, J. Y., Kang, S. S. and Hyun, S. H.: Influence of swabbing solution and swab type on DNA recovery from rigid environmental surfaces. J. Microbiol. Methods, 161, 12–17, doi:10.1016/j.mimet.2019.04.011, 2019.

Yu, P., Zaleski, A., Li, Q., He, Y., Mapili, K., Pruden, A., Alvarez, P. J. J. and Stadler, L. B.: Elevated levels of pathogenic indicator bacteria and antibiotic resistance genes after Hurricane Harvey's flooding in Houston. Environ. Sci. Technol. Lett., 5, 481–486, doi:10.1021/acs.estlett.8b00329, 2018.

Zhong, R. and Wille, K.: Material design and characterization of high performance pervious concrete. Constr. Build. Mater., 98, 51–60, doi:10.1016/j.conbuildmat.2015.08.027, 2015.

LIST OF ACRONYMS

AWI	Air Water Interface
CCA	Chromocult Coliform Agar
CFU	Colony Forming Units
CSO	Combined Sewer Overflow
DO	Dissolved Oxygen
FIB	Faecal Indicator Bacterium
MDGs	Millennium Development Goals
PBS	Phosphate-saline Buffer Solution
PC	Plate Count
PFU	Plaque Forming Units
QMRA	Quantitative Microbial Risk Assessment
ROS	Reactive Oxygen Species
SDGs	Sustainable Development Goals
TPB	Triple Phase Boundary
TYGA	Tryptone Yeast Glucose Agar
TYGB	Tryptone Yeast Glucose Broth
UV	Ultra Violet
2D	Two Dimensional

LIST OF TABLES

LIST OF FIGURES

ABOUT THE AUTHOR

Iosif Marios Scoullos, born in Marousi, Athens, Greece, is a Chemical Engineer, graduate of the National Technical University of Athens, Greece. With a strong interest for the protection of the environment and its resources, his thesis in Biotechnology and the Environment was related to the production of second generation biofuels, entitled "Contribution of the inhibitors which are produced during the hydrothermal pre-treatment of biomass on the growth and ethanol production by the fungus *Fusarium oxysporum*".

Having understood that in order to achieve great changes it is not only technological means which are necessary, but also appropriate management tools and interdisciplinary cooperation, he studied MPhil in Technology Policy at the University of Cambridge, UK, focusing on environment and sustainable development issues.

Since the problem of water is one of the most difficult, though needed, issues to be addressed in the beginning of the 21st century he decided to focus on this with a PhD in Environmetal Biotechnology and Sanitary Engineering at the Delft University of Technology and the IHE Delft Institute for Water Education. He received an Academy of Athens Scholarship from the Vasiliki Bekiari-Vekri Bequest and obtained an Alexander S. Onassis Public Benefit Foundation Research Grant.

Journals publications

Scoullos, I.M., Lopez Vazquez, C.M., van de Vossenberg, J., Hammond, M. and Brdjanovic, D.: Effect of artificial solar radiation on the die-off of pathogen indicator organisms in urban floods. Int. J. Environ. Res., 13, 107-116, doi:10.1007/s41742-018-0160-5, 2019.

Scoullos, I.M., Lopez Vazquez, C.M., van de Vossenberg, J. and Brdjanovic, D.: Die-off of *E. coli* as fecal indicator organism on different surfaces after urban floods. J. Environ. Manage., 250, 109516, doi:10.1016/j.jenvman.2019.109516, 2019.

Scoullos, I.M., Adhikari, S., Lopez Vazquez, C.M., van de Vossenberg, J. and Brdjanovic, D.: Inactivation of indicator organisms on different surfaces after urban floods. Sci. Total Environ., 704, 135456, doi:10.1016/j.scitotenv.2019.135456, 2020.

Scoullos, I.M., van de Vossenberg, J., Lopez Vazquez, C.M. and Brdjanovic, D.: Assessment of microbial sampling methods for flood-prone urban surfaces. In preparation.

Conference proceedings

Economides, S.B., Han, C.L., Orowitsch, S., Scoullos, I.M. and Nuttall, W.J.: Paradigm shift for future mobility: A cross country analysis of behavioural policies, in: Procedia – Soc. Behav. Sci., 48, Transport Research Arena 2012 on Sustainable Mobility through Innovation, Athens, Greece, 23-26 April 2012, 2588-2596, doi:10.1016/j.sbspro.2012.06.1229, 2012.

Skoullos, I.M., Lopez Vazquez, C.M., Hammond, M., Vojinovic, Z. and Brdjanovic, D.: Pathogen occurrence caused by flooding of sewerage networks, in: Proceedings of the IWA Balkan Young Water Professionals Conference, Thessaloniki, Greece, 10-12 May 2015, 362-367, 2015.

Netherlands Research School for the
Socio-Economic and Natural Sciences of the Environment

D I P L O M A

For specialised PhD training

The Netherlands Research School for the
Socio-Economic and Natural Sciences of the Environment
(SENSE) declares that

Iosif Marios Scoullos

born on 17 September 1986, Marousi, Greece

has successfully fulfilled all requirements of the
Educational Programme of SENSE.

Delft, 13 May 2020

The Chairman of the SENSE board

Prof. dr. Martin Wassen

the SENSE Director of Education

Dr. Ad van Dommelen

K O N I N K L I J K E   N E D E R L A N D S E
A K A D E M I E   V A N   W E T E N S C H A P P E N

The SENSE Research School declares that Iosif Marios Scoullos has successfully fulfilled all requirements of the Educational PhD Programme of SENSE with a work load of 36.2 EC, including the following activities:

SENSE PhD Courses

- Environmental research in context (2014)
- SENSE Summer academy (2014)
- Research in context activity: 'Representative of IHE PhD Fellows' Association Board (PAB) in IHE PhD Programme Committee (June 2015-March 2016)'

Other PhD and Advanced MSc Courses

- PhD start-up course, TU Delft (2014)
- Hydrology and hydraulics, IHE Delft (2013)
- Urban drainage and sewerage, IHE Delft (2014)
- Conventional wastewater treatment, IHE Delft (2014)
- Modelling of wastewater treatment processes and plants, IHE Delft (2014)

Management and Didactic Skills Training

- Member and Chair of IHE PhD Fellows' Association Board (2014-2016)
- Supervising MSc student with thesis entitled 'Inactivation of indicator organisms on different urban surfaces after urban floods' (2019)

Oral Presentations

- *Pathogen occurrence caused by flooding of sewerage networks.* IWA Balkan Young Water Professionals Conference 2015, 10-12 May 2015, Thessaloniki, Greece
- *Effect of simulated solar radiation on the die-off of indicator organisms.* Soil contamination symposium: scope, advances and challenges, 1 September 2017, Wageningen, The Netherlands

SENSE Coordinator PhD Education

Dr. ir. Peter Vermeulen

9 780367 556921